THE BATTLE
FOR THE
BUFFALO RIVER

THE BATTLE FOR
THE BUFFALO RIVER

THE STORY OF
AMERICA'S FIRST
NATIONAL RIVER

Neil Compton

The OZARK
SOCIETY
FOUNDATION

The University of Arkansas Press
Fayetteville   1992

Frontispiece: Remains of the great sweet gum, beech, sycamore, and birch trees that bordered the Buffalo above Roark Bluff, destroyed by the management of Valley-Y Ranch, April 1965.

This book was designed by Chiquita Babb using the Minion and Optima typefaces.

⊖ The paper used in this publication meets the minimum requirements of the American National Standard for Permanence of Paper for Printed Library Materials Z39.48-1984.

The Library of Congress has cataloged the hardback edition as follows:

Compton, Neil, 1912–
      The battle for the Buffalo River : a twentieth-century conservation crisis
in the Ozarks / Neil Compton.
            p.   cm.
      Includes index.
      ISBN 1-55728-235-8. —ISBN 1-55728-236-6 (pbk.)
      1. Nature conservation—Political aspects—Arkansas—Buffalo River.
I. Title.
QH76.5.A8C65   1992
333.91'6216'097668—dc20

                                        91-27540
                                        CIP

TO THE MEMORY OF MY DEAR WIFE, LAURENE,
WHO FOR FIFTY-FIVE YEARS
FLOATED THE RIVER OF LIFE WITH ME

Laurene Compton in Indian Creek, 1968.
*Courtesy of Special Collections, University of Arkansas
Libraries, Neil Compton Papers.*

It is with sincere thanks that I acknowledge the designation of money donated by many individuals to the Ozark Society as a memorial to my wife, Laurene, to aid in the publication of this work.

ACKNOWLEDGMENTS

In assembling material for this book the most important source of information was the Special Collections Division of the University of Arkansas Library. In addition to my own papers, which they had acquired in 1975, was valuable material from the files of such personages as Senator J. William Fulbright, Congressman Jim Trimble, Congressman Clyde T. Ellis, Governor Orval Faubus, Congressman John Paul Hammerschmidt, and Senator John L. McClellan. I was especially welcomed in Special Collections by coincidence. My daughter Ellen Shipley was their field archivist. She, Michael Dabrishus, the director, and her coworkers were of invaluable assistance during the two years I spent reviewing those lengthy documents.

Kenneth L. Smith, who figures throughout the text, had given his extensive papers to Special Collections and had persuaded me to do the same. As the author of *The Buffalo River Country,* Ken was qualified by experience as a critic for this book. No one else has been so intense a participant in the Buffalo River battle. He has known all contestants both pro and con. Without his corrections I would have made many errors great and small.

Ellen Shipley and Mike Dabrishus knowing of my dilemma in finding a publisher suggested that I approach Miller Williams of the University of Arkansas Press. In an interview with him that matter was happily settled.

Ellen also knew of an expert typist, Jeanie Wyant, who produced the first readable manuscript.

Brenda Crities, current vice-president of the Ozark Society, came from Cape Girardeau, Missouri, to proofread the book for spelling and punctuation.

Tom Dearmore, a native of Mountain Home and an early decision maker in favor of the Buffalo National River (but at the same time a grandson of that apostle of the Norfork Dam, Tom Shiras), was another valuable proofreader. Tom Dearmore, recently retired as editor of the editorial pages of the *San Francisco Examiner,* had also been an editorial writer for the *Washington Star* and the *Arkansas Gazette,* and with his cousin, Pete Shiras, published the Mountain Home *Baxter Bulletin* in the 1960s.

May I also commend most highly the Ozark Society Foundation for assuming the expense of providing the original electronic manuscript and the index. The Ozark Society Foundation is a publishing enterprise on its own and an outgrowth of Ken Smith's original Ozark Society publishing fund.

To retired architect Paul Young of Fayetteville especial thanks are due for providing information concerning the early day conservation efforts of Dr. Thomas Hardison that led to the establishment of Petit Jean State Park, an accomplishment which Arkansas citizens should not forget.

Recognition of those who furnished some of the photographs is also in order. They are Harold and Margaret Hedges, John Heuston, Nancy Jack, Jeannie Rush, John Swearingen, Everett Bowman, Jack Atkins of the Arkansas Parks Commission, and Joe Clark.

We are also indebted to the various newspapers and magazine whose running account of this contest enabled the author to set down events in proper order without dependency on memory alone.

And last but not least thanks to the staff of the University of Arkansas Press for their interest in this episode of Arkansas history and their hard work in putting it all together.

NEIL COMPTON

CONTENTS

AUTHOR'S FOREWORD

The struggle to save the Buffalo River in the Arkansas Ozarks brought to the fore manifestations of a worldwide plague generated by the hand and mind of man. If we in our great wisdom cannot develop insight enough to control that affliction, we might well become the principal agents in the ruination of our only possible home in the universe.

It is not likely that we shall ever find sustenance on any other celestial body beyond the earth. Some may provide temporary abode, but none other than the earth can harbor us on into the future. And here we have only a hairline two-dimensional surface upon which to place our feet. There will never be habitation for us down in the grinding superheated interior. We are not at ease and can never remain for long in the air above, and in its waters we drown. Where the earth's gaseous envelope meets dry land is our natural domain, and today we stand dominant upon it. We now exercise the power to change and mutilate it in ways undreamed of a generation ago. With that power we now course the skies like angels and live like gods on terra firma. If we have at last become as gods, it is now past time to extend to the earth and all of its creatures the compassion and understanding that we have hitherto assigned to the gods.

But whatever we may think of ourselves, our mission here, and whatever we may imagine our origins to be, the fact remains that we are of this earth, part and parcel of its elements, evolved in and irrevocably adapted to its oxidizing atmosphere, to the radiation it receives from the sun and beyond, and to its waters borne on wind and cloud, contained in lakes and seas, glorifying the land in ever-flowing rivers and streams. Any reasonable man cannot look upon these marvels in their pristine state without feeling an innermost sense of awe and humility. To realize that we are a part of this grand combination of natural forces and basic particles woven on the loom of time comforts the soul and restores our often jaded spirits. With that comes the realization that if we are to survive the unpredictable future, we must not upset this wonderful balance on planet earth.

What we have accomplished has most often been for our own good, but a significant part of our impact on the world is not so good. With our superior and ever-expanding base of knowledge we should by rights be able to discern and avoid problems relative to our presence here; but, too often, emotion, runaway enthusiasm and blind faith cancel out the best of reason and even everyday common sense.

Thus it is that we overexercise every new technical discovery, so enamored do we become of it. Let us then consider one of these unquestioned advances and its undesirable side effects.

FOREWORD FOR THE OZARK SOCIETY PRINTING

The Battle for the Buffalo River tells a story that ended in March 1972 when Congress authorized the creation of the Buffalo National River, essentially ending the ten-year-long controversy of park versus dam. When thinking about the history described in this book, I see a fortunate coincidence of the right man and the right time.

The Right Man

Neil Compton, who led the campaign to protect the free-flowing river, was the one person—the *only* person—who could assume that role. He was, first of all, a serious person, serious of purpose, with a creative bent. He was intensely curious about and in love with the world of nature. In college he studied geology and biology and in 1932 first visited the Buffalo River. During later years, he ventured out at every opportunity to learn more about the Ozarks backcountry.

When it fell his lot to lead the river campaign, he came as a small-town doctor, his medical practice established, his social credentials well regarded, his daughters and son grown or soon to be. His wife, Laurene, was supportive. His mother-in-law, Edna Putman, who shared the Compton home, would become a behind-the-scenes positive influence.

All of that, plus his skill in photography, would help in the campaign. At times through the 1960s Neil would show me the latest movie he had made to promote saving the river. He would use a hand-held 8mm camera and put his own voice on the soundtrack as well as music, nineteenth-century classical music (no need there to get permissions or pay royalties, Neil told me). So he added melodious flowing music to the river's canoeing sequences, and grand operatic passages for the Buffalo's majestic scenery.

Technically, Neil's home movies were a bit rough, but the photos were telling and the narrator obviously sincere. These qualities together had power to influence people.

Neil spoke and showed his movies to any group that would listen. Month after month, year after year throughout the sixties and into the seventies, Dr. Compton labored for the river, devoting not only his leisure time but also time from his medical practice. Through his efforts he attracted others to the cause—many from home in Arkansas, many also from nearby Missouri, still others from farther away. Neil conceived, and then led, a new organization—the Ozark Society—whose first purpose was to save the Buffalo River from being dammed.

This campaign was a great grassroots effort to create a new national park. Many joined as leaders and foot soldiers; together they gave the campaign more visibility, a sense of rightness, and, very importantly, critical mass. Beyond anyone else, however, Neil provided a long-term, personal commitment. No one else ever came forth to match his own combination of passion, creativity, and stamina.

The Right Time

The 1960s became favorable for promoting a park for the Buffalo, the Ozarks' most spectacularly beautiful river. The Corps of Engineers had already taken the region's best dam sites, but now they were being stopped from damming the Current River in the Missouri Ozarks, where in 1964 the Current and its Jacks Fork became the Ozark National Scenic Riverways.

Other battles were being fought and won in the sixties to create or expand national parks—in Arizona's Grand Canyon, among California's redwoods, in Washington's North Cascades. In each case—the Buffalo River and all the rest— the battle was between those who saw the area's natural resources to be used for material gain (often to benefit local interests) and those who saw the resources as having intangible, even spiritual, benefits (with the park advocates usually living outside the immediate area). In simplest terms, locals versus outsiders, exploiters versus preservationists.

The 1960s economy also permitted protecting the river. There was war in Vietnam, but then-president Lyndon Johnson had a full domestic agenda, and there was widespread confidence that all was possible. Park advocates suggested that a Buffalo National River would cost less than a Corps of Engineers dam, or not much more than the cost of a single combat airplane.

And so, as this book describes, after ten years of struggle Congress in 1972 authorized creation of the Buffalo National River. Neil Compton, as leader of the conservation forces, would become known as the park's father, joining other leaders for creating national parks, even John Muir for Yosemite.

The National River Today

Since 1972 the National Park Service has acquired the 95,000-plus acres within the park boundary authorized by Congress. The service has installed facilities such as campgrounds and boat landings, visitor centers and walking trails; provided for the display of historic sites;, and put in place a corps of park rangers and maintenance workers. The service has also developed working relationships with other public agencies such as the Arkansas Game and Fish Commission, the park's privately owned canoe-rental concessioners; and people owning homes and farms within park boundaries. And now and for the future, the service welcomes instructors and students to use the national river as a natural, living laboratory for teaching and learning.

More recently, businesses in the river area are renting out vacation cabins and opening bed-and-breakfast retreats, group lodges, and restaurants. People are buying acreage for cabins and retirement homes. Realtors are finding it lucrative to promote sale properties as being close to the Buffalo River. The park has become accepted by locals who may have formerly opposed it and is known and visited by outsiders from near and far. The Buffalo has become ever more appreciated. Loved, actually. Revered.

The Buffalo National River, however, constitutes only eleven percent of the Buffalo River's watershed. It is a strip of parkland, often narrow, with a long boundary abutting private property. The park has unwelcome intruders: bands of destructive feral hogs, invasions of non-native plants that overwhelm native species, farm wastes and eroded topsoil coming into the river from tributaries. The park service does whatever it can to deal with such problems.

As I write in 2009, the condition of our national economy hinders public spending for parks. The park service at the Buffalo River looks for more help from volunteers. A new organization of volunteers, Buffalo National River Partners, is to assist with park service programs. Another new entity, the Buffalo River Foundation, stands ready to help protect natural and historic sites within the river's watershed. And the Ozark Society continues to actively work for the river's protection.

But far greater, overarching challenges are also in view. Human-caused climate change is now visible. The world's ever-increasing human population makes unsustainable demands on the world's natural gifts: clean air and water, fertile soil, timber and minerals, living space. There are too many people using up nature's capital instead of only its interest.

As time passes *everybody* will suffer the effects. We face difficult challenges in dealing with shortages of natural resources as well as stabilizing the population and coping with climate disruptions. Considering this unsettling prospect for the future, what role can the Buffalo River fill?

It is simply that the Buffalo National River and its surrounding natural landscape can be our sanctuary, a refuge—at least from time to time—for physical, mental, and spiritual renewal. A place to be enjoyed, cherished, treasured, and defended. Get to know this river, get informed about anything threatening its integrity, and then do whatever you can to protect and improve it.

But work to protect Nature and her gifts, anywhere you can.

KEN SMITH

THE BATTLE
FOR THE
BUFFALO RIVER

ONE

THE COMING
OF CONCRETE

The discovery of how to manufacture construction adhesives was a great step in our advancing technology. Probably, observant pottery workers in ancient times came up with plaster and cement to hold tile, brick, and stone walls and floors in place. After that came concrete, the ideal material for big dam builders.

The Dam-Building Brotherhood

Dam-building technology and engineering know-how peaked in the years following the Second World War. However, significant strides had been made before that. Easily approved federal deficit spending for almost any new project had placed federal agencies in the forefront, although private enterprise had funded some of the first projects. The Bureau of Reclamation in the west, the Tennessee Valley Authority in the east, and the Corps of Army Engineers across the board were the federal agencies involved. They were favored arms of government because any congressman astute enough to obtain a big construction project for his district would be practically assured of reelection. Local promoters clamored for these developments, sensing the financial gain to be made during construction and in real-estate speculation thereafter.

With the discovery of electromagnetic force and how to generate and transmit it in the late nineteenth century, a new dimension was added to comfortable living. Falling water tumbling down from the mountains of Norway was used to drive turbines leading to the electrification of that country. Americans were envious of that luxury enjoyed by a small nation, but we had few suitable natural waterfalls beyond Niagara. The eastern half of the nation was, however, laced with many rivers with a fair gradient flowing out of the Appalachian highlands. Dams containing penstocks to deliver water to dynamos could be built on those rivers. In the years immediately after the First World War an initial step was taken. Wilson Dam was built on the Tennessee River at Mussel Shoals after some congressional squabbling. It was a special project, with no further developments being undertaken until after the Great Depression. Then, with an administration and Congress in office sworn to set things aright, the Tennessee Valley Authority (TVA) was born. Backward Appalachia would be rejuvenated. All of her rivers would be dammed, and a utopian society would emerge, its economy based upon this new technology. The TVA proceeded apace. The dams were built on every visibly flowing waterway, except, curiously enough, the Buffalo in middle Tennessee. In the years that followed, hydropower proved to be not nearly enough, its potential limited by the inventory of river systems that could be dammed. The TVA is now a federal power cartel, deriving its electricity mainly from coal and oil generating plants.

Private power companies were not nonexistent during these times. But they did operate under the constraint of having to make a profit. Consequently, their undertakings would not be as unnecessarily large, as numerous, or built in such inopportune places as federal projects were later on.

Powersite and Its Progeny

In 1911 there arrived on the banks of White River at Branson, Missouri, men looking for a good place to build a hydroelectric power dam. One site was found about three miles upstream from Forsyth, Missouri. Funds were provided by the Henry L. Doherty firm of New York, and work on Powersite Dam began that year. It was a prodigious undertaking for that time and place.

Local laborers were hired, and newly arrived Italian immigrants were brought in from New York. Along with mule teams they toiled from dawn to dusk, moving dirt, gravel, and rock blasted loose from the bluffs and riverbed. Coffer dams were built and concrete poured until the top of the barrage stood forty feet above the normal river level.

In March 1913 Lake Taneycomo (for Taney County, Missouri) filled and water flowed over Powersite Dam. The day of concrete barriers had arrived for Ozark rivers. But Lake Taneycomo was only eighteen miles long, a puddle compared to what was to follow. For the most part, it was contained within the natural banks of the river, and its shoreline was relatively stable, fluctuating only with the natural rise and fall of the river. It was not a multipurpose project, having no flood control pool; and at that point on White River there was enough water flow to take care of power demand at all times. In other words, there was no drawdown. Tailwaters below the dam were not cold like those downstream from the deep

Powersite Dam was the first dam ever for an Ozark stream. Powersite (now called Ozark Beach Dam) on White River, finished in 1913, is operated by the Empire District Electric Company. Water from the great Bull Shoals Reservoir stands on its forty-foot face.

reservoirs that were to follow, and native fish populations were not decimated therefore.

Lake Taneycomo is, in fact, a lake and not a reservoir. Its impact upon the environment was negligible except for what came in its wake. It was a great curiosity for those days and times. The Ozark Highlands are an old peneplain eroded by rivers and indented by their valleys. Here there were never any glaciers to gouge out lakebeds, no rifting and faulting to blockade the waterways. As a result there were no natural lakes anywhere. They would have set off the beauty of the hills to great advantage. Local citizens and visitors were delighted with Lake Taneycomo. Rockaway Beach and much, much more sprang up around Branson, Forsyth, and Hollister. In a short while it was a tourist mecca of significance in the whole country and remains so today.

The next river in the Ozarks to be contained behind the concrete was a smaller one, the clear and sparkling Spavinaw, down in the old Cherokee Nation, now Oklahoma, not far west of the Arkansas line. It too was no federal pork-barrel job but a public project nevertheless. Tulsa, Oklahoma, by 1919 had awakened to the fact that it was the Oil Capital of the World. The city needed a good water supply, and Spavinaw was only sixty miles away. As soon as the First World War was over, work was started on the Spavinaw Dam by the Tulsa Water

Department; by 1920 it was finished. It was very similar to the one at Powersite, but the lake flooded the valley from hill to hill. The scenic effect was charming, and people came from all around to boat ride and fish. After the dam at Spavinaw Tulsa built another at Eucha on the same stream, and now we are left to imagine what wonderful floating waters for canoeists were lost to thirsty Tulsa. Meanwhile, the upper Spavinaw in Arkansas now receives more than its proper share of runoff from hog, cattle, and chicken farms. We have been forced to look elsewhere for unspoiled Ozark rivers.

The next dam on an Ozark stream was to be no little forty-foot pile of concrete. It would create what some have called "the Missouri Dragon," a system of drowned valleys sprawling across the landscape with rampant fore and hind limbs, tail and claws clutching the adjacent hills. It would flood 129 miles of the Great Osage River in north central Missouri, covering sixty thousand acres of land. The headwaters of that drainage lie in the Flint Hills of Kansas. In its course eastward, it crosses the state line into Missouri, and in those reaches is known as the *Marias des Cygnes* (the River of Swans), named by the first French trappers and explorers to see it. As the Osage, it flows on to join the Missouri River near Jefferson City.

Bagnell Dam was built by the Union Electric Company of St. Louis and was completed in 1931. It was a hydroelectric power project with little provision for flood control. Immediate tourist development and real-estate activity were inhibited for a time by the Great Depression and then by the Second World War. But today the Lake of the Ozarks is built up around most of its shoreline.

In a reservoir of that size the environmental impact would be profound because of the loss of bottom land, forest, and farms and the displacement of native wildlife.

During the days of the dinosaurs, 120 million years ago, the western border of the Ozarks was a seacoast facing the Kansas Inland Sea. But the shallow sea filled with sediment, and the continent was elevated. The Ozark hills still rise, a low headland seen from out on the prairies of Oklahoma; their western margin is still demarcated by water. The Neosho, or more properly the Grand River, flows south out of Kansas and across Oklahoma along that old coastline. But recently the Grand River has become notably wider and deeper than any river ought to be, the result of what might seem to some an attempt by engineers to reestablish the inland sea.

Next after the Great Osage, the Grand River succumbed to the high concrete. In 1939 private funding by the Oklahoma Power Commission succeeded in completing another regional hydropower installation, the Pensacola Dam at Disney, Oklahoma. The reservoir rivaled in size the Lake of the Ozarks and would be given the sentimental name of the Lake o' the Cherokees. Prosaic natives call it Grand Lake. It, like those so far mentioned, was not built with government money, was not a multipurpose undertaking, and did not alter the regional balance much more than did the Lake of the Ozarks.

These works of man on the Missouri's Osage River and Oklahoma's Grand River were, up to the late 1930s, within reasonable bounds of economic venture. Their builders did not wish to be involved in construction or land acquisition costs that could not be recovered by the sale of the product, the energy that would be generated.

Takeover by the Corps

That was not the case with the next generation of dam builders to appear on the scene. They were the U.S. Army Corps of Engineers, highly regarded as doers of difficult tasks, especially when it came to moving the troops about and building military works of defense and offense. Their organization and execution of orders was as good or better than that of those highly efficient engineers who attended the Legions of Ancient Rome. Their people were some of the best we had. In his youth Robert E. Lee was one of them, fresh out of West Point. Their personnel to this day remain in general no less honorable, and it is not intended here to assault or demean them, either as a group or as individuals. If we judge them to be wrong in their undertakings, we must remember that the mistake has been inspired by our political system and by Congress. The Corps' reason for existence is to perform big engineering feats. It is only natural that they devise plans for them, but such plans can only be activated by legislation. In this country the responsibility ultimately falls back upon us—the voters who elect those congressmen who order the big dams to be built.

During and after the Civil War the Corps of Army Engineers gained experience in building canals and cut-offs and gradually assumed responsibility for the nations riverine problems, all outside of the military field. They snagged obstructing logs and dredged bottoms in designated navigable streams. They built levees to hold back floods up and down the Mississippi and its tributaries. Undertaking enterprises even further removed from military beginnings, they dug drainage ditches across the Everglades south of Lake Okeechobee for the benefit of truck farms and sugar cane growers. In northern Florida, they even began construction on the ill-advised Trans-Florida Ship Canal.

The formal entry of the Army Corps of Engineers into matters of civilian import came on June 13, 1902. On that date was created, under Section 3 of the Rivers and Harbors Act, the Board of Engineers for Rivers and Harbors. That board would hear pros and cons for projects pertaining to the nation's rivers and harbors and would then approve or reject them. The Corps had leeway to attend to levee construction, canal digging, and harbor dredging.

But that authority proved to be not enough. They then were not in the dam-building business. The Bureau of Reclamation, the TVA, and private power had stolen the show. The benefactors of the Corps in Congress and the officers and staff could see the many advantages of getting into the act. Thus in 1938 a bill was submitted in Congress, passed and signed into law by President Franklin Roosevelt. It was the Flood Control Act of 1938. It gave the Corps of Engineers authority to build high dams on virtually every remaining free-flowing stream of any consequence in the United States. In a very short time plans existed for every one of these. Each of these plans would remain on file in the Corps of Engineers central office throughout the years to come. If serious objection did arise to any one of them, the policy of the Corps would be to shelve the matter until favorable public officials came to power. The plan could then be brought out, polished up, and added to, its basic authority being based on the Flood Control Act of 1938.

Anyone considering this law should not overlook its basic tenet: flood control. The logic of the method was that a high dam across any running stream would transform the valley above into a receptacle to catch and hold excess water in time of flood. To be 100 percent effective, the holding basin in dry times should be empty. There should, by rights, be no conservation or power pool. Any standing water above the dam would detract from the flood control potential of the undertaking. It meant also that any good alluvial bottom farm-land above the dam would be in jeopardy in the event of any rain of consequence. In the Ozarks, the premium farmland consisted of such bottom land, the steep slopes and hilltops roundabout being generally poor and rocky. In the more stable multipurpose reservoirs with conservation and power pools, this bottom land loss was continuous and total, a permanent flood offered up to the catechism of flood control, hydropower, and recreation.

The Corps of Engineers, their champions in Congress, and supporters in the hinterland soon realized that flood control alone was an inadequate, unreasonable, and even a ridiculous purpose for such massive undertakings. Surely other features could be added to guarantee some benefit to mankind. Hydroelectric power generation was already a primary working factor with the early dam builders. It was readily adapted to the Corps of Engineers' plans for flood control projects already conceived. Municipal water supply was another "reason" added to flood control and close on its heels came recreation (fishing, camping, boating, and water skiing). Recreation was, in the end, to become the principal use, along with real-estate development around the big impoundments. Irrigation and navigation would be reasons included for special projects like the Arkansas River Seaway.

Clyde Ellis and the REA

In the Arkansas Ozarks, one such champion, Clyde Ellis, was not long in coming. To begin with, he was a country schoolteacher from Garfield in eastern Benton County. He entered into politics at an early age and was elected to the state legislature from Benton County in his twenties. In 1933 he ran against Claude Fuller, the "Backwoods Baron," then entrenched in Congress for many years. From Eureka Springs, Claude Fuller was an institution: a seasoned campaigner and master of political mores of the times. With Works Progress Administration (WPA) funds, he had been successful in getting two small recreation lakes built in Arkansas's Third Congressional District, one at Wedington in Washington County and one in Leatherwood Hollow near Eureka Springs in Carroll County. These small projects proved to be no great political advantage to Fuller. He was sure of reelection and was careless and sometimes uncouth in his campaigning.

Clyde Ellis, on the contrary, entered the contest with youthful energy and imagination. The main plank in his campaign was rural electrification. In those days everyone out in the country used kerosene lamps or Coleman lanterns, and some had small Delco systems for generating electricity. Ellis's appeal to the people was great. The New Deal was going strong. It would do just about everything for everybody, and Ellis argued that if he were elected dams would be built to generate electricity. From new hydroelectric dams, power lines would march out into the hills to distribute electrical power for one and all. In the end Fuller was defeated by only one hundred votes in the large Third District, and Clyde Ellis went to Washington at the age of thirty-three. We then called him the Boy Wonder of the Ozarks in an effort to be facetious, but it was no joke. Ellis would in due time become the godfather of the Rural Electrification Administration (REA), which had been established by the New Deal in 1936.

Ellis immediately became a key figure in the Corps of Engineers' plan to dam the North Fork of White River, called in early days the Great North Fork and by latter day abbreviationists, Norfork.

That program went off so well that Clyde Ellis, propelled by political enthusiasm, ran at the end of his second term against John L. McClellan for a seat in the U.S. Senate and was defeated. With World War Two in progress, he joined the navy and at the end of that conflict was appointed executive secretary of the REA, which important position he was to occupy for many years thereafter.

Despite protestations to the contrary, the REA is another deficit generating federal agency, a part of the Department of Agriculture. It is not analogous to private power companies, although it strives to create that impression. In the

Clyde T. Ellis preaching the big dam religion at Bentonville. Orval Faubus, *far right.*
(Courtesy of the University of Arkansas Library, Special Collections. Clyde T. Ellis Papers)

beginning, the REA derived funding directly from the treasury department. In 1973 Congress established what is called the Rural Electrification Revolving Fund, administered by the REA itself. The different units of the agency could borrow from this fund in the event of shortfall, which was frequent. The original source of money, however, continued to be the U.S. Treasury. The loans in many cases could not be paid back, and a substantial deficit ensued.

The *Arkansas Gazette* in an editorial comment made on the subject on July 24, 1984, had this to say about the then long-standing situation:

> The task of providing electric service to sparsely settled communities was as difficult as training a dog to walk on its hind legs and there are those who would argue that it was not done well. The surprise was that it was done at all.

By 1984 the REA was supplying service for twenty-five million people. They had built and were maintaining many thousands of miles of power lines and like the TVA were involved in the construction and operation of coal-fired power plants around the country. By that time the Revolving Fund owed the federal treasury 9.7 billion dollars. A few congressmen, like Ed Bethune of Arkansas's Second District, who wished to stabilize the federal debt, sought to require the REA to repay their obligation; but in 1985 that measure was voted down in Congress, and the 9.7 billion was officially forgiven. This episode is described because it illustrates the way in which such agencies are perpetuated.

The rough mountainous country of much of the Ozarks and the Boston

Mountains had failed to sustain the initial population of white settlers. The numbers of white pioneers had peaked about 1900 after which an exodus began. Between 1950 and 1960, Arkansas lost 250,000 people and one congressman as a result, but that was not necessarily a calamity. Cut over, eroded, and impoverished uplands were going back to nature. Opportunity for productive, long-term silviculture and wildlife restoration existed.

But the introduction of federally financed projects designed to bring urban amenities to the most remote hollows and faraway mountaintops has reversed that depopulation. Federally financed, paved roads and now even government-funded rural water systems have laid upon the land an urban transformation that can only intensify as the years go on.

The REA became ever more efficient in its effort to make the land conform. The clearing of power line right-of-way had always been a matter of hard labor and expense. At first, these swaths through forest and field were not too obvious, but during recent years we have seen new ways of clearing the timber. Spraying with 2,4,5–T will kill everything that intrudes under those lines, but better yet are new mechanical methods that will lay bare wide, unsightly gashes up and over the steepest slopes and in any and all directions, literally ruining once lovely landscapes. The instrument used in that case is a bulldozer with a brush-hog attached behind.

With the defeat of Clyde Ellis in the 1942 election there stepped into his shoes in Congress a gracious low-key politician, Circuit Judge James Trimble from Berryville. Ellis's Norfork Dam was abuilding, and its political popularity was not overlooked by Judge Trimble, who was destined to make history as a federal dam builder.

The Great North Fork

The North Fork of the White River was in many respects the finest of our Ozark streams. Its headwaters lie in the Salem plateau south of Cabool, Missouri. It flows on through the Mark Twain National Forest, emptying into the mainstream of White River at Norfork, a short distance above Calico Rock, Arkansas. The dolomite base of the region is soluble in ground water, and consequently many caves large and small along with extensive solution channels have been dissolved out over the ages. In those underground spaces water in large quantities accumulates, and wherever one of these conduits intersects the surface a spring is born. The numerous waterways so formed are all cool and clear. Their flow is relatively stable compared to other Ozark streams. They decline but little during sometimes prolonged summer drought in the Midwest.

They are the Current, the Jacks Fork, Meramec, Piney, Eleven Point, Black, St. Francis, and the North Fork. The mountain-born Buffalo in Arkansas is the most spectacular Ozark river because of the tall palisades and rugged tributary gorges, but its upper reaches do decline sharply in summer. In those seasons the North Fork flows on diminished but little.

The North Fork was, in addition, the best brown bass fishing stream in the country. That was proven by the regard that sports fishermen held for it. One of these was my father-in-law, W. H. Putman, whose greatest pleasure was to fish the North Fork. It was 150 miles away on all gravel roads, but when he could he didn't hesitate to make the long trip.

Tom Shiras and the Norfork Dam

In 1901 there arrived on the banks of the North Fork River at Red Bank Ford a young man, Tom Shiras, with his stepfather, Percy Gehr, a prospector. They were on their way to Mountain Home, Arkansas, to surmise the possibility of industrial development in the area. At their camp on the riverbank the next morning there was a beautiful sunrise; a flock of white cranes winged up the river; deer played around the camp; and the stream was alive with big bass. It was so entrancing that Tom told "Captain" Gehr that he was staying in the Ozarks forever. Shortly thereafter he was able to buy the *Baxter Bulletin* in Mountain Home of which he was the editor with his brother, Emmet, as co-owner. Years later his grandson, Tom Dearmore, and nephew, Pete Shiras, took over the paper and became key decision makers in Mountain Home in favor of the Buffalo National River. A simple incident that took place while the North Fork lay dying may have influenced their decision.

With the signing of the Flood Control Act of 1938 the Corps of Engineers had undertaken the impoundment of the North Fork with determination. The first requirement was positive support of the congressman in whose district the project was located. The next was the approval of the governor of the state. The next requirement was a show of support from local citizens. The local chamber of commerce would be indoctrinated by a submission of benefits, but most important was the enlistment of the support of the editor or editors of the local newspapers.

Tom Shiras was approached and was convinced that the plan for the North Fork would bring great benefits to Mountain Home and Baxter County. He espoused it in the *Baxter Bulletin,* and when the time came he went to Washington to testify before the congressional committee hearing the proposal. He is reported to have said to that body: "Down where I come from there are

Tom Shiras, the original big dam promoter, fishing the North Fork River before inundation. *(Courtesy of Tom Dearmore)*

100 thousand wild horses running loose!" A most effective statement in regard to the power in that running river.

With nothing to block its progress—no one, no individual, no group in opposition—the impounding of the North Fork of White River went on without pause.

In spite of our involvement in World War II, this fundamentally unnecessary, nonmilitary, big federal program rolled on without a hitch. The site was excavated, the abutments laid, the concrete poured, the valley behind it all laid bare, and the downed forest burned.

It was spring 1944. The dam was built, and it was time to close the gates and let the trapped waters rise. Editor Shiras, his grandson, his nephew, and their friend, Raymond Seward, had been busy through the winter with mundane things. They hadn't paid much attention to what was going on down along the river. The editor decided that the time had come to go down to the North Fork, before the gates of the dam were closed, for one last fishing and picnicking trip. With their gear aboard his car, the four passengers set forth. Shortly they came to a hill that led down into the river valley. There they were in the midst of desolation and hardly knew where they were. They stopped the car and got out to survey the ravaged landscape. The two older men shed tears, and Raymond Seward said, "I'm leavin' this country and am never coming back." That he did, moving in a few months to Muskogee, Oklahoma.

Whatever remorse that Tom Shiras may have felt at what he had seen that day was not evident in his continuing support of the North Fork Dam and Lake. He had been its champion and could not change. But we would like to think that somewhere within there was a burden of regret. Grandson Tom Dearmore said of him years later: "He liked the lake and saw it as a fine victory for economic progress, but I'm sure he missed the old river also, the original ambience of nature that lured him away from the city years before."

Aftermath of the Norfork Dam

Only three or four years before the contract for the North Fork dam had been let, the Arkansas Highway Department had completed a beautiful multiple-arched concrete bridge over the North Fork on U.S. Highway 62. It was similar to the still-existent bridge across White River at Cotter and which has now been declared a historic structure.

The presence of this then new submerged multimillion-dollar bridge was at first ignored by the Corps of Engineers, and the Arkansas Highway Department was left holding the bag. Traffic was not too heavy in those days, but something had to be done. The "Arkansas Navy" turned out to be the answer. A flotilla of ferry boats would carry passengers back and forth on highways 62 and 101. For forty years they did the job well but slowly, too slowly for the ever-increasing flow of traffic in fast-growing Baxter and Marion counties. In the very beginning some had clamored for a bridge across that lake. To have to wait for ferryboats was intolerable in our fast-moving society. But the times were not right. Although the highway department had been awarded almost four million dollars as "just compensation" by the Corps of Engineers for the drowned bridge on the North Fork, that state agency could not possibly finance a new bridge across the lake. Federal money would be required, but efforts to obtain it by the

An engineering extravaganza—the Highway 62 (now 412) bridge over North Fork Reservoir.

Arkansas delegation were halfhearted to nonexistent. But as the years went by the idea picked up momentum. The late Senator John L. McClellan got into the act along with Congressman John Paul Hammerschmidt in the House. The rest of the Arkansas delegation joined in, and finally on a nice day in November 1983 a grand ceremony took place on the bridge over bridges on Lake North Fork.

Federal, state, and local lawmakers joined residents and flag-carrying veterans during the dedication of the forty-million-dollar twin Veterans Memorial Bridges. Also present were Congressman Robert Roe of New Jersey, chairman of the House Subcommittee on Water Resources; Congressman Bill Alexander of Arkansas; Senators Dale Bumpers and David Pryor; Governor Bill Clinton; and various Army Corps of Engineers officials.

Senator Dale Bumpers said with amusing candor that the magnitude of the event, attended by two thousand north Arkansas residents, could be judged by the "many politicians gathered," a timely acknowledgment of the basic reason behind that and almost all big governmental construction jobs. In this case one led to another, a double political bonanza for concurrent officeholders over a period of forty years.

Today the Veterans Memorial Bridges take off from the hills bordering the valley of the North Fork, leap into the sky, and soar across the lake like arterials in a big city traffic stack. The Highway 62 bridge is 3,460 feet long and 44 feet wide. The 101 bridge is 2,880 feet long and 32 feet wide; its piers reaching down to the foundation rock were screw sealed into the lake bottom at depths of 180 feet.

All of this has been begotten, in the first place, in the name of flood control, now a minor factor in the operation of Norfork. Hydroelectric power is also a secondary factor. Recreation and real-estate development are far and away the most important activities emerging from what has been done to the North Fork River.

The preeminence of the Corps in federal funding notwithstanding, there occurred at the big bridge dedication in 1983 a curious gesture. Henry Gray, director of the Arkansas Highway Department, presented the Corps with a check for four million dollars. That was re-reimbursement for the "just compensation" given by the Corps to the highway department for the original bridge now on the lake bottom.

Congressman John Paul Hammerschmidt, who has often expressed concern over the national debt and needless federal spending, stated that $4 million was too much. He recommended that the highway department refund only $1,700,000. An ordinary observer of these complications might wonder what difference it made. Wasn't it just a game of musical chairs played with public money? Both the Corps of Engineers and the Arkansas Highway Department operate on federal grants on projects as big as this except for, in some cases, a small percentage of state funds.

This monetary maneuvering is related here only to illustrate how complex the web of governmental finance can become once its weavers go into motion.

We are left now to wonder about the future. To what use will the next generation put this immense revision of land and water? We wonder who will traverse that skyway and for what purposes? Will the shoreline below become the city limits for those embryonic urbanities now abuilding? Will there be any recollection of what was there before the day of inundation? Who will have ever heard of Bill Putman, Tom Shiras, or Tom Dearmore and the happy hours spent along the shady banks and on the lively rapids of the Great North Fork?

White River, The Queen of Ozark Waters

The Queen of Ozark Waters, the White River, arises in three prongs on the high backbone above two thousand feet in Arkansas's Boston Mountains. Through deep gorges it flows north out onto the Boone Stripped Plain, where it cuts down through the limestone bottom of the primordial Ozarkian Sea, now uplifted to thirteen hundred feet above current sea level.

From there it curves eastward in a grand arc through southern Missouri, receiving numerous interesting clearwater tributaries. Then it turns southeast-ward back into Arkansas on the Salem Plateau at nine hundred feet above sea level. At Newport, on the shoreline of the ancient Mississippi Embayment (the ancestral Gulf of Mexico) it flows out onto the broad coastal plain or delta three hundred feet above sea level. From there it meanders southward to ultimately join the Arkansas and Mississippi rivers below St. Charles.

It was a majestic stream, its waters clear all the way through that lowland country of muddy bayous. Its name, given no doubt by the first European pio-neers to see it, attested its clarity.

Henry Rowe Schoolcraft

In 1818 Henry Rowe Schoolcraft traveled the White River and left us a few words to tell us what it was like in those days. Schoolcraft and his companion, Levi Pettibone, traveling with a pack horse, had made their way down the valley of the North Fork to a point where it was necessary to cross that stream:

> It is so clear, white and transparent that the stones and pebbles on its bottom at a depth of 8 or 10 feet are reflected through it with the most perfect accuracy and at the same time appear as if within two or three feet of the surface of the water. Its depth cannot therefore be judged by the eye ... In attempting to ford the river where the water appeared to be two or three feet deep the horse suddenly plunged in below his depth and was compelled to swim across, by which our baggage got completely wetted. ...

Further on, in order to better descend White River itself, Schoolcraft pur-chased a dugout canoe at a hunter's camp. His narrative continues:

> The men, women and children followed us down to the shore and after giving us many directions and precautions we bid them adieu and shoving our canoe into the stream found ourselves, with little exertion of paddles, flowing at the rate of three to four miles an hour down one of the most beautiful and enchanting rivers which dis-charge their waters into the Mississippi. ...
>
> Its shores are composed of smooth spherical and angular pieces of opaque, red and white gravel, consisting of water-worn fragments of carbonate of lime, hornstone, quartz, and jasper. Every pebble rock, fish or floating body is seen while passing over it with the most perfect accuracy; and our canoe often seemed as if suspended in the air. ...
>
> Sometimes the river for many miles washed the base of a wall of calcareous rock, rising to an enormous height and terminating in spiral, broken and miniform masses, in which the oak and cedar had forced their crooked roots. Perched upon these the eagle, hawk, turkey and heron surveyed our approach. ...

Here the paths of the deer and buffalo where they daily came down to drink were numerous. The duck brant and goose continually rose in flocks before us and alighting on the stream a short distance below, were soon again aroused. ...

Very serpentine in its course, the river carried us toward every point of the compass in the course of the day. Sometimes rocks skirted one shore, sometimes the other, rock and alluvion generally alternating from one side to the other, the bluffs being much variegated giving perpetual novelty to the scenery, which ever excited fresh interest and renewed gratification, so that we saw the sun sink gradually in the west without being tired of viewing the mingled beauty, grandeur, barrenness and fertility as displayed by the earth, rocks, air, water, light, trees, sky and animated nature. They form the everwidening, diversified and enchanting banks of White River.

We in this day of the world transformed have cause to wonder about the past. What words would that eloquent observer, Schoolcraft, have used in regard to the fate of White River? Today there is no such thing; as a river it does not exist.

The Panorama Remembered

For those of us who were not avid fishermen or johnboaters, White River was a thing of sentiment. It was not considered unusual, since this land was well supplied with similar water courses from border to border. It was a place where we went with our true love on moonlit summer evenings in those youthful years. From up high on the Panorama, near Monte Ne, you could survey the White River Valley stretching away to the south until it merged with the distant hills out of which issued the river itself, a silvery current, alive and flowing in its deep meanders down below.

On the car radio there was wafted to us then the strains of the original Moon River program. It was Kreisler's *Caprice Viennois,* played softly on the organ, with a hypnotic spoken voice superimposed.

> Moon River ... Moon River
> Beautiful white river ...
> Flowing onward, ever onward
> To the sea ...

May we be forgiven our sentimentality for that lost scene, by the engineers, politicians, assorted developers, and speed-boat salesmen for whom it now means something else.

Congressman Jim Trimble, *left,* and constituent at Waldron, Arkansas. (*Courtesy of the University of Arkansas Library, Special Collections. Trimble Family Papers*)

Judge Trimble in Washington

The next congressman, after Clyde Ellis, from Arkansas's Third District, was a gentleman politician, gracious and considerate. He honestly wanted to accommodate every constituent with whom he came in contact. He had no harsh words for those who disagreed with him. He was a quiet campaigner, no source of spellbinding or blood-and-thunder oratory, but he was firm. If he was convinced of a thing, he held his ground, and for that he was to be commended and chastened.

Jim Trimble was born up at Possum Trot in the hills above Osage in a remote corner of Carroll County. There he accidentally lost an eye in a country boy reenactment of the Battle of Pea Ridge, but it did not slow him down. After graduating from the University of Arkansas, he was elected prosecuting attorney and then circuit judge in Arkansas's Northwest District. He would be from then on known as "Judge" Trimble even in the halls of Congress. He was elected to that body in 1942, succeeding Clyde Ellis before the North Fork Dam was completed. Judge Trimble was not unaware of the political advantage of that undertaking. Many citizens up and down the main stem of White River clamored for more of the same.

Then in the spring of 1948 floods plagued the area. The Corps of Engineers flew Congressman Trimble over the worst of it, the Kaw River, near Kansas City, explaining how they could control such disasters. Judge Trimble was convinced that it was the answer to high water in the White River Valley and never swerved from then on in his determination to dam the White and all of its tributaries. Before his term came to an end, Jim Trimble would become the principal champion of federal dam building in Congress.

The first great concrete barrier on the main stem would rise at Bull Shoals, 256 feet above the stream bed, creating a reservoir of 71,240 acres at flood control level and 45,440 acres at conservation pool level. Waters driving the generating units would come off of the bottom at fifty-eight degrees summer and winter. That chill would prevail all the way down to the Mississippi, a temperature in which native fish could not survive. The largest fresh water commercial fishery in the United States was over 300 miles down river at St. Charles. A few years after the gates were closed at Bull Shoals in 1951 the St. Charles fishery closed down its operation for lack of native fish in White River.

Table Rock

After Bull Shoals came Table Rock, a few miles above Branson, Missouri, and just above the head of Lake Taneycomo. The dam stretched 6,423 feet across the valley and stood 252 feet above the stream bed. It backed water all the way up to the village of Beaver in Arkansas. The James River, a major Missouri tributary and a fishing stream almost as good as the North Fork had been, was now a backwater all the way to Galena, Missouri.

In the years just before and after the Second World War float fishing had come to be a major recreational activity on the White and its larger tributaries. With Table Rock completed in 1959 that activity necessarily came to an end. With that, grumbling and words of dissatisfaction began to be heard from those who liked to catch our native fish out of their natural free-flowing habitat.

Float fishing on White River had been a sizable commercial enterprise. Its

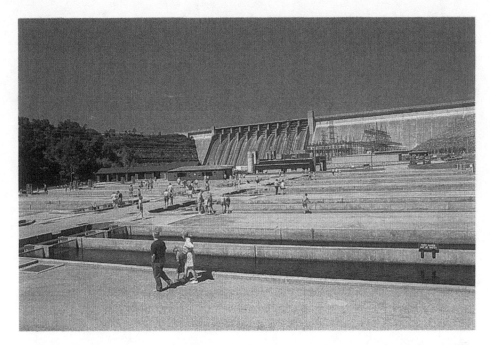

The concrete encasement of White River's valley floor, Table Rock Dam, and the Shepherd of the Hills Fish Hatchery stand as monuments to man's effort to make up for his destruction of native warm-water fish by their replacement with trout, which sometimes also perish due to surges of oxygen-depleted bottom water below the dam.

best-known operator—a raconteur, an oracle, and an unparalleled Ozarks character—was Jim Owen, owner of the Jim Owen Fishing Service in Branson, Missouri. In 1958 he was obliged to submit, via the outdoor section of area newspapers, a final word on White River. It was a general letter to his former customers. It said in part:

> Dear Fisherman Friend:
>
> This letter is in regard to a change in our plan of outfitting our float trips. We are moving this operation to Cotter, Arkansas, and our new base of operation will be the Miller Trout Dock.
>
> Naturally, I hate to turn the management over to someone else but there are two big reasons for this—
>
> The building of Bull Shoals Dam took away all of my wonderful floating water on the lower White River and the just completed Table Rock Dam has done the same thing to the floating water on the upper White River.
>
> To all of you good friends who have floated with me for the last 25 years I want to again, thank you. ...

Trout, of course, are not native to Ozark streams. They could survive, but not breed, in the cold tailwaters below the big dams. As soon as the Corps of

Engineers and their supporters became aware of this, an expensive ongoing remedy came into being. It was put into practice to mollify increasing numbers of hard-core float fishermen. The government would build fish hatcheries below Norfork and Table Rock dams from whence to replenish the trout population in the cold tailwaters in the river and in some instances in the deeper waters of the lake. Other exotic species such as white bass and striped bass did better in the impounded waters above the dams and were also introduced.

This laborious and complicated method of providing fish for fishermen has helped to decrease criticism of the loss of the native fishery, but one wonders what will happen if the system should fail, which is not improbable with the passage of time and the vicissitudes of government.

After the completion of Table Rock there still remained significant reaches of the river above it and below Bull Shoals. For them the Corps had plans.

The most convenient and least controversial of these was the upper one. It was not named for a dam-building rodent, as some might think, but for a family whose last name it was and who settled on the banks of White River early in this century at the point where the now-vanished Missouri and North Arkansas Railroad crossed the river. There the Beavers operated a country store for many years. The original plans for the dam located it at the hamlet of Beaver, but later the site was moved upstream about six miles. The village of Beaver was already on a lake, the head of Table Rock. When completed in 1964 Beaver Dam stood 228 feet above the stream below and backed water as far south as Fayetteville.

Of the three reservoirs on the main stream of White River, Beaver is the most spectacular, since it lies in close proximity to outliers of the Boston Mountains. One of these, Whitney Mountain, is eighteen hundred feet above sea level and rises precipitously seven hundred feet from the western shore of Beaver Lake, providing tremendous views either from out on the lake or from the mountain's top.

The Ozark Rivers, Dammed and Undammed

Greers Ferry Dam on the Little Red River, which joins the White east of Bald Knob, was built concurrently with the Beaver project, being finished a couple of years earlier in 1962. It was similar in size and expense, being 243 feet above the stream bed, creating a lake between thirty thousand and forty thousand acres in extent. It is said to have cost over forty-six million dollars.

These two would prove to be the last dams built on White River and its tributaries, but that was not because of a lack of plans for more.

Those on the drawing boards at headquarters in Washington were Doniphan on the Current and Galena on the James River in Missouri; in Arkansas, Water Valley on the Eleven Point; Lone Rock, Gilbert, Pruitt, Compton, and Point Peter (the last two are pump-back projects) all on the Buffalo; Hardy on the Spring River; and Wolf Bayou on the main stem of the White River near Batesville.

Wolf Bayou in its original plan would have bisected north Arkansas with a lake eighty miles long. It would have put water forty feet deep on the face of Norfork Dam and would have backed water seven miles up the Buffalo. It would have covered up forty miles of the Missouri Pacific Railroad between Kansas City and Memphis and because of that drew the ire of many previous big dam enthusiasts.

Wolf Bayou was held in abeyance for several years; but in 1978, after a series of setbacks on other rivers, the Army Engineers revived Wolf Bayou, announcing that they would spend two hundred thousand dollars on a new study of the project. Members of the Batesville business community came out for the new plan, and much publicity was generated. The revised version was much smaller, backing water only up to Calico Rock, but it still covered up the railroad. After a number of convulsions and some changes in the political scene, the clamor for Wolf Bayou died down. But was it really dead? It only proved something we knew all along: All old dam plans never die, they are just put on hold.

If they can't get a dam, then a big ditch will do. In February 1987 the Memphis District of the Army Engineers held a meeting at Newport, Arkansas, to publicize their latest program for Henry Rowe Schoolcraft's enchanting White River.

They would construct and maintain a two-hundred-foot-wide, nine-foot-deep, navigation channel from mile 10 at Arkansas Post Canal, to mile 254 at Newport; the total estimated cost, $29.3 million.

What has been said here in criticism of the overall policy of the Corps of Army Engineers is not one man's opinion. Thousands share that view, and of those there was one who deserves a quote. Leonard Hall was one of the principals in the long contest to stop the damnation of the Current and Jacks Fork rivers in Missouri. He was outdoor and environmental editor for the *St. Louis Globe Democrat* and the author of *Stars Upstream*, a moving plea for the preservation of the Current. In one of his columns dated July 1, 1964, he stated his opinion of the Corps:

> Most powerful of all the bureaus, because of the vast wealth of its many faceted lobby, is the Corps of Engineers. Starting as an agency which had river and harbor improvement under its care, the Corps got into the dam-building business during the years when we were coming out of the Great Depression and the Dust Bowl. Almost at once it saw the possibilities for building political power in damming every river in America. Here was the greatest pork barrel opportunity in history. Here was the

chance to line up the support of big contractors, cement manufacturers, makers of dirt moving equipment, real estate speculators, operators of subsidized barge lines. The Corps built its influence until today it holds power of political life and death over countless elected public officials. And it proceeds on its way in the dam-building business with no regard for the public, the facts about such supplementary values as recreation and wildlife, or sound knowledge of modern water technology.

This listing of unrealized plans is not complete. There were others, but it is proper now to enumerate those that have been completed by the year 1987. That is necessary, for anyone so interested, to understand the enormity of what has been done, not just to the White River Basin but to all of the Ozarks and the Ouachitas to the south.

In addition to the non-Corps projects at Powersite, Spavinaw, Bagnell, Disney, Remmel, and Carpenter, completed Corps projects include Norfork, Bull Shoals, Table Rock, Beaver, Greers Ferry, Stockton, Pomme de Terre, Clearwater, Wappapello, Trumann, Nimrod, Blue Mountain, Greeson, Gillham, DeGray, Blakely Mountain, Millwood, Broken Bow, Tenkiller, Fort Gibson, DeQueen, Dierks, and last, but not least, the Arkansas Seaway with twenty major dams, complete with locks, to take care of all that water-borne traffic which will, like the Cargo Cult of the South Seas, surely arrive some day.

T W O

THE CONSERVATION IDEA
TAKES WING

There have always been those who grieve at the loss of the earth's original endowment, whether from natural calamity or from our own doing. Most primitive tribes had taboos against the killing of certain species and the despoliation of sacred places. More recently, emperors, kings, and noblemen sought to stay the denudation of their countries and the decimation of wild creatures by the creation of royal preserves. That may have been based upon the enthusiasm of nobles and princes for the chase, but there were no doubt many commoners as well as aristocrats who were sympathetic and took pride in such parks.

But it would be in America that the comprehension of that sometimes delicate relationship that we call ecology would reach its apex. Unfortunately, ecology has become a buzzword in the lexicon of over-enthusiastic activists. It has been so overused as to render its true meaning obscure. It is simply the study of all living things in relation to each other and to the physical environment in which they exist.

Another related term now much in vogue and also corrupted is conservation. It has been appropriated by various agencies and organizations who use it for their own ends in spite of their lack of intent to conserve anything. Those of us who have been so intensively involved in problems of conservation would like to believe that we abide by its true meaning, which by rights should be the salvaging from unreasonable development the best of our remaining forests, prairies, rivers, lakes, seashores, wetlands, mountains, valleys, and wildlife of all varieties.

The nineteenth century, which has been called by some the century of slaughter, also saw the coming of a national conscience in the United States in regard to the subject of conservation. That was, no doubt, inspired by a series of shocks experienced by our citizenry upon the realization of what we had done, or were doing, to our marvelous new world.

One of the first voices heard was that of Henry David Thoreau from that already largely despoiled east coast in those years before the Civil War. Although he generated no organized group of followers, his observations have become increasingly pertinent with the passage of time.

A tireless investigator in the field of botanical and natural science early in the eighteenth century was John Bartram, whose travels up and down the eastern seaboard left us priceless information about the plant communities as they were in those days.

Somewhat later came John James Audubon, whose accomplishments as an artist, traveler, and recorder of avian science were incredible in his time and in ours as well. Not a conservationist exactly, his name now most properly nevertheless is applied to one of our most prestigious nonfederal conservation organizations.

Later in the nineteenth century there came to prominence on the west coast another monumental figure in conservation, John Muir. From him came the inspiration to preserve such magnificent features of the land as Yosemite and many another western park. The Sierra Club, which he organized and led, still functions as one of our most influential, nationwide, private conservation agencies.

In 1935 the Wilderness Society was founded and remains today the most knowledgeable private conservation agency on deck in Washington, D.C., to diagnose, report, and advise the citizenry of the progress, or the lack of it, of conservation measures in Congress. They have been behind the initiation of legislation to create game refuges, public park lands, and wilderness areas.

In 1936 the National Wildlife Federation came into being at a convention called by President Franklin D. Roosevelt at the urging of cartoonist Ding Darling. Now the largest conservation organization in the nation, the National Wildlife Federation is perhaps the most influential in the political arena where votes count. Its basic appeal is to those millions whose favorite outdoor activities are sport hunting and fishing. It was they or their early-day predecessors, along with allied groups such as the Izaak Walton League, who first mounted a concerted drive to stop the slaughter through the means of public law.

On the political front it would be a big game hunter who did the most for the conservation movement in the U.S.A. Theodore Roosevelt was the man, our president. It was he who set in motion the political machinery that would result in the establishment of our national parks, forests, rivers, sea- and lakeshores, botanical reserves, monuments, historic sites, and game refuges.

The shocks that inspired such action on the part of these men and the general public came throughout the span of the nineteenth century and on into the twentieth.

The first may have been felt by Daniel Boone when in 1810 he returned to

Kentucky where forty years before he had seen: "wild beasts of all kinds and buffalo had been more abundant than cattle in the settlements." During that visit Boone would wonder where it had all gone: "a few signs of deer were to be seen, but as to the deer itself, I saw none."

In that century many species, like the deer in our own eastern states, would go on into extinction; the heath hen on Martha's Vineyard, the Carolina parakeet in the southern states, the jaguar in South Arkansas, and the grizzly bear in Missouri and Kansas. A few others, the mountain lion, the eastern black bear, the southern red wolf, and the timber wolf, would linger on into the next century.

PUBLIC PARK LANDS, A BEGINNING

The environmental movement that began in the nineteenth century was not limited to wildlife management, however. The land, its forests, and its waters were matters of concern as well. But it would be one of the most fantastic displays of the dynamism of the earth's crust and mantle that would attract the attention of one and all. Once this presence was known, it would be imperative that it should be set aside as a natural preserve never to be commercialized and ruined.

The Great Western Parks

John Coulter, a veteran of the Revolutionary War and a far-ranging explorer, was the first white man to see it. He was not believed back east when his tale was told. His listeners, in wry derision, called the place he had seen "Coulter's Hell."

It was Yellowstone.

The significance of Yellowstone as a stupendous example of the earth's natural forces was not lost on the earliest white settlers in the region. One night in September 1870 a group of them camped in that region of thermal display. They discussed its best possible utility to mankind. Cornelius Hedges, from nearby Montana Territory, proposed that it should be a national pleasuring ground for all the people, a federal preserve never to be commercialized by private investors. That there should be such places in the nation had first been suggested by Henry Thoreau and Ralph Waldo Emerson. It was an idea whose time had arrived, and two years later Congress passed a bill to make Yellowstone a national preserve. The bill was signed on March 1, 1872, by President Ulysses S. Grant, creating the first national park in the world.

At first Yellowstone was administered by the U.S. Army. Then the Department of Interior took over, attempting to manage Yellowstone and recently established western parks (Sequoia, Yosemite, and Mount Ranier) from Washington, D.C. That task was difficult, and in 1916 Congress created the National Park Service to oversee these areas of national interest under the direction of one of our outstanding conservationists, Stephen Mather, for whom Mather Lodge at our own Petit Jean State Park is named.

A few years after the creation of the National Park Service many of the states instituted park systems of their own. In 1927 Arkansas became one of these. In those years there lived in Morrilton, Arkansas, a physician, Dr. Thomas W. Hardison, who knew the country thereabout. Some of it he was enamored of, especially the mesa-like Mount Petit Jean south of the Arkansas River. Dr. Hardison tried diligently to get Petit Jean with its spectacular cliffs, canyons, and waterfalls included in the national park system. It was not of sufficient size, but in 1927 the state legislature finally voted instead to make it a state park, with Sam Davies as the first director. With the arrival of the Civilian Conservation Corps (CCC) on the scene a few years later the handsome old lodge and the original sturdy stone and log cabins were built at Petit Jean, all under the supervision of national park architects.

Following that, other outstanding state parks were established in the 1930s such as Devil's Den, Mount Nebo, Crowley's Ridge, and one of especial interest to us here, the Buffalo River State Park, established in 1936.

Dr. Hardison and Petit Jean

The importance of the efforts of Dr. Hardison and the influence of the CCC program has been recorded in a booklet entitled *The CCC in Arkansas* by Frank Burggraf and Karen Rollet and furnished me by architect Paul Young of Fayetteville, who was employed on the Devil's Den project.

Its relevance to our subject is such that a summary of Dr. Hardison's report is here offered:

> The idea of a great area for public recreation in Arkansas, which at the same time would conserve one of the few remaining primeval forests of our State, had its beginnings back in 1907; but at that time we knew nothing about State parks. I simply knew that near my home was an area of surpassing natural beauty that ought to be conserved for the benefit and enjoyment of future generations. In this I had the support of the Fort Smith Lumber Company, with whom I had come to this country as physician, and who owned most of the area that it was proposed to conserve. But after thinking and talking about it more or less constantly for fourteen years, during which time I received no helpful suggestion from any source, I was as much at a loss as to how to proceed as in the beginning. Finally, knowing that it was not the proper

step to take, I asked our representative in Congress to introduce a bill providing for the acceptance of this small area—it was only 1,540 acres as I recall—as a national park.

In this account of the events, Hardison goes on to describe his meeting with Stephen T. Mather:

It happened that, as the bill was introduced, Stephen T. Mather, Director of the National Park Service, was en route to Hot Springs National Park, and our representative wired me to see Mr. Mather and try to enlist his support of the bill. After an hour's conversation with that kindliest and most courteous of men, and with his assistant, Arno B. Cammerer, now Director of the National Park Service, I was told that our area was not suitable for a national park and advised to turn my attention to bringing about its acceptance as a state park.

I had never heard of a state park, and I doubt that anybody in my State had; but I took Mr. Mather's advice, and sustained by his constant encouragement, in a comparatively short time I had the satisfaction of seeing Petit Jean State Park created by a special act of the Arkansas Legislature. ...

Though we had one state park as early as 1923, we did not have an adequate state park law until a few months ago. When Petit Jean State Park was formed, responsibility for its administration, for want of a better agency, was placed in the reluctant hands of the State Highway Commissioner; and from one agency to another, mostly reluctant and altogether incompetent, the interests of the state park movement in Arkansas were shifted with changing political administrations until two years ago, when by sheer political accident we acquired a state park leader in the person of Attorney General Carl E. Bailey, ex officio Chairman of the State Park Commission, who saw something else in state parks than material for a new sort of campaign oratory.

During his two years as chairman of the commission, Carl Bailey took more than a perfunctory interest in state parks. He visited and became acquainted with the natural advantages and needs of each park. He envisioned the twofold scope of an intelligently planned and wisely administered system of parks in Arkansas, and when he became Governor last January, one of his first recommendations to the Legislature was the passage of an act that would create authority and the means to undertake a constructive work in Arkansas.

For his significant effort in establishing Arkansas's state park system, Hardison Hall at Petit Jean State Park is named in Dr. Hardison's memory.

Thus there came to be park lands set aside for the sake of their natural beauty, their scientific significance, and sometimes their historical importance, instead of for game and fish management purposes. In addition to influential hunters and fishermen, there was now added a substantial host of birdwatchers, hikers and backpackers, campers, photographers, botanists, zoologists, geologists, artists, canoe paddlers, and plain and simple nature lovers. There now existed in the court of politics (where all matters relative to conservation are eventually decided) a quorum of sufficient numbers and influence to question and even to contest the headlong transformation of the environment by engineers, developers, sawmillers, ranchers, and plowmen.

But organized and sustained action in regard to any environmental problem in the twenties and thirties was poor. The Sierra Club was essentially a western agency; the Audubon Society was stronger in the east; the Wildlife Federation was nationwide but still in the process of growth and organization; and the Wilderness Society had not yet been formed. In the Central Highlands there was no specific citizens group to give comprehensive consideration to the many unnecessary and undesirable activities being enacted upon the land. People in the employ of federal and state conservation agencies were (and are for the most part) under compulsion not to enter into controversy that might embarrass their elected or appointed supervisors. The instigation and continuing surveillance of a good conservation program in a government such as ours must come from John Q. Public. That is what would happen, almost too late, in the case of the Buffalo River.

Eastern Parks

The establishment of the great western national parks around the turn of the century attracted the attention of the nation. We were proud of these superlative natural features and were glad that they enjoyed the recognition and protection of the federal government. America had set an example for the rest of the world on the proper way to harbor the best that nature had bestowed. The mechanics of their creation had not been difficult. In those western states were vast expanses of federally owned land. No private landholders had to be considered in Yellowstone. Presidential decree or an uncontested act of Congress was sufficient to beget most western parks in those early days and even to the present time.

But there were many citizens in the east who felt that their region had been overlooked. If we were to have national parks, the east should not have been left out with its incomparable seashores, mixed hardwood and evergreen forests, lush wetlands and distinctive ranges of hills and mountains, great freshwater inland seas, and running rivers without number. But here there was no limitless expanse of public domain. Different methods would have to be devised if national parks were to be designated in the east.

The first of them was created without fanfare in 1919, the Acadia National Park on the rugged coast of Maine, the land donated by prominent citizens of the northeastern states.

Then in 1921 Hot Springs, Arkansas, achieved national park status. It had been declared a national reservation as early as 1832, but there were no significant guidelines for its administration, no specific laws or legislation for true conservation; and Hot Springs grew up as a commercial spa, the presence of its considerable thermal emission being obliterated by intensive urban development.

The first concerted effort to establish a very large national park in the east began in 1923. Businessmen in Knoxville, Tennessee, and Asheville, North Carolina, conducted an organized effort to have a significant part of the Great Smoky Mountains included in the national park system. The land was in private ownership, and the U.S. Park Service was not in the land-buying business in those days. The legislatures in the two states appropriated large amounts, private donations (some even from school children) were solicited, and finally the Rockefeller family declared a grant of five million dollars to make up the final half of ten million dollars, the needed purchase price. After congressional approval, the Great Smoky Mountains National Park came into being in 1934.

It was followed the next year by the Shenandoah National Park in Virginia, Isle Royale in Michigan in 1940, Mammoth Cave in Kentucky in 1941, and Everglades in Florida in 1947.

The Ozarks and the Ouachitas

Those of us who lived in the Ozark-Ouachita uplands noted these developments with interest. Our Interior Highlands were extensive, over 500 miles from northeast to southwest and 250 to 300 miles across. Nowhere did its crests exceed twenty-nine hundred feet above sea level. We were inured to our mediocrity, both in regard to our people and the land. We had nothing to compare with Pike's Peak, the Grand Canyon, or the Great Smokies. We were country folks and always would be. The rest of the nation had let us know that, and we accepted it with laconic, hill-country humor. We would go our way, making what we could from the thin soil of the flinty hilltops and the sandy ridges. In the narrow valleys there was some good alluvium, and there were the prosperous farmers of some of our citizenry.

But it was as it should be. Wild game in any quantity had long since been eliminated. The great forests of hardwood and pine had been cut over, and the remaining sawmillers scrounged the second and third growth for a two-by-four. Agriculture had gone through cycles: first, the production of grain—wheat, oats and corn—out on the flats; then the big boom in tobacco right after the Civil War. Next was the production of fruit: apples, strawberries, grapes, and peaches. But it was a hard and unpredictable way. People were leaving the Ozarks country by the thousands. Texas, California, and Oregon beckoned; and there they went, leaving the old hills vacant and lonesome. Then nature laid her healing hand upon the Ozarks.

The tide of empire had flowed around this low, deeply dissected plateau. Nowhere in its reaches were there great cities. Springfield, Missouri, was the titular capital with seventy-five thousand people in the 1930s. In those days the Ozark-Ouachita region was a curiosity of sorts. It was an arrested frontier, recalled with nostalgia by many of us who knew it well. To us the exodus that lasted from 1900 to 1964 was no disaster. It was a natural adjustment to early-day exploitation, a trend that should have been allowed to proceed without governmental intervention.

My father, David Compton, summed it up in a sentence. "That's a place that should never have been cleared," he would say when we passed some worn-out, abandoned farmstead back in the Spavinaw hills.

In the days immediately after the end of World War II there were those long, winding, dusty roads that led us back over the hills and mountains through Dongola, Snowball, and down to Eula on the Richland. There along the old wagon road was a scattered assembly of pioneer dwelling places behind their rail and picket fences, with the yellow rose of those early comers and the burning bush copses out in front. Some were of log construction showing the meticulous axe work, reflecting long hours of labor that went into the dovetailing and precise fitting that only those old-time builders knew how to do. Some were sawed lumber and all in the dogtrot style with rusty but enduring metal roofs. Inside of a few there would be mementos of yesteryear; old coffee grinders, glass mason jars, hickory-bottom chairs, and sometimes even a spinning wheel.

For me it was a fascinating, not-to-be-repeated experience to stand there sensing the human melodrama that had transpired here on the lower Richland that joined the Buffalo a short distance to the north. What were their names, their hopes and fears, their trials and tribulations that had caused them to pick up and leave? What had their fortunes been in Harrison, Springdale, Odessa, or Bakersfield? Time and tide, the storms of spring and summer, would level those old houses, barns, and fences in a few years and little would remain.

The land was vacant and ready for some other use for man to impose upon it.

The first move to establish a national park in the interior highlands came not in the Ozark region but down in the Ouachitas. In the 1920s citizens in the town of Mena in Polk County, Arkansas, being aware of the ongoing effort to establish the Great Smoky Mountains National Park, organized a similar proposal to be located in western sections of the Ouachita National Forest. A. W. Dodson, a leader of the move, I came to know well. He was superintendent of schools in Mena and was active in the chamber of commerce there.

He and his coworkers had ready support of the local papers and the state news media in Little Rock as well. Guy Amsler and members of the Arkansas

The Bill Henderson cabin on the Center Point Road, Upper Buffalo, as it was in 1949, now long since fallen in.

Game and Fish Commission lent support, and the Arkansas legislature endorsed the proposal. It had the unanimous backing of the Arkansas delegation in Congress where a bill for the creation of the Ouachita National Park was passed without difficulty. But the National Park Service was not too eager for such an acquisition. The Ouachitas resembled—but just weren't as big or as impressive as—the Great Smokies. Besides that, President Calvin Coolidge knew little or nothing about the Ouachitas and cared less. He pocket vetoed the bill, and the U.S. Forest Service offered as a consolation measure recreation areas at Bard Springs and Shady Lake and Albert Pike campground.

However, a small segment of the proposed park did achieve national protection under the Wilderness Bill of 1974. The Caney Creek Wilderness area in the Cossatot Mountains was created by Congress in that year with its supervision in the hands of the U.S. Forest Service.

A COUNTRY BOY COMES OF AGE

In those distant days before the Great Depression, those of us in late grade and early high school in the flatland town of Bentonville began to hear from newly arrived companions tales of their mountainous homeland to the south and east. The Mayeses, Joneses, Yateses, Rushes, Cornetts, and Barbers had departed that rugged country for various reasons. A great exodus was in progress from that highest range of the Ozarks—the Boston Mountains.

Boys like me from Falling Springs Flats were intrigued by the tales that Rex

Barber from St. Paul and Audrey Yates from Mount Judea (pronounced Judy) could tell about the high mountains, rushing rivers, great cliffs, and hidden waterfalls that abounded in Newton and Madison counties. In this boy there was born an abiding desire to see all that if it was really there.

A first view of that mysterious range was not long in the offing. It was to be from the train. In the spring of 1924 my father, David Compton, newly elected county judge of Benton County, was obliged to go to Memphis, Tennessee, to look at road machinery for the county; I was delighted to accept an invitation to go along. We boarded the Frisco and headed south to pass through the tunnel under the Boston Mountains' crest at Winslow. The weather had been cloudy and wet, and rumbling on down along the banks of Frog Bayou we could look up and see those clouds impaled on the wooded mountain tops a thousand feet above. What an exciting revelation it was to a one-time dweller on Coon Creek, whose bordering hills, scarcely higher than one hundred feet, never wore such cloud caps.

Then there would soon be a view of that water-carved table land from another vantage point—up on top.

That summer there was to be a big barbecue down at Winslow. Its purpose was to get together those people involved in plans then underway to pave Highway 71 from Bentonville to Fort Smith. County judges were a factor, and again I was to accompany my dad on a day-long expedition to Mount Gaylor. This time we mounted the county car, a wide-open Model T roadster, and chugged off down the then all-gravel Arkansas 71 to the big jamboree up and above everything in the country around. While there it was possible to wander over to the rim of Mount Gaylor where one could see out and down into Winfrey Valley to the east. It was that day a blue abyss that seemed to have no bottom, a breathtaking sight for one from our modest Sugar Creek hills. Off on the distant horizon were the wrinkled lineaments of Pilot Knob and White Rock Mountain, although I knew not their names that day. What an exciting thing it was to realize that in Arkansas we had such mountains. If they were not as immense as the Rockies, no matter; their impact upon one's vision was the same.

Yet another phase of our Boston Mountain grandeur was to be revealed a few years later. We were on our way to the Arkansas State Fair in Little Rock in our Model A, cloth-topped roadster. From its rumble seat one was afforded a full view of everything. In the White River Valley south of Elkins there was cloud and drizzle all the way up to Fly Gap on Highway 23. But there in that pass we emerged above the clouds and beheld a stunning panorama. It was as if we were on an eminence overlooking an arm of the ocean. The cloud bank filling the valley of the Mulberry below stretched away southward to cover the wider

Arkansas River Basin. It was as if the land below had been engulfed by the sea— a vision of Eden to be remembered for a lifetime.

Within a few years it became possible to study that landscape in 3-D miniature scale. My inclination as a student in the University of Arkansas was toward the natural sciences. The fundamental subject was geology, and there on the fourth floor of Old Main was a fascinating seven-by-seven-foot relief map of Arkansas. There you could stand and survey that surface assigned to us in all its detail. There were the ancient and wrinkled Boston Mountains stretching from Bald Knob, Arkansas, to the Oklahoma line and sixty miles across at the waist between Harrison and Clarksville.

There were the river gorges eroded into that old plateau, where born on its divide were the three prongs of White River, the War Eagle, Kings River, the Buffalo, the Richland, and the Little Red, all on the north slope. On the steeper south slope coursed Lee Creek, Frog Bayou, the Mulberry, the Big Piney, and Illinois Bayou.

On that map there was one deep cleft that attracted the attention of this curious freshman more than the others. It was the gorge of the Buffalo River. Its course north through the Boston Mountains was clearly shown, and its exit from that range between two high ridges was there to see.

To survey that map was to be filled with a desire to go out and see it all firsthand. But in those days there were no bands of outdoorsmen to join, tramping the distant hills or cruising those rapid streams. Roads were rough and difficult, hard on those old Model Ts without steel-belted tires. To go by horse or mule was not my thing. We had no kinfolk in places like Erbie, Iceledo, or Dongola with whom to visit. For a lone teenager to go awandering in that land in those times would have been foolhardy. Being no adventurer, I stayed with the boys in Ma Cornett's boarding house, attended classes, and awaited the day.

On the Buffalo with Stan Hayden

That day came surprisingly soon—the next summer, August 1932.

We had then in Bentonville a Methodist minister, Stan Hayden, a slight, tousle-headed, intense fisher of men and fish as well. The Haydens had originated over at Boswell on White River, and Stan Hayden knew that country well. One of the tributaries out of the White River in that region was noted for its smallmouth bass. It was the Buffalo, and there Stan Hayden was determined to go and try his luck. He would do something unusual for those days. He would float the Buffalo in a homely johnboat for several days on end, but he needed a couple of crewmen. Owen Hayden, whose father had recently died in Detroit,

Neil Compton, *left,* with the Rev. Stan Hayden, on the Buffalo, 1932, amid the gear of the day and time: tin buckets, bushel baskets, lard cans, cigar boxes, and canvas army cots.

Neil Compton at Dillard's Ferry (Highway 14 crossing), at the end of the johnboat float with Stan Hayden, August 1932.

had come to live with his Uncle Stan and had enrolled in the U of A. Along with me he was a denizen of Ma Cornett's boarding house. When Reverend Hayden got the call to go fishing that August, he asked Owen and me to go along as boat handlers. Thus came my introduction to that mysterious Buffalo River, deep in the Ozarks.

Stan and Owen drove up in Stan's green Chevy touring car that summer morning in 1932, and how we managed to load all our gear aboard I do not recall. There were boxes, baskets, tubs, old beat-up suitcases, and folding canvas cots. It was worse than what was yet to come with wives and children during the Battle of the Buffalo in the 1960s.

Stan headed east to Rogers and then north on 62, which was being black-topped to Gateway. From there it was already surfaced into Eureka Springs. But from Eureka Springs it was mostly gravel upon which we ground along through Alpena, Harrison, Bellefonte, Western Grove, and St. Joe. Finally, we descended a long hill into a valley through which flowed a pretty, clear river. It was spanned by a new, gracefully arched concrete bridge. Crossing the bridge, Stan pulled up in front of a shanty perched on the east embankment above the river. Its occupant was resoundingly named Claude Rainbolt; he was a manufacturer of johnboats, which proves that the Buffalo must have had some utility as a recreation stream even in those days. From him Stan purchased a newly put-together johnboat for five dollars. Its lumber wasn't fully seasoned, and it was as heavy as lead, but it was twenty feet long and would hold all of us and all of our gear without trouble. For propulsion there were two homemade paddles and a long hickory pole to "pole" it with if we got in a hurry. Here we left Stan's Chevrolet to be moved down river to Dillard's Ferry, which somebody said was fifty miles. By river it was twenty-five or thirty miles, but it seemed like fifty by the time we got there.

That evening we camped a short distance down river within sight of the bridge on a gravel bar across from a fifty-foot bluff that overlooked a big, deep hole of water, a perfect place for swimming as we were to find out. After camp was made and supper finished, we bedded down for the night in the open on our canvas cots under the stars. Along about midnight we were awakened by loud voices from the big rock across the river. That night our man of the Lord had to listen to a vocabulary that was not from on high but from St. Joe, which was as low as you can get, so we learned. That evil recitation has echoed in memory down through the years and refuses to be cast out. It was arranged like an old-time Burma Shave ad with one line following another from different participants, the first of whom ascended the bluff and announced for all to hear— which was us.

"St. Joe, Arkansas!" and splash! into the river he went. His successor added a line.

"The asshole of creation!" and kersplash! in he went. The third ascended the rock and delivered the punch line.

"Lowlier than whale shit nestling on the bottom of the ocean!" and kersplash! into the dark river he went.

With that as an introduction, we set out the next morning down the beautiful Buffalo River, alternately poling and paddling along. We passed under the Missouri and North Arkansas Railroad bridge which was still in use then and upon which was perched a couple of citizens who waved us good-bye. Except for one other, they were the last humans we were to see on the rest of the trip. The other one startled us that afternoon while we coasted along the forested shore. He was an old man with a long beard down to his waist who came running down the mountain with a tow sack, which he dipped in the river, and then ran urgently back again. It seemed obvious that something was afire up there and that he was trying to extinguish it. We wondered what, a still perhaps, but it was never revealed; and we drifted on down the river that was to become the subject of such controversy for me thirty years later.

Stan Hayden fished and fished, from the boat, from the gravelly banks, and from out in the water up to his armpits, but in the whole time he caught but one small brown bass. Meanwhile, fish swam under our boat, easily visible in the clear water, by the millions it seemed, mostly brown bass but all of small size. Since one of the days we were out was on Sunday, I suggested that there might have been divine disapproval for a preacher to be so engaged.

We drifted on for two and one-half more days, camping two more nights, shooting pleasantly through some riffles, and hanging up in the gravel on others. Then Owen and I would perform our appointed tasks, pushing and pulling mightily to get the heavy slab-sided johnboat on through, while Stan fished away to no avail. This was the way that those early adventurers, French fur traders, and English pioneers explored this country, in pirogues and dugouts, coming up the rivers so laboriously. It just wasn't the way to travel for such tenderfooted fellows as Owen and I. At the end we were tired, cross, sore, and sunburned and bereft of any ambition to repeat such a journey.

At last we drifted around a bend, and there on the left bank was a welcome sight, Stan's Chevrolet waiting to carry us on the next leg of this tour of the deep Ozarks. We were at Dillard's Ferry, exactly where the high bridge on Arkansas 14 stands now. One of the Dillard boys was there with the ferry boat ready to go. It was attached to a cable across the river and received its power from the force of the current, which was manipulated by the angle of the ferry boat against the flowing water. We loaded what was left of our stuff and headed on east toward the ancestral Hayden home. The road wound around and about through the forested hills. We went through such improbable hamlets as Harriett, Big Flat, and Fifty-Six, finally coming down off the ridge into a valley with the melodious

The Hale Hayden barn near Boswell was still in use in 1973, forty-three years after Neil Compton's first visit there.

name of Sylamore. Here we turned up the south bank of White River and in a few miles arrived at Boswell, which place I do not recollect. We were at the home of Hale Hayden, Stan's brother, where we were to stay for a couple of days.

Hale was a sort of overseer for big property owners down in Batesville. They operated big cotton plantations this far up White River, which at this point has a fairly wide valley. White sharecroppers lived out on the plantation with Hale Hayden serving as a supervisor. The area was in a condition once applying to the whole of the Ozarks but which had not been true of Benton and Washington counties since before the turn of the century; the building of rail fences and log cabins, the manufacture of homespun clothes, and freedom from shoes were things that had disappeared from our part of the country. The truth of the survival of the pioneer way was inscribed upon my awareness by a homely episode, the recollection of which still brings back the harsh memory of how rough it was.

On the morning of the first day at Hayden's, with necessity making itself known, I approached Hale Hayden and made respectful inquiry as to the where-abouts of that private sanctum that all farmsteads in my part of the country were provided with. He appeared taken aback and nonplussed. Finally he said: "Well, thurs one out behind the garden, but we don't use it much, hit was built for the womenfolks."

After prowling around in the garden for a while through an increasingly impenetrable jungle of big ragweeds and pokeberry bushes, I found to my dismay the object of the search. It was surrounded by a thicket of blackberry bushes and covered with poison ivy vines. If it ever had been of any utility, there was no evidence. It was no safe place for man or beast and could only harbor such things as wasps, spiders, and copperhead snakes. Not wishing to present anything whatever to these venomous creatures, I retreated with nervous anxiety from the garden to the front porch where Mr. H was taking the morning air in a hickory-bottom chair. After having been offered a description of conditions behind the garden, he looked a way off down White River and said thoughtfully:

"Well, we allus go to the barn, they's plenty of cobs."

To the barn then I did retire, watching every step, for it served as a facility for dogs, hogs, cats, horses, and cows as well as humans. That place was alive with high jumping fleas, which made the experience almost unbearable. Furthermore, one was overcome with misgivings about the condition of the cobs. Were they new or used ones? On account of their shade and texture you really couldn't tell. There was no box or bin to indicate which was which, and they lay scattered about around the barn for one to pick and choose with the outcome being left to blind fate or lady luck. My folks had often said that Neil was a finicky kid, something that I had always felt to be unfair, until this—the moment of truth. Having taken my chances on recycling something and having borne up like a man to the last scratch, I returned to the house trying to display that quality once ascribed to me as a compliment by Oscar Cornett, Jack's dad. He said in a discourse on the various boys in the Ma Cornett boarding house: "Neil has poise."

Arriving home sunburned, tired, and sore from all that paddling, poling, and wrestling the heavy johnboat, there was no abiding desire to go back and do it again … no yen to save anything that we had seen. It had all been there forever and would continue as it was on until the end of the world. We weren't proud of the rude image ascribed to us here in Arkansas by the outside world back then. But I now knew that in the heart of our country it was true, no matter if we did live in the Land of the Big Red Apple and the Land of a Million Smiles with a new paved road down the middle and electric lights long since. Now I knew that back yonder in the hills some folks didn't wear shoes.

This journey to the unknown Buffalo did, however, implant the seed of interest in our native land so that for me from then on honest comparison was sought between our rivers, forests, mountains, and prairies and those in other parts of America and the world. And for our people a feeling of not so much pride but of sympathy and understanding for their unpretentious manner, their honest approach to the uncertainties of life, their wry and whimsical method of expression, and their tried-and-true moral values.

The View from White Rock

Further realization that the Ozarks do possess charm and grandeur of significant magnitude came in 1941–1942 when serving as county health officer for Washington County. My duties lay within the western province of the Boston Mountains around Winslow, Weedy Rough, and Evansville. Dora Deen, one of our staff nurses, could not say enough in praise of the scenery from White Rock Mountain, a principal peak down in the Ozark National Forest. The Second World War was on by then, and my application for a commission in the Navy Medical Corps was in process. A visit to White Rock was a must before departure for parts unknown.

One Sunday in early May my wife, Laurene, and I made the journey over precipitous forest roads across the narrow isthmus connecting Pilot Knob with White Rock Mountain. The latter eminence, rising to twenty-three hundred feet above sea level and nineteen hundred feet above the Arkansas River Valley to the south, is surrounded by tall, vertical, sandstone cliffs. From its southwest overlook one could see Mount Magazine across the Arkansas River to the southeast, the peaks of the Ouachitas on the southern horizon, and Poteau Mountain out in Oklahoma. It was as grand a landscape as one could hope to see anywhere.

As we started to depart for home in Fayetteville, Laurene looked out at the bluff line and declared that there were azaleas growing among the rocks. My response to her observation was that it couldn't be. They didn't grow in Arkansas. In order to verify that fact, I got out to inspect the strange plant at close range. It was indeed an azalea, beautiful rosy pink, with a delightful clove-like fragrance, a wild, hardy native whose presence in the Ozarks I had, until then, never suspected. They grew all around the rim of that soaring mountain top, some in the woods, but the best and brightest perched in crevices and crannies on the towering cliffs, a fitting diadem for this Olympus of the Boston Mountains.

Shortly thereafter I came again for a final look at this heavenly eminence, and with me came Willis Campbell, that paragon of exuberance and enthusiasm and the then secretary of the Fayetteville Chamber of Commerce. We stood on the breathtaking overlook and wondered what other thrilling sights might lay unknown to us beyond Potato Knob, the Black Mountains, and those distant eastern ridges.

Willis Campbell could only say with a great flourish of both arms: "It just goes and goes … and goes."

It was a landscape to be etched in memory, a vision to be carried on through the war, a place to return to if fate would permit.

HOME FROM THE WAR

Out on the island of Guadalcanal, ensconced as lord and master of a small navy sick bay at Koli Point, I awaited the day when, if spared by fate, there would be a homecoming. One thing that I would do then would be to drive up that road to White Rock and then on to Cass, Oark, Mount Judea, and beyond to see what mysteries lay beyond White Rock and Weedy Rough.

That opportunity would arrive almost before we in the military were ready for it. Suddenly, it was April 1946, the war was over, and I was out of the navy. Laurene and I were in Little Rock after the arrival of a son and heir in those days when all maternity cases stayed in the hospital for two weeks. It was necessary to return to Bentonville alone to attend to business, and upon arriving at Clarksville, I suffered an urge to cross the Boston Mountain range at its widest girth. I turned the blue 1940 Ford north on steep and winding, all-gravel and dirt #21.

It was a trip through yesterday: rail fences, log cabins, one-room country school and churches, ploughmen working fields and gardens, and little country stores at Ozone and Fallsville. Beyond the Mossville church the road plunged downward endlessly from the oak and pinewoods into a forest of great trees, some of them unknown to me. Near the bottom there were wild azaleas in bloom, like the ones on White Rock Mountain four years before, a reason to stop for a moment of admiration. There, working on the rutted road, was Jim Lewallen, who sought to relieve my ignorance. "They was honeysuckles," according to him.

Growing nearby in a damp cleft was a small spreading tree with leaves two feet long and big, saucer-sized, white flowers. That was a "cowcumber tree" he had me know, one of our magnolias, I would later learn.

At the bottom I had come to the narrow and sheltered valley of the upper Buffalo. The old road crossed the stream on a rusty iron bridge and turned to the north passing a few fields behind sandstone fences and bordered by ancient beech and sweet gum trees in places. There were thickets of redbud, dogwood, and wild plum along the way. To the west loomed the red sandstone ramparts of Moore Creek Bluff and Winding Stair Mountain, shutting out the world beyond so recently disrupted by the great war.

In the midst of that peaceful scene, in plain, white-painted elegance, the Boxley community building and church stood guard by the cemetery.

Below Boxley a grove of tall pines marched down to the road, from which the waters of the brawling, newborn Buffalo could be seen in a few places. Along the way were scattered well-tended farmsteads with their assorted dwellings and outbuildings, some of more recent 1920–1930 bungalow style, and others of

Dick Murray unloading the International Scout at the Lost Valley Lodge in Ponca, October 1967.

older, two-story, end-of-the-1880s vernacular. Those homes, along with their functional barns and outbuildings blending well with the landscape, attested to the presence of the self-sufficient yeoman farmers whose ancestors had settled here over one hundred years before.

For me it was fulfillment of those dreams of homecoming during the long months on Guadalcanal. I had come to as lovely a vale as could be found on earth, a place that Willis Campbell and I knew must surely be there somewhere beyond the Black Mountains. It was Shangri-La, the Shalimar, the Vale of Tralee. It was a place to restore the troubled spirit of any man. For it, great soaring passages of music had been written. For such a scene in the Alpine hinterland Beethoven had set down the score for the *Pastoral* Symphony. For such music Boxley Valley was that day no less an inspiration, and the strains of that music were in my ear on down the road to Ponca.

For me it was a time and place for sincere thanksgiving. For so many of my friends and classmates there would be no homecoming. …

For the Lincoln boys, Tommy and Jack (distantly related to Abraham), lost in the air force flying the Hump, and on the ranges of New Guinea …

For Iggy Knott, snuffed out in his parachute harness in Normandy on D-Day …

For my wife Laurene's uncle, Carl Swift, who looked like his dad, Dr. Charles Swift, down in a B-24 off Rabaul …

For Dickie Knott, a young genius from Bentonville, shot down by his own men one dark night in Italy by mistake while returning from a reconnaissance assignment …

For A. J. Yates, a Bentonville Tiger and later an All-Star Razorback, killed in a wreck in North Africa …

And for Dr. H. C. Baker, Jr., Lt. MC, USNR, a class behind me in medical school (and a brother-in-law of Congressman Clyde Ellis), now on the bottom of the Pacific Ocean after the great typhoon.

Ponca was a minuscule village hidden deep in the hills and devoted then to the business of mining whatever zinc that could be gouged out of the mountains roundabout. There was McCarthy's store and lodge where I stopped for a cold soda pop and a visit with Mr. McCarthy, who had arrived there in 1916 during the First World War when zinc was at a premium and mining activity intensified. McCarthy had built the store and lodge of rounded, river-worn cobbles. The store, a low one-story structure, was full of country store merchandise, many antiques, and other artifacts. The two-story cobblestone lodge sat back in a grove of tall cedars and had been built as a hotel for miners and transients. There was a distinctive charm and atmosphere about this place named Ponca by McCarthy, who had come from Ponca City, Oklahoma, to begin with. That the lodge would become a principal rallying point for us in the long struggle for the fate of the Buffalo River or a pleasant refuge in the hills after the building's purchase by my son-in-law and daughter, Curtis and Ellen Shipley, and good friends, Roy and Butch Clinton, was beyond comprehension in 1946.

THREE

THE LEGACY
OF LOST VALLEY

At about that same time the charm and beauty of Boxley Valley had drawn the attention of others, and one of them would implant the seeds of a great and impossible idea which would nevertheless one day bear fruit. His name was Glenn Green, also known as Avantus or Bud Green, and he was publicity director for the Arkansas Publicity and Parks Commission.

In early 1945 the *National Geographic,* wishing to do an article on Arkansas, contacted Green for assistance. He in turn called upon Tom Millard, soil conservationist of Harrison, who knew rugged Newton County very well. They went to Cob Cave in Boxley Valley, a great yawning, open-mouthed cavern that could only be reached by crawling through a water-carved opening in a limestone bluff.

Upon arriving there, Avantus Green was overwhelmed by the spectacular bluff, by Cob Cave (named for the many thumb-sized corncobs left on its floor by bluffdweller Indians one thousand years before), by the glorious waterfalls, and by the magnificent mixed hardwood forest that filled the valley.

Green wrote several articles describing it for various newspapers and renamed the place Lost Valley. His reports on Lost Valley were published over the next few months in the Sunday magazine sections of the *Arkansas Gazette* and the *Arkansas Democrat.*

Later in the year Green led newspaper reporters and even two contingents of army personnel into Lost Valley. As a result, in 1946 the *St. Louis Globe Democrat* did a full-page feature story on Lost Valley.

After having seen Lost Valley, Fitton's Cavern, Hemmed-In Hollow, and other scenic areas, there came from the mind of Glenn Green the idea that some of Newton County's scenery was worthy of national park status. He was the first to say so and deserves the credit for striking the spark for what was to follow.

Having read his newspaper description, and having that same year driven all

Glenn Avantus (Bud) Green was
the first to publicize the idea that
the scenery in Newton County was
worthy of national park designation.
(Courtesy of Susan Green Hartsfield)

too hurriedly through Boxley Valley, I determined to go back to see what I
might have missed. Along with my father and a couple of friends, a first pilgrimage was made into Lost Valley, through the tunnel cave, under the great
precipice, and into the yawning overhang. Sure enough, there were still the tiny
corncobs of the ancients scattered about. That day I realized that we had traversed a forest unlike the familiar oak/hickory elsewhere in the Ozarks. The
great smooth-barked giants were beech trees and mingled in were tall cucumber
magnolias. It was a relict stand of Appalachian forest hidden in this Ozark
recess. Lost Valley was better than Green had been able to describe, its reality an
inspiration to any visitor, native son or not.

A Word with Jim Trimble

Being by now convinced that Green's idea was a good one, I decided to
take some action if opportunity presented. That came about a month or two
later. Congressman Jim Trimble, who was just getting started as a big dam
builder, came to our Rotary Club in Bentonville for a talk, and after it was over
we had a discussion in Applegate's drugstore. Judge Trimble was very sympathetic with the national park idea.

But not wishing to be a serious party to such an unorthodox proposal, he sent me up a blind alley. He recommended that I go to see Jerome (Jerry) Dahl, the superintendent of the Ozark National Forest. That I did in all innocence on my next trip through Russellville. The National Forest Service and the National Park Service were not at war with each other, but their interagency relations were, at best, indifferent. Jerry Dahl listened politely and dismissed me without any suggestion as to where to go from there. He had to deal with eager beavers who didn't know the score from time to time and tolerated the interruption as best he could.

I went on to Little Rock to serve a two-year residency in order to refurbish my rusty talent as a physician after those years in the South Pacific.

Surely Avantus Green, from his important position as publicity director, could generate enough help for the national park proposal to bring it off if it was at all possible.

The Return to White Rock

Summer came and it was time for a return to White Rock to see how it had fared during the war. Laurene and I took the road through Locke, coming in from the southwest. That approach when last seen as one neared the mountain passed through a lovely arcade of white oaks. But that day in 1946 we were sickened by the sight of their dismembered tops lying in chaotic heaps along the road, left there by loggers who could care less about the beauty of the mountain now that they had their due in whiskey barrel staves. In our hearts were deep misgivings about the National Forest Service, whose domain it was, and who had made the timber sale. How could they, as custodians of this scenic treasure, have been party to what in our minds amounted to no less than senseless depredation, done for a pittance, leaving no one enriched?

Driving home that day there was in mind but one thought: Why doesn't someone do something about such an outrage?

Not realized at the time was an obvious fact: the thinker of such thoughts must be the doer if he is true to his convictions.

The Buffalo River State Park

In Little Rock, doing a residency in obstetrics, I renewed my acquaintance with dapper Dr. Alan Cazort, an allergist, who would try to rectify a hay fever problem for me. He was from Ozone up in the mountains and loved that country as well as I. He described his many visits to the Buffalo River State Park, a place that I should have known about but had overlooked. It was just downstream from Dillard's Ferry where Owen and I had come off the river with Stan Hayden in 1932.

Dr. Cazort's description of the place could not be resisted, and one day in 1948 Laurene and I and our three children loaded up the family sedan and repaired to that far realm on the lower Buffalo.

The Buffalo River State Park was then run by a Mr. Hill, a thin and nervous man. His anxiety was especially related to his neighbors, some of whom did not wish a park near their domain. To demonstrate that, they made use of an ancient weapon, fire. Some local landowners would set the nearby woods ablaze almost every spring when the wind was from the west so as to carry it on into the park lands.

That did not happen in 1948, and we were shown to our quarters, one of a series of beautifully designed cabins constructed of sawed oak lumber, covered with white oak shingles, and each equipped with a fine stone fireplace and chimney made from local pink limestone full of crinoid fossils. In those days visitors were few, and we settled down in our cabin to enjoy the solitude and listen to the night sounds. Off somewhere there would be the tinkling of a cow bell or the baying of an old fox hound, but that night we heard a back country sound that was different. For the first time, we heard the rapid, ringing, precise call of the real whippoorwill. In our Benton County countryside we had the chuck-will's-widow, a related night-crying bird that was to disappear as intensive agriculture compromised their nesting ground. But the whippoorwill was still there along the Buffalo with his evening and nighttime concert to make a trip to his realm worthwhile.

Like Alan Cazort, we were taken with the place and would come back many times over the years, bringing our Bentonville friends to hear the whippoorwill, to fish, to swim, and to explore the hiking trails. After Grumman made their sturdy canoe available, we learned to steer the craft and float the lower Buffalo to our heart's content.

The Buffalo River State Park would become a meeting ground, like Ponca on the headwaters, for us in the oncoming, long-drawn-out contest to prevent the obliteration of the Buffalo.

The Shivaree

For a time Glen Green continued as publicity director for the Arkansas Parks Department, and in the spring of 1946 he was called upon to arrange a trip on the upper Buffalo for the Standard Oil Company, who wished to do a travelogue movie of the area, entitled "Invitation to Arkansas." Green had arranged for boatmen from the Jim Owen's boat line to transport these visitors down the river. Most of this crew was camped out near the Ponca low-water bridge and had bedded down for the night when a first-class disruption took place.

A couple of cars came down the steep, tortuous highway from the north at breathtaking speed. Their drivers and passengers gave forth with loud shouts uncomplimentary to the foreigners from Missouri and even discharged what sounded like pistols and shotguns. To cap it off, terrific explosions of dynamite erupted along the road. The thoroughly upset Missouri boatmen took refuge in the woods, spending an uneasy night, but managed to fulfill their mission the next day.

This was an unnerving development for Green and his national park idea. It was believed that there was much local resentment against further publicity and against the employment of out-of-state johnboat operators. It might be best not to stir up further animosity.

Then Green had also met with discouragement elsewhere. He had approached Governor Ben Laney with a proposal to make Lost Valley a state park. Laney was interested, but the state legislature failed to endorse it, due mainly to a lack of funds.

Feeling that his efforts were being wasted, Avantus Green resigned from the Parks Department. Green would spend several years as a representative of the National Education Program in Washington, D.C., and would contribute nothing further to the conservation effort here in Arkansas.

Twenty years after the night of the dynamiting at Ponca we were to learn that it was not what it seemed. Dexter Curtis, an old friend and native of Ponca, knew the facts. That night there had been a shivaree up on the mountain near Compton. All old-timers know that that post-nuptial send-off requires the use of shotguns, dynamite, and moonshine. Some of the boys that night had a lot of each left over and decided that it would be great sport to scare the devil out of that bunch down at the low-water bridge; it was just a sporting thing to do—no hard feelings toward anyone.

Bud Green, like many another personality interested in conservation of our scenic and outdoor recreational inventory, had made a contribution and then faded away.

But there were two who would stay the course.

Ken Smith and Harold Alexander

Ken Smith, from Hot Springs, was a young student at the University of Arkansas in the early 1950s. In 1953 he became a member of a small hiking club founded by Dr. Hugh Iltis and frequently led by Dr. Kurt Stern. Lost Valley and

Kenneth L. Smith in Boxley Valley, 1957, with one of seven iron rendering kettles used by Confederates to make gunpowder at Bat Cave on Cave Mountain. The iron kettle was broken by Union army cavalry in 1862 during the Battle of Boxley. *(Courtesy of Kenneth L. Smith)*

the region around it was a favorite with them, and Ken maintained his interest after graduation in mechanical engineering in 1956, returning to Lost Valley many times. He demonstrated a proclivity for writing and in 1958 had published several articles about Lost Valley, Hemmed-In Hollow, and Big Bluff in the Sunday magazine sections of the *Arkansas Gazette* and *Arkansas Democrat*. In

Harold Alexander with his wife, Virginia, and two daughters on the Buffalo, April 1963.

the process, he became deeply concerned for the integrity of Lost Valley, realizing that it would need protection of some kind by either governmental or private means. In due time he prepared a voluminous report on Lost Valley for the
Nature Conservancy, representing untold hours of field and home work.

Ken Smith befriended the owner of the key parcel of land controlling
entrance to Lost Valley, Mrs. Harry Primrose, and her son, Gene. Although he
gained her good will and confidence, Ken had difficulty in persuading her to sell
her tract to the Nature Conservancy. Mrs. Primrose valued the scenic canyon as
much as anyone, but she trusted no one but herself as guardian.

At the same time another voice was being heard in Arkansas, keeping those of
us interested in conservation abreast of developments and telling us the proper
way to go. Harold Alexander is still with us to this day (1992), a sort of Aldo
Leopold for Arkansas. He was educated as a wildlife biologist, coming here from
Kansas at the end of the war and entering the employ of the Arkansas Game and
Fish Commission. He was a rumpled but straightforward, clear-headed fellow
who laid it on the line.

People in the service of state and federal agencies are not exactly free to stand
in the forefront in regard to measures that are potentially controversial. They
may find themselves at the mercy of various judges, public executives, assorted
senators, legislators, and overwrought citizens. Harold Alexander was to have
that problem, but he was able to adapt to the situation after a certain amount of

trauma. He soon acquainted himself with such conservation organizations as were on stage at the time, such as: the Arkansas Audubon Society, the Arkansas Wildlife Federation, and the embryonic Nature Conservancy. He joined those groups, offering his expertise and authoring some of their documents in reference to conservation problems in Arkansas.

Harold also was able to have published in the state newspapers reports on conservation and wildlife for the edification of us out in the state who weren't always acquainted with the latest news. He also made it a point to become acquainted with those citizens with a known conservationist background and to encourage them to take positive action in reference to burgeoning environmental crises in our state.

In 1956 I had read one of Harold Alexander's reports in the *Arkansas Gazette*. It concerned our free-flowing, recreational, and good fishing streams, a timely inventory for those who weren't sure just what we did yet possess in the way of rivers and streams.

I was program chairman for the Bentonville Rotary Club at the time and had Harold Alexander give us his talk on stream preservation. We had a long visit afterwards and have been friends and collaborators ever since.

Harold's voice was of prime importance to the developing conservation movement. Many of us had not realized the magnitude of the big dam program for our Ozark rivers. That every last one of them was to be subjected to major engineering alterations could scarcely be believed; surely not all of them, surely not the Buffalo. But it was true, and Harold Alexander was the man who made us realize it. We read the newspapers and paid some attention to the radio and the TV but were just too busy to become involved. We knew that President Eisenhower vetoed the bill for the Lone Rock and Gilbert dams on the Buffalo. But we thought surely that that was the end of it. It took Harold Alexander to prompt us to write to President Eisenhower thanking him for the veto and begging that any other such bill receive the same treatment in the future. Actually that did happen three times in the late 1950s, and it took Harold Alexander to point out that this was only a short breathing spell, that economy-minded President Eisenhower would soon leave office, and his successor might be a horse of a different color. Then the big dam plan for the Buffalo would roll on. Thus by a quirk of fate the Buffalo River was saved from damnation in the 1950s.

The big dam program had been so popular; it had been such a panacea for politicians, a guarantee of tenure for anyone in Congress who could get such a job done in his district, that it was difficult to believe some of the dissent that was beginning to be heard.

Gus Albright and Other Dam Critics

In 1955 a book appeared that would receive some national attention, its author—Elmer Peterson, an Oklahoma newspaperman. It had a catchy title, *Big Dam Foolishness,* and described most of the objections that dam opponents would use in the future.

In April 1956 the Sunday magazine section of the *Arkansas Gazette* ran a double-page feature by Dorothy Wyre, a free-lance writer from Conway, extolling the beauty of the Buffalo and condemning the Lone Rock and Gilbert dams. In it she quoted some of Peterson's remarks, causing many readers to rethink the big dam foolishness.

The same paper in 1957 printed another lengthy article on the new bridge then abuilding on state Highway 14 across the Buffalo. The author, Bill Lewis, listed objections to the Lone Rock reservoir that would cover the new bridge, with that fact being a principal one. According to Lewis, the highway department wasn't worried; if the new bridge should wind up under water, the Corps of Engineers would reimburse the highway department, and they could build a higher bridge elsewhere. It had all happened before, you will recollect, to the unending frustration of the American taxpayer.

On February 24, 1957, Elizabeth Carpenter, the *Arkansas Gazette*'s reporter in Washington, submitted an item that verified Harold Alexander's cause for concern ... a new big dam bill had been conceived in Congress:

Protests Mount against Move to Dam Buffalo

By Elizabeth Carpenter

Nature lovers in Arkansas are worried about the proposed damming of Buffalo River. Last week they wrote their protests to Arkansas members of Congress who have been working hard to get the economy-minded Eisenhower administration to approve the two power dams in Searcy County—Lone Rock and Gilbert. ...

Now there is a mild flurry of protest by the Arkansas Audubon Society which was organized in May of 1955 by sportsmen and nature lovers.

Letters to each of the Arkansas members of Congress were signed by Douglas James, corresponding secretary of the Society's conservation committee.

They feel that with the trend under way to "destroy" most of the nation's swift-water streams, the Buffalo River will have even more value to the state's resort program in its present state.

Representative James Trimble of Berryville maintains the dams can be built "without destroying the scenic beauty."

After that we began to hear from Gus Albright, a sturdy outdoor writer, then

Gus Albright, one of the first
Arkansas journalists to favor a free-
flowing Buffalo, was a writer for the
Arkansas Gazette and the Arkansas
Game and Fish Commission.
(Courtesy of Gus Albright)

the news editor for the Arkansas Game and Fish Commission. Gus also authored a column in the *Arkansas Gazette,* which became steadily more hard-hitting as the threat to the Buffalo increased.

In April 1957 Gus had taken note of action by the Federated Garden Clubs of Arkansas:

Buffalo River Backers

CONGRATULATIONS to the Federated Garden Clubs of Arkansas—336 clubs with a total membership of 3,435—for its renewed stand, in resolution form, for the permanent preservation of the Buffalo River in its natural state. Copies of the resolution were mailed to each Arkansas congressman and to each appropriate Congressional committee.

As these women have so capably stated, "We believe that the preservation of a 'variety' of recreational features and scenic qualities in Arkansas will enhance and add to the scope and variety of recreational opportunities, and will 'increase' the value of the Ozark region as a playground for the Central United States.

"Because of its unusual quality the Buffalo River is particularly worthy of preservation, and thousands of Arkansas citizens oppose its alteration and destruction by high dams."

He summed up the situation very well in the Sunday *Arkansas Gazette* for July 14, 1957:

Now the Buffalo—Nature's Last Stand

Should the beautiful Buffalo river be damned? Certain interests say "yes"—others say it would be an "unforgettable" mistake.

Well, we have always recognized that "there are two sides to everything," and the proposed Buffalo river projects are certainly not exceptions to this rule. So, let's consider for a few paragraphs some of the arguments for and some of those against. But first—a brief history.

Two dams have been proposed—one near Gilbert just west of highway 65, the other at Lone Rock near the mouth of the Buffalo where it runs into White river. A bill was introduced in Congress last year, proposing Lone Rock for flood control and Gilbert for flood control and power.

The Corps of Engineers study revealed that cost-benefit ratios did not justify the projects, and the Corps made a negative report on their construction. However, Congress passed the bill anyway, only to have President Eisenhower veto it.

Now this year the proposal has been revised, whereby Lone Rock would be expanded to include power facilities—an additional cost of more than $57 million. At this writing, the two projects have been tentatively approved by the House Public Works Committee for inclusion in an omnibus authorization bill. The total cost of the two dams has been upped to 86 1/2 million dollars.

And where is the strong Congressional push coming from?

Arkansas Third District Congressman Jim Trimble is the Washington force behind the projects. ...

Arkansas is well stocked with large reservoirs—Norfork, Bull Shoals, Ouachita, Hamilton, Catherine, Nimrod, Blue Mountain, and Greeson. There is under construction—Table Rock in Missouri just across the North Arkansas line.

Authorized—Greers Ferry on the Little Red River. Now in the planning stages—Beaver Dam, near the Benton-Carroll county line.

Gilbert and Lone Rock would simply be two more of the same. But there are no more Buffalo rivers! Dam this one and it is gone forever.

Dam the Buffalo? In this writer's opinion—"it would be a grave injustice, a calamity to the people of Arkansas. To the outdoor public of the nation."

What do you think?

The Arkansas Audubon Society had been organized only two years before, but they were able to take action by approving a timely resolution drafted by Harold Alexander and signed by H. H. Daniel, president, voicing opposition to the Buffalo River dams.

This position paper was placed in the hands of public officials on the state and federal levels and was published in many newspapers.

A year later Gus Albright would lament the dearth of effective action in the Sunday, April 13, 1958, edition of the *Arkansas Gazette*:

The Beginning of Buffalo's End

It appears now that 1958 will mark the "beginning of the end" of the beautiful Buffalo River—thanks or "be damned" to what the politicians tell us is a temporary recession.

Congress recently passed the Omnibus Bill—the "catch all" for billions of dollars worth of federal water projects—which provides that two dams be constructed on the Buffalo, one near its mouth at Lone Rock and the other upstream near Gilbert.

Omnibus—the swap out gimmick that forces the bad in order to get the good. So what? It's only the tax payers' money.

Well, the damming of Buffalo River will put one of the state's better and most popular parks under about 40 feet of water. Need more parks? Sure! But what about losing this one? There isn't enough money in Fort Knox to find and buy in Arkansas a more beautiful site than the Buffalo State Park.

Audubon Society Speaks Up

The real sad part about this near loss of such a beautiful stream is the fact that there are so many responsible people in Arkansas who, in their hearts, deplore the impounding of these waters but who have been reluctant to do anything about it.

As far as I know, the only state organization or agency that has this year expressed opposition to the dams, both publicly and to our congressional delegation in Washington, is the Audubon Society. A few garden club chapters have done the same. More recently these people have asked for the president's veto. My hat is off to them.

But it was good to know that there were some in the area willing to stick their necks out, even though there was no definitive plan to combat the big dams.

The Decision Makers: Fulbright and Trimble

On June 25, 1958, the *Arkansas Gazette* carried a brief news item of ominous portent for the free-flowing Buffalo River:

Gilbert, Lone Rock Restudy Ordered

Army Engineers have been directed by the Senate Public Works Committee to restudy the proposed Lone Rock and Gilbert Reservoirs in the White River Basin in Northwest Arkansas.

Committee Chairman Dennis Chavez (Dem., N.M.) advised Senator J. William Fulbright (Dem., Ark.) today of the Committee action.

Mr. Eisenhower and Republicans contended that the projects were not economically justified as shown in a report from the Army Engineers.

Senator J. William Fulbright. *(Courtesy of the University of Arkansas Library, Special Collections. J. W. Fulbright Papers)*

Representative James W. Trimble (Dem., Ark.), and Fulbright who championed the projects, contended that they had been justified by the Army Engineers in 1951 before Republicans took over the White House and changed the Engineers' justification standards. …

Fulbright said the restudy of the projects would keep the projects alive so that they might be included in the next authorization bill, possibly in 1959 or 1960.

That information increased the growing discomfiture of those citizens who were following developments on the Buffalo.

But there may be recourse, hopeless though the case may seem. In this case Congressman Trimble and Senator Fulbright both clearly stood in support of the Gilbert and Lone Rock dams. Other members of the Arkansas delegation could be counted on to go along with them. The principal obstacle, President Eisenhower, would soon be out of the way. The chance of an economy-minded, veto-wielding successor was not good, and the omnibus bill would then gain full approval.

The personalities of these two influential representatives, however, would prove to be a very important factor. Trimble, as we have said, displayed gentlemanly determination. Once his course was set, he would not swerve, no matter how much we conservationists might plead for him to change his tack.

Senator Fulbright was different, though no less a gentleman. He was pragmatic. Although a liberal in politics, he was capable of confounding his warmest supporters in that field. An example was his signing of the Southern Manifesto during the early integration brouhaha. When his hand was called by many of his compatriots, he said simply that a majority of the voters in his state approved of what he had done and that if he was to stay in office it was a necessary act.

No matter how important his influence was to be later on in the struggle for the Buffalo, he was no conservationist, as we well knew in 1958. In July of 1959 we were to have from the senator a real shocker, news of which appeared in the *Arkansas Gazette*:

Brush Control

Senator Fulbright has proposed a project through which millions of acres of Ozark uplands, now covered with brush and scrub timber, could be used for pasturage or the growing of short leaf pine. Under the plan the federal Forest Service Experiment Station in Northwest Arkansas and the University would undertake a six-year research program to determine the best means of eradicating the brush and utilizing the reclaimed acreage.

... Both the Soil Conservation Service and the University's Experiment Station at Fayetteville have been testing sprays to determine the best way to kill off the brush. ...

Senator Fulbright has urged Secretary of Agriculture Benson to recommend a $100,000-a-year appropriation in his next budget to accelerate and coordinate these studies under his proposed federal-state project. A reclamation program such as this would benefit not only Arkansas but every state whose forests are plagued with choking underbrush.

Several of us in the towns of northwest Arkansas had become aware of a mutual interest in general conservation problems, as well as Lost Valley and the Buffalo River. No formal meetings were involved, but there were chance get-togethers at Professor Sam Dellinger's home or at meetings of Harry McPherson's Northwest Arkansas Archeological Society.

We were all upset to read of Senator Fulbright's plan to have three million acres of northwest Arkansas sprayed with 2,4,5-T.

One of our number, Evangeline Archer, was an old friend of Senator Fulbright. She never spoke of or to him except as "Bill," as if he were a family member. But she was horrified by the brush control proposal and dispatched a letter taking him to task in no uncertain terms. Others followed suit, and he was the recipient of a significant stack of anti-spray mail. After that the big brush-killing program for the Ozarks died a quiet death, a good example of the importance of writing your representative regardless of election time. We would not forget Evangeline Archer's good example.

The Nature Conservancy in Lost Valley

Our increasing interest at the time lay with Nature Conservancy. It was no federal agency, but a strictly private conservation group modeled after a similar organization in England. Its financing depends upon private donations alone, with the donors being eligible for tax exemption, a strong inducement to give in this case. Any effort to influence legislation by lobbying or attempt to determine political decisions by the Conservancy would terminate its tax advantage. This limits its maneuverability in the case of extensive controversy such as was developing along the Buffalo River. Nature Conservancy generally concentrated on smaller areas of prime scientific or scenic interest available from willing sellers and in need of protection by either private or public agencies. The Conservancy does not engage in managerial activity but upon purchase of a tract turns it over, after proper recompense, to some desirable organization such as a state parks commission or wildlife management group. Thus the Nature Conservancy would have been a good organization to initiate a protection plan for Lost Valley.

That idea to protect Lost Valley was first suggested by Dr. Hugh H. Iltis, professor of botany at the University of Arkansas, in 1955; but Dr. Iltis left the area and nothing came of it until three years later. Then Ken Smith, in a communication from Dr. Iltis, was urged to enlist the aid of Nature Conservancy. On July 1, 1958, Ken contacted George B. Fell, the executive director of the Conservancy, to see if Lost Valley could be placed on their agenda. Harold Alexander was active in the organization trying to recruit members.

A reply to Smith's inquiry was furnished by Paul Bruce Dowling, ecologist and field representative for Nature Conservancy, who suggested that Ken contact Harold Alexander and me in order to organize a project committee for Lost Valley. This was not accomplished immediately, but Paul Bruce Dowling did come to Arkansas to expedite the matter.

Ken Smith's proposal was accepted, and the Conservancy earmarked twelve thousand dollars as a purchase price for Mrs. Primrose's 160 acres. It would now be up to Ken to close the deal.

But all of this afforded no solution to the problem of the fate of the Buffalo River.

Dowling called our attention to ongoing activity in Missouri where big dams on the Current and Eleven Point rivers had been proposed by the Corps of Engineers. He pointed out that Missouri's conservation agencies opposed those dams and had requested that the National Park Service survey those streams for possible inclusion in the national park system.

Dowling realized that such an effort would be beyond the bounds of Nature Conservancy but that it would be a proper course for us to pursue after the Lost Valley problem was settled as a Conservancy enterprise.

Dowling's suggestions gained considerable dissemination in the local press, with Gus Albright featuring those ideas in his column. It would have been a workable way to go except that we were to have no help from timorous state agencies in Arkansas. The Game and Fish Commission would like to have supported it but was afraid of the legislature. The State Parks Commission assumed a stand-off position altogether and was later manipulated extensively by the Buffalo River Improvement (big dam) Association.

The Arkansas Wildlife Federation

Bill Apple of Little Rock was a seasoned campaigner in the conservation field. A short, dark man with the appearance of a Spanish grandee, but with the best down-south accent in Arkansas, Bill was a decisive plotter of plots and head of the Arkansas Wildlife Federation. He was no kin to Bob Apple of Dardanelle, also the head of the same organization later on.

Bill Apple remembered Avantus Green's national park proposal and had some ideas of his own as to how best to keep the big dams off the Buffalo. He would have the Corps of Engineers build a series of small dams on that river's tributaries. A lot had been said about the upper Buffalo falling to levels too low to float during the frequent summertime dry spells. From those small reservoirs water could be released from time to time to maintain useable river levels … that in addition to flood control. It was said that Andy Hulsey, head of the Arkansas Game and Fish Commission, had originally proposed it, and even Gus Albright came out for it. Most of the membership of the Wildlife Federation, however, were against that program.

The rest of us concerned conservationists, to whom the idea was most outrageous, kept our cool and said little. We needed the good will of those little dam people in the long haul ahead and didn't want to alienate them. We knew that some of the most outstanding scenery in the Ozarks lay in those deep recesses whose waters fed the Buffalo. To alter them in a manner after the style of the Corps of Engineers' flood control, big bathtubs would be a dispersed but comparable evil.

As time went on and the going got rougher, the little dam plan was dropped except for an occasional proponent still in need of therapy.

A Developing Stalemate

It was now becoming more evident that Ken Smith's exhaustive efforts to place Lost Valley in the hands of a protective agency were approaching an impasse. In a letter to Paul Bruce Dowling, dated May 28, 1959, Ken had this to say:

> Today I had a letter from Mrs. Primrose's son Gene … "Mom said she was not interested in selling 160 acres, at any price. I am sure she will not sell this summer. She might decide to sell after I get through school. I just don't know."

Whether or not the stalemate on Lost Valley was a factor or whether Ken was dissatisfied with his life as an engineer in the employ of the Crossett Company is not known. What was certain is that he, sometime during the next several months, decided to leave Arkansas and start life anew in California.

The Timber Thieves

In the first week of October 1960 a real calamity befell Lost Valley. Some of Mrs. Primrose's neighbors purposely talked her into going into Jasper to spend the day. As soon as she departed, a local sawmiller moved in with a bulldozer, without permission, and scraped out a road across her land and up over the hill behind the natural bridge and into the heart of the canyon. It should be noted that the core of the area, the big bluff and the waterfall, was owned by a Pendergrass family then living in Little Rock.

The loggers cut indiscriminately whatever was in reach of their winch truck, about one hundred feet on either side of the bulldozed road: big beech, sweet gum, black gum, and red oak. The usual jumble of downed tops and defective logs were left to rot away over the years.

Mrs. Primrose was outraged, but there was nothing she could do. Tactics of that sort were regarded proper by local sawmillers. They had a tradition of cutting timber wherever it was to be found, whether on their property or not, whether properly paid for or not, and oftentimes on National Forest land, all without permission.

Ken Smith's shock was now compounded. He submitted a short editorial to all the local newspapers entitled "We Are Losing Lost Valley." All who read it were angry at such sorry doings, but they were, like Mrs. Primrose, without recourse.

Those of us in the embryonic Arkansas Nature Conservancy group around

Fayetteville realized how helpless we were in the face of such exploitative tradition. We realized that it was time that somebody DID something to counter such abuse.

Lost Valley in My Lap

Nature Conservancy officials were disturbed at this turn of events. The Lost Valley project was one of their important programs at the time.

But Ken Smith's plans for departure proceeded without pause. It was imperative that he find a successor to take his place. He had not met me at the time, but, having heard of my interest and limited activity in Lost Valley from Harold Alexander, turned my name in to the Conservancy as a possible chairman for the project. As a result, I received a letter November 26, 1960, from Paul Bruce Dowling with this proposition:

> Dear Dr. Compton:
> We have tried unsuccessfully to buy the major tract at Lost Valley from a Mrs. Primrose. There is still a possibility there but someone must be able to really convince Mrs. Primrose to sell or else the state must be convinced to condemn it for a park preserve. We really need someone to act as project chairman. ... Would you take on the job?

I was interested in the problem but really did not want to assume that responsibility. It was not something that a busy physician should be embroiled in; but hoping that a turn of events might change things for the better or that we might come up with a more proper leadership before long, I made this reply to Bruce Dowling on December 1, 1960:

> I received your letter of November 26 and it is needless to say that I am surprised at your wanting me to assume the chairmanship of the Lost Valley project.
> I have been visiting the Lost Valley area since 1946 and have done a great deal of study on its botanical and geological features and I feel that it is unique in this part of the nation. If you think no one else is better qualified, I would be happy to do whatever I can to help in its preservation.
> ... I do have a good many colored slides made over the years and also movies which can be shown with lectures on the subject. As you know, I am engaged in the practice of medicine which is sometimes a demanding profession; however, I would give this project first priority over other extra-professional activities in which I am engaged.

In a letter to me dated December 3, 1960, Ken mentions his reasons for leaving:

Rather than phone you as Bruce Dowling suggested in his letter of November 28, I am writing this letter to explain in more detail what the Lost Valley chairmanship would involve, and the ways in which you could receive help in carrying out the project. First, though, I'll explain why I've resigned the chairmanship. ... I want to go back to school for graduate work in business administration—it appears that I'll enter the University of California (Berkeley) next September. The reluctance of Mrs. Primrose at Lost Valley to sell, plus my having to clear up a number of personal duties before leaving Arkansas, caused me to relinquish the chairmanship October 1. Had I not planned to leave the state, I would never have done so. ...

In a further communication Ken Smith proposed to furnish me with his extensive report on Lost Valley, all past correspondence on the subject, and a list of all past contacts interested in Lost Valley. He would come to Bentonville for a visit, and we would go to Lost Valley to meet Mrs. Primrose. He would continue to promote the project as long as he remained in Arkansas.

On December 17, 1960, Ken Smith did come to our home in Bentonville, on which date Sam Dellinger, my old zoology professor at the U of A, and Evangeline Archer, an elderly conservation-minded lady, came up from Fayetteville to discuss the Lost Valley problem. We decided to journey to Lost Valley the next day for a firsthand look and to interview Mrs. Primrose.

Ken prepared an exhaustive report on what we saw, along with photographs showing what had been done by the loggers. Evangeline and Mrs. Primrose struck up a cordial relationship, and we went home hopeful that this might lead to something positive for the Lost Valley program. These doings, although not momentous, did generate some publicity in the local papers. We continued to be aware of its importance from then on.

At about this time Evangeline Archer volunteered to devote her efforts full time as a secretary to our proposed chapter of Nature Conservancy. This was a welcome offer since she could type as fast as one could talk.

Nineteen sixty-one arrived with Lost Valley more or less in my lap, a responsibility that required a better knowledge of the place and the problem. As soon as possible another inspection trip was planned. The findings were reported to Paul Bruce Dowling in a letter dated January 26, 1961:

> ... We feel that what has been done was nothing more than vandalism. Some large sections of tree trunk had rolled clear to the bottom of the canyon and down into the mouth of Cob Cave. One large treetop had fallen into the canyon above Eden Falls where it is hanging top down. Another large white oak had fallen across the ravine above this and was left in that position, thus forming a log bridge. Altogether, damage had been not so extensive as to render it unfit for consideration as a preserve, but a clean-up job is certainly in order. ...

While this attention was being given to Lost Valley, other significant activity was going on to the east in Searcy County. The Corps of Engineers, pursuing their tried-and-true policy of enlisting the support of the local editors and businessmen, had laid their plans for the Buffalo River before James Tudor, editor of the *Marshall Mountain Wave* in Marshall, and were received with enthusiasm.

The *Wave* carried frequent and lengthy glowing accounts of the many benefits that would come from the building of the mammoth dams, now a probability since John Kennedy had been sworn in as president.

There would be no more disastrous floods, all of the area towns would experience rapid growth, land values would take a sharp upturn, millions of tourists would come to go boating on the lakes, hunting and fishing would be greatly improved, and the Buffalo Valley would be comparable to the TVA project.

Tom Dearmore and Pete Shiras, grandson and nephew of Tom Shiras, at the *Baxter Bulletin* in Mountain Home were converted to the same gospel as were also the editors of the *Yellville Mountain Echo* and the *Harrison Daily Times*.

If any of us wanted to stop the big dam train now, our work was cut out for us.

The situation during those first weeks of 1961 is best revealed in a letter to Paul Bruce Dowling (February 8, 1961):

> The last meeting of those interested in the Lost Valley project was held at the home of Mrs. Laird Archer in Fayetteville on Saturday Feb. 4th.
>
> Present at the meeting was Mr. Clayton Little of Bentonville whom I have mentioned before. Mr. Little is going to prepare options for all the property in the Clark Creek watershed. He is very interested in Lost Valley and inasmuch as he lives in Bentonville and would be able to handle many of the details for which I will not have the time, I am going to entrust him with some of the work, especially the legal aspect of it.
>
> Professor S. C. Dellinger, emeritus professor of Zoology at the University of Arkansas, whom you have no doubt heard of, was also present. He is on good terms with Gov. Orval Faubus and has volunteered to see him after the current session of the legislature is over to discuss the Lost Valley project in detail. In so doing he will attempt to make contact with the Pendergrasses who live in Little Rock and own the property where the recent logging was done and who are related to the Governor's wife.
>
> We felt that it would not be wise to bring up this subject during this session because of other legislation and a general shortage of funds. I do not believe there will be any effort on the part of the state to purchase the property at this time.
>
> Also in regard to the Governor, I would like to say that I discovered further evidence of his interest in Lost Valley the next day while visiting with Mr. John Williams who is head of the department of architecture at the University of Arkansas. Mr. Williams has long been familiar with the Buffalo River area and told me that a short time ago during an interview with Governor Faubus that the subject of Lost Valley

was brought up. He said that the Governor informed him that he had heard of the logging operation and that he had used his influence to put a stop to the cutting. Ken Smith and I had heard rumors of this already, but this was the first real verification of the Governor's action that I discovered. ...

At this stage of the game I was beset by increasing misgivings as to the effectiveness of Nature Conservancy as the primary organization to accomplish what we had in mind. The Buffalo River was the real problem, not Lost Valley. There were many Lost Valleys along its course. We were in need of a comprehensive, hard-hitting group that could take the bull by the horns and not by the tail. We needed people who would work hard to save the whole Buffalo, Lost Valley along with it.

The Audubon Society, the Federated Women's Garden Clubs, and the Arkansas Wildlife Federation were all in the proper alignment and had done much already, but their primary interest, in those days at least, lay elsewhere— hunting, fishing, bird watching, gardening.

The Sierra Club was a proper example. They were experienced in the rough and tumble of politics, publicity, and lobbying and would come out fighting hard for a good cause. But it had a strong Western flavor, not then as nation-wide as now. What would they be willing to do for us yokels down here in Arkansas with our hills less than mountains and our unknown back-country rivers? To find out more about that I joined the Sierra Club but was never able to attend any of their functions. Their membership in Arkansas was then near to nothing.

On March 1, 1961, I wrapped the idea up in a letter to Ken Smith, who was still in Crossett:

> I have some extra information for you which I didn't want to mention in the letter.
> For one thing I am considering the possibility of starting a region-wide organiza-
> tion to engage in all matters relating to ecological problems such as we now have in
> Lost Valley. We discussed the idea at this last meeting but only briefly. It needs a
> period of gestation right now. We wouldn't want to appear to be in competition to
> Nature Conservancy. In fact the first step would be to organize a chapter of the latter
> group in N.W. Arkansas and this is now underway. But we do need a sensible,
> influential and dedicated group of people who will work for and give priority to the
> Ozark Region. We are going to need a name. It should be descriptive but not too
> cumbersome. "The Ozark Conservation Society" for instance. Can you think of a
> better one? Its activities would involve economics, politics, education, research, pub-
> licity and any other factors that might affect the balance of nature in the Ozarks. ...

This in brief marked the genesis of the Ozark Society.

The outlook for success of any movement to "save the Buffalo" was rendered evermore unlikely by a news item appearing in the *Arkansas Gazette,* February 2, 1961, in which the word conservation is used in its doublespeak connotation:

Conservation Prospects Good

Chances are good that Arkansas's water, forest and park projects may get additional funds because of President Kennedy's step-up in economic spending.

The President's economic message asked cabinet and agency heads to go over the list of public works projects, roads, recreation and national forest and park expenditures to see if any of the projects could be speeded up. A list of those which can be speeded up is requested by the President by February 17.

Senator J. W. Fulbright has made a request of the secretary of the Army to ask for $2 million more for bank stabilization on the Arkansas River. He also requested the secretary to put on his list $125,000 for a survey of the feasibility of Lone Rock and Gilbert Dams, and $250,000 for planning Ozark Dam. If these get secretarial and presidential approval, then Arkansas will fall heir to some extra money on the federal project front soon.

With Senator Fulbright, Congressman Trimble, President Kennedy, and all the rest of the Arkansas delegation all lined up on the big dam team, there was little hope for the Buffalo.

I came in that evening after getting the message off to Ken and told my wife, Laurene, that we were going to put together a bunch of people to try to stop the United States Army Corps of Engineers from damming up the Buffalo River.

She looked at me with pity, if not sympathy, and said: "Neil, are you out of your mind? Don't you know that you haven't a chance of ever stopping them from building those dams?"

I could only make a feeble reply: "Well, somebody has got to try. Whether we win or lose it would be a crime not to."

And so it would have been.

Whatever her misgivings might have been, the very next day Laurene sat down and wrote Jim Trimble a long letter.

March 4, 1961

Mr. Jim Trimble
Washington, D.C.

Dear Mr. Trimble:

I am a housewife and mother who belongs to that legion of "silent" voters who, though interested in the welfare and development of our state of Arkansas, has never before taken the time to "write a letter to your Congressman" about matters that have been of concern to me.

My own eldest daughter was graduated from our state University last June and is teaching in Denver, Colorado this year. She, too, is most "homesick" and is looking forward to returning this summer to work on her Master's degree. She once made this remark. "If I could establish my own home in one of the small towns in this part of Arkansas and rear my children here and they could be as happy in their childhood and schooldays as I have been in mine, I could ask for little else." We love our state and have faith in it and its natural beauty has been a constant joy to us and played a very important part in our family life.

Now, certain rumors have come to my attention about which I feel I can no longer be silent but must somehow convey to you my feelings on the subject and my real concern. I have heard that there is now proposed the building of at least two dams on the beautiful Buffalo River and that these dams are already in the planning stage. This part of Arkansas through which the Buffalo flows has long been a favorite retreat of my family and of many of our friends. I am sure this is true of many, many other people. We have always found a peace and satisfaction in our explorations of the scenic and the many unusual wonders that abound in the areas surrounding the Buffalo River that we shall always be grateful for and that we like to share with friends at home and with visitors to our state. We sincerely believe that in this grand and awesome section the state of Arkansas has a natural attraction comparable to more famous places elsewhere that have already been set aside and preserved for the con- tinued enjoyment not only of these present generations but of the generations yet to come. We strongly feel that before the Buffalo River and its surroundings are drowned by man made lakes we should all seriously consider what the loss to our state would be. And it would be an irretrievable loss. The thought of this possibility saddens the hearts of more of our people than you might realize. Shouldn't we be far sighted enough to keep at least one of our fresh water streams in its natural state? And is not the Buffalo River our most beautiful and the one to deserve our every pro- tection? We are fast becoming surrounded with our man made lakes. Wouldn't it be well to consider the possibility, in this instance, of trying to preserve this particular section—at least a major part of it—and all the unique features it boasts, as a state or even a national park? I think future generations would thank us for it.

We need flood control programs and power for industry—yes. It is nice to have the large lakes for the sports of fishing and speed boating. But there are other needs of our people that cannot be discounted. We have the great and rare privilege to live in one of the most beautiful natural sections of the United States. Let us be grateful for this privilege and not misuse our God given gifts but proceed slowly and with all deliberation before we make any drastic change of what has already been created so perfectly. Drowning this river can surely not mean progress but only a shameful destruction.

So this is my too long letter to my very busy congressman who, I am sure, hasn't much time to give to the wandering thoughts that disturb this female mind. But, Mr. Jim, you are so important where matters of this sort are concerned and I know your influence is great. And this thinking is from my heart and I must share it with you and ask your careful consideration on this matter. Your help could be invaluable in preventing the ruin of one of our state's great natural assets.

Yours very truly,
Laurene Compton

Bad News from Marshall

While we were busy organizing an anti-dam faction in Fayetteville, its antithesis was being born in Marshall, the county seat of Searcy County. That event was recorded by the *Marshall Mountain Wave*, March 20, 1961:

Organization to Push for Buffalo Dams

At a public meeting held in the County Courthouse in Marshall last Friday evening, an organization for the development and construction of the Lone Rock and Gilbert dams on the Buffalo river was formed by the unanimous consent of the more than 150 interested citizens and property owners attending.

James R. Tudor, editor of the *Marshall Mountain Wave,* opened the meeting and explained as closely as possible his understanding of the status of the Buffalo river projects at the present time. He pointed out that the people of the area have been plugging for the construction of the Buffalo river dams for a period of several years, and that at the present time it seemed that the possibility of securing these projects were brighter than ever.

He also pointed out that during the past few months members of the National Wildlife Federation, The Arkansas Garden Clubs, Arkansas Chapter of the Audubon Society, and others who have in the past opposed the construction of the Buffalo River Dams have caused to appear in newspapers of the state and in the April issue of *Field and Stream* magazine articles opposing the construction of the dams on the Buffalo.

A majority of the people attending the meeting thought that the time for action was at hand, and that an organization should be formed to combat the unfavorable publicity, and also to let members of Congress and the Senate, as well as the U.S. Corps of Engineers know that the people of the area traversed by the Buffalo river highly and almost unanimously wished the ultimate completion of the proposed dams.

Gibson Walsh, manager of the Searcy County Title Co., of Marshall, was named chairman of the Association, and Mrs. Kate Ruff, co-owner and manager of the Sunset Court in Marshall, was named secretary and treasurer of the organization. ...

It was voted that a small membership fee of $1.00 per year be charged and several present turned in their fee. Mrs. Ruff and Mr. Walsh have continued to receive memberships through the mail since Friday, and by Tuesday morning approximately 60 members had paid in the $1.00 membership dues.

Officers and directors of the Association plan to contact leaders in Newton, Marion, and other adjoining counties and solicit the help of these people to secure the Buffalo projects. ...

The same paper, in its close monitoring of the situation, was forced on April 6 to report that in spite of Congressman Trimble's request for $125,000 from the Bureau of the Budget as planning money for the Gilbert and Lone Rock dams, the Bureau had failed to include the request.

James R. Tudor, publisher of the
Marshall Mountain Wave during the
"battle of the Buffalo." (*Courtesy of
Jim Tilley, 1991 publisher of the*
Marshall Mountain Wave)

The April 13 issue of the *Wave* contained more adverse news for the pro-dam advocates. In urging readers to write letters favoring the dams to Congressman Trimble, the *Wave* quoted him as saying that he had received few such letters; but, on the other hand, he had received numerous letters opposing the projects.

In the same issue the editor took *Field and Stream* magazine to task for printing an article critical of the Gilbert and Lone Rock dams. He accused the Arkansas Game and Fish Commission of fostering the article, which they denied. In view of what was to follow, these reports might be construed as evil omens for the Buffalo River "Improvement" Association, which the big dam group had modestly chosen to call themselves.

A Night in the Trees

By May 1961 Ken Smith was still in Crossett, preparing for flight, but before he left he wanted to take me on a field trip around Boxley and Ponca to show me other features, like Leatherwood Cove, Running Creek, Big Bluff, and the Goat Trail. But nature, not man, would interfere in a most devastating way.

I was to join Ken and his friend Bill Beall at the Cedar Crest Lodge (later the Lost Valley Lodge) on the night of May 5, and we would go out the next morning for the first day's sightseeing.

The weather had been rainy and damp, but the forecast did not include anything unusual. That evening I got into my Chevy station wagon and started out for Newton County. But a couple of miles south of Bentonville, I beheld a cloud bank of unbelievable proportions filling all of the southeastern horizon, the exact direction of my anticipated journey. Whether or not it reached ten or twelve miles up into the stratosphere I didn't know, but it looked it. I got out of the car and beheld that monsoon for several minutes, then chickened out, went back home, and went to bed.

It was a rainy night in Bentonville but nothing to compare with what happened over Gaither Mountain in Newton County. That included Lost Valley, Boxley Valley, Hemmed-In Hollow, and the headwaters of Crooked Creek that flows on east through Harrison, Arkansas. In that area it rained twelve inches between 11:00 P.M. and 3:00 A.M. Ken and Bill on the second floor of the lodge couldn't sleep because of the thunder and lightning. They were entertained by the sight of a wall of water that came down Ads Creek from the north through the hamlet of Ponca. It surged from hill to hill and rose five feet deep in the lower floor of the Lodge. The next morning the telephone lines were out, power lines were down, the roads into Harrison and Jasper were gone in places, and Ken and Bill had to walk out.

In Lost Valley that night some old friends of mine were camped out: John Swearingen of Rogers, his wife, and their children. They were on a knoll across Clark Creek from Mrs. Primrose's house.

Let us relive their ordeal in the words of Mrs. Swearingen as recorded in the *Rogers Daily News* for May 11, 1961:

Rogers Family Spends Four Terrifying Hours in Trees above Floodwaters

A Rogers mother recounted today how she and members of her family spent four terrifying hours early Sunday in treetops above raging floodwaters, with each not knowing in the murky darkness whether the others were safe.

The frightening experience happened to Mr. and Mrs. John Swearingen when they were caught in a flash flood while on a camping trip at Lost Valley in Newton County over the weekend.

Even after the five members of the family were able to climb down out of the trees, they found they were marooned by the flood-waters.

It was more than two days later—on Tuesday—before they were able to get out of the area and return home.

The young mother said she and her husband and their three sons, David, 14, Jimmy, 9, and Tommy, 6, arrived in the area early Saturday afternoon, and that they set up camp on a little knoll about 200 feet from Clark Creek after first checking with Mrs. Harry Primrose, who lives nearby, to see whether it was safe.

"She told us later that she had lived there 35 years," Mrs. Swearingen said, "and that the water had never been that high before."

Mrs. Swearingen said it was only about an hour after a deluge of rain began at midnight that the water began to rise.

"My husband and the two older boys were sleeping in a tent," she said, "and the little boy and I were sleeping in the car.

"Then Jimmy, the nine-year-old, climbed into the car with us while John and David began to patrol and inspect the area. They decided it was unwise for us to try to get out.

"It was sometime around one o'clock when John and David put ropes in the trees because there were no low-hanging limbs, and the trees could not be climbed easily.

"That was only a precautionary measure," she said, "but it was raining terribly hard.

"About two o'clock," she said, "the water began to rise, and it rose several feet in just a few minutes.

"The ground where we were standing was covered with water, and the hard rushing water was only about 20 feet from us.

"We decided to take to the trees!"

Mrs. Swearingen said her husband put young Tommy into one tree, and then helped her into another tree, using the ropes to help get them up.

Fourteen-year-old David helped young Jimmy into another tree.

Then, Mrs. Swearingen said, her husband went back to the tree where he had put Tommy.

"Now the rest of this story I didn't know until morning," she said, "because the noise was so loud and the rain was so heavy that we couldn't see or hear each other except during flashes of lightning.

"But as John went back up into the tree with Tommy he got hold of a limb that was not very strong, and he fell out and broke his collar bone."

Young Tommy held a light for his father, Mrs. Swearingen said, and in spite of the broken collar bone, he struggled into the tree, where he clung to a branch beside Tommy for the rest of the night.

"Then the water got up to the tree I was in," Mrs. Swearingen recalled. "My tree was nearer the edge of the water."

She said she caught glimpses of the rising water during flashes of lightning.

"At about 4 A.M. the storm subsided," she said, "and the water went down quickly."

They climbed out of the trees, still not aware all roads of the area had been washed away, and that they were stranded.

"The boys didn't even want to come out of the trees then," she laughed—"they were having a ball!"

By 11 o'clock Sunday morning the water in the creek had gone down enough so that the family could walk safely across the creek, so they walked to the Primrose home, about 400 feet from their campsite.

They soon received word from Kenneth Smith, an author who has written many stories about Lost Valley, and Bill Beall, both of whom had walked into the area from the Lost Valley Hotel, that they were completely surrounded by water.

Smith and Beall, who left their car and walked out, told the family that they would let Rogers friends know that they were safe.

"I understand that the message which they gave a local forester to relay by radio was picked up by some ham operator in Oklahoma," she added.

"At about five o'clock (Monday morning)," she said, "John decided he wanted to try to get the car out in case another rain came. He and the boys, and Mrs. Primrose's son, Gene, had to build a new crossing to the creek by rolling in big stones."

She said that the route which they had taken across the stream was an old wagon trail which had never been used by a car.

"To get out he had to go between two trees," she said, "but I don't know what happened. Maybe he didn't get enough traction or maybe his arm wasn't functioning too well, but the car slipped and lodged against a tree.

"So then David and John walked to the nearest farmhouse—the Keeton farmhouse—where they got the help of a boy and a tractor."

They cut the tree, and with the help of the tractor, pulled the automobile across the stream.

"It was about 11 o'clock Monday morning, while we were surveying the damage, that we saw the plane," she said.

The plane to which she referred was carrying Rogers businessman F. G. Larimore, Swearingen's employer Jim Shofner, and Attorney Eli Leflar.

Mrs. Swearingen said that David, who is a Boy Scout, told them that the circling plane was trying to signal them.

The boy told them that the pilot had dipped his wings to notify them that they had been spotted.

The three men returned to Rogers, bringing back word that they had spotted the Swearingen car.

"We had various plans to get out of the area," Mrs. Swearingen explained, "but we decided that since the Buffalo River was just barely running over the bridge, that we would try to go by that route."

The Primrose youth and David guided Mr. Swearingen over what was left of the road, she said, explaining that in some places there was only a six or eight inch ridge on which to put the wheels, and that the rest of the road was washed out.

"The rest of us walked," she said.

In a conversation with the Swearingens shortly afterward they said that the scariest thing about all of it was the rumbling of the big boulders being rolled down the bed of Clark Creek by the force of the torrent.

That night Crooked Creek went on a rampage and washed away the whole south side of Harrison, which was located on the stream bank. A couple of citizens were drowned there that night. One would have thought that after that the town would simply be rebuilt further back on higher ground. But in the end the federal government appropriated ten or twelve million dollars to rebuild it where it had been. The next proposal was to build a high flood control dam on Crooked Creek. That would have covered up a lot of the best farmland in Boone County, and hot opposition against it was generated. Thereupon, the Corps of Engineers was authorized to build a levee along the south side of Harrison, a seawall, if you will, around one side of this Ozark upland town.

All of this left us to ponder an important problem, the forces of nature versus the forces of man and his engineers. What should be done about such things?

Those scenic recesses, Hemmed-In Hollow, Lost Valley, and Leatherwood Cove, had been reamed out. They weren't the same, but they were not destroyed. Time would refit the jumble into the natural scheme of things. After all, this was the means by which they had all been created, the incessant and overwhelming forces of erosion carving up the ancient elevated seabed into its present arrangement of canyons, gorges, valleys, and hills.

A strange thing about it was that it was a localized event, over Gaither Mountain for the most part. The Buffalo River didn't get up afterwards at Buffalo River State Park, seventy-five miles downstream.

It seemed to have been a fitting end for the unrewarded efforts of Ken Smith in Lost Valley.

Meanwhile, those of us now responsible for the Lost Valley project got on with the activation of that program and the establishment of an Arkansas chapter of Nature Conservancy. We published a list of aims and objectives which was overly ambitious, to say the least, for such a small and inexperienced membership.

A Letter to Senator Fulbright

During the late spring and early summer of 1961, the newly born Buffalo River Improvement Association (henceforth to be referred to as the BRIA) redoubled their efforts to gain wide support for Lone Rock and Gilbert dams.

The great rain over Gaither Mountain was played up as a cause of a flood of huge proportions at Batesville far down on White River, which flood could have been prevented if the Buffalo had had its dams in place, all of that in spite of the fact that the Buffalo didn't get up noticeably below Buffalo River State Park.

The "Batesville Flood" argument would be repeated from time to time over the next several years. It actually originated in the Little Rock office of the Corps of Engineers and came complete with a flood of irrefutable statistics, graphs, charts, and engineering linguistics.

The BRIA ran large ads in the *Arkansas Gazette,* the *Arkansas Democrat,* and other state papers boosting the dams. They instituted a letter-writing campaign to various state and federal officials.

Harsh commentary appeared in the *Marshall Mountain Wave* directed toward the Arkansas Wildlife Federation, the Audubon Society, and Bill Apple and Gus Albright, in particular. The BRIA was bold enough to state that at least 90 percent of their opposition had been silenced if not erased. In that assumption they were in for some surprises.

Governor Faubus had made an address in Bentonville in late June and had stated his interest in the state park system. It seemed a good opportunity to suggest the addition of Lost Valley to the system, and in a communication to him on July 3, 1961, I took that liberty, saying in part:

> As is usually the case after a gathering such as we had at the Town House, it is difficult to discuss matters and I am taking this means to express my sincere approval of your speech. I was especially interested in what you said about our parks in Arkansas and am enclosing a copy of the objectives and purposes of the Ark. Nature Conservancy.
>
> So far we have been stalled in our efforts to purchase Lost Valley by the reluctance of the principal land owner to sell, but are continuing as tactfully as possible to further this project. We have heard of your interest in Lost Valley and that you did have something to do with the prevention of further timber cutting operations there, for which we are most grateful.
>
> Once again I want to express my gratification in the interest you stated the other night in the park system in Arkansas. It isn't too often that those of us born and raised here realize that the natural beauty of our state constitutes one of our very greatest resources.

The governor's reply was encouraging but was not a final statement. Some would say that it was only politically motivated, but subsequent positive and nonpolitical action on his part would disprove that.

Many of us felt in 1961 that we had a friend in Governor Faubus, but none of us knew him personally. None of us had direct access to him or his advisors, but from his behavior and such statements as are here recorded we took heart and felt free to communicate whenever indicated.

His reply July 14, 1961:

> Thank you for your fine letter of July 3rd in reference to the park program and development of recreational areas in Arkansas.

One of my dreams and ambitions for the state is to develop a fine recreational program with parks and areas for the enjoyment of our people and our many tourists. If the program presented to the people by the legislature had succeeded, we would have had a fine park development program for the next two years.

However, now it will be difficult and I can foresee that only a meager amount of funds will be available for such a program in the immediate future.

I assure you that I shall be glad to work with you in any way and I hope that together we can find some means of developing the many scenic and potential recreational areas of Arkansas. Certainly Arkansas can be one of the finest states in the union in this respect.

Most sincerely,
Orval E. Faubus

[handwritten] You might speak to your representatives in the interest of the park program. It would help.

Orval Faubus was not the only decision maker on our agenda in the summer of '61. To use an old navy term, we had become aware of some vague scuttlebutt to the effect that Senator Fulbright might be interested in "saving the Buffalo" (Harold Alexander being the medium of transmission of that too-good-to-be-true news). We were urged to write letters to him pleading the case.

My letter to him revealed the uncertainty of our program at the time. We didn't know exactly what should be included nor how it should be managed, but at least we specified the entire river:

July 3, 1961

Dear Senator Fulbright:

… Evidence of your interest in the possibilities that exist in Arkansas with regard to the establishment of a national park or wilderness area has been brought to our attention and I wish to express the sincere approval of our organization. You can be sure that we will support you and wish to work with you in every way possible. We intend to coordinate the efforts of other conservation minded groups in the state so that we may present a definite program to the proper authorities.

We understand that there is some possibility the Secretary of the Interior may be invited to come here or that an investigation may be made by the National Park Service. If this is to be done we have suggestions to offer as to what to show them and we will consult all of the other allied organizations to make certain that they see the very best.

As you will see in the outline enclosed, we hope to have the Buffalo River maintained in its natural state from its origin to its mouth on White River. The part of it not in a park should be maintained as a wilderness area and we feel that it would be worth far more to Arkansas thus protected than dammed for other purposes. It has already passed from a state of relative isolation to one of considerable popularity for float fishermen.

We hope when you are again in Arkansas that our group and others interested will have the opportunity to discuss these exciting possibilities with you.

Sincerely yours,
Neil Compton

Senator Fulbright's reply was electrifying. We could scarcely believe that we had come by such a favorable decision so soon. Certainly the tocsin had been sounded, and the big dam people, the BRIA, the Corps of Engineers, the REA, and the fast-buck real estaters had viable opponents in the contest about to begin.

July 8, 1961

Dear Dr. Compton:

Thank you for your letter of July 3 concerning the plan for a national recreational area along the Buffalo River.

I am very interested in getting the Buffalo River area included in the National Park System. This would not only preserve this wonderful stream but would be a fine economic asset to one of the poorer sections of the State. I plan to make an effort to get an appropriation which would finance a survey by the National Park Service of this area. This will be the first step and if the results of the survey are favorable, I will introduce a bill to authorize the development of a park.

I am very glad to know that you are interested in this project.

With kind regards, I am

Sincerely yours,
J. W. Fulbright

FOUR

THE BATTLE JOINED

But that was not all in that fast-breaking year of 1961; the *Time* magazine for July 14 was devoted to outdoor recreation in America and in it was a beautiful, full-page color photograph of a canoeing party camped under Big Bluff on the upper Buffalo. It was an absolute eye-opener, better than anything else in that issue.

We were absolutely dumbfounded. Who were these people who knew all that much more about the Buffalo River than we did? What in the world was the Ozark Wilderness Waterways Club (the O.W.W.C.)?

At our meeting that month we talked about nothing else. Evangeline Archer stepped forth and said that she would leave no stone unturned until we found out. She wrote to *Time* magazine and was referred to the *Kansas City Star*, who identified the photographer, A. Y. Owen, who then named the people in charge of the O.W.W.C.

They were Harold and Margaret Hedges who lived in Lake Quivira, Kansas. We later learned that they had begun their outdoor experiences in the Ozarks right after the end of the Second World War. They were bored with the city and for kicks first bought a johnboat in which they paddled around on the Missouri River. But that wasn't much. Soon the Grumman Aircraft Company came out with their sturdy, lightweight, aluminum canoe, and the Hedges quickly learned canoe handling, becoming experts at that.

They ventured on the nearby Big Blue River and then on to better ones: the Gasconade, the Current, the Jacks Fork, Swan Creek, Beaver Creek, Big Sugar, the Buffalo, the Mulberry, and Big Piney. In a few years they had traveled all of the canoeable streams in the Ozarks, accumulating an entourage of avid out-doorsmen who would travel all night from Kansas City just to get to float the Buffalo the next day.

Compared to them, us Arkansawyers born and bred, were pikers and green-horns, but we were going to learn from them everything we could.

Harold and Margaret Hedges would become indispensable members of the anti-dam coalition.

The BRIA Prepares for the Contest

On August 3, 1961, the *Yellville Mountain Echo* published an interesting account of a meeting of the BRIA that had been held in the Marion County courthouse a few days before.

Chairmen for each county concerned were selected: Luther Cavaness, Marion County; Tom Dearmore, Baxter County; James R. Tudor, Searcy County; Arl Jones, Newton County; and J. C. Sutton, Boone County. This was to be a subdivision of the BRIA to foster pro-dam activities in those counties and was to be known as the Buffalo Dams Association, with Tudor as overall chairman; but it apparently died aborning for we heard nothing more of it. The BRIA would continue on as the only viable proponent for "improvement" of the Buffalo River.

Arthur Wood, an attorney at the meeting, suggested that they should opt for only one high dam to be located at the Highway 14 bridge, saying, "If we had only lakes to sustain us now we would be hurting."

That summer, Evangeline Archer kept busy at her typewriter. There was a frequent exchange of letters with the Hedges, and through them contacts were made with several O.W.W.C. members, including Ray Heady, outdoor editor for the *Kansas City Star,* who would play an important role in the pro-park publicity later on.

One communication was dispatched to Justice William O. Douglas of the United States Supreme Court, requesting that he come to Arkansas to see the Buffalo. He was an important proponent of conservation measures, and we felt an endorsement from him would be very desirable.

What we didn't know was that wheels were already in motion to bring that about.

Justice Douglas had seen the photograph in *Time* magazine and was quite struck by it. He wanted to see it firsthand and had one of his aides call a close friend, Charles Whitaker, in Kansas City to see if a trip to the Buffalo could be arranged. Whitaker then contacted the Hedges, and the expertise of the O.W.W.C. came into play. They would gladly escort Justice Douglas down the Buffalo the next spring.

Evangeline Archer in one of her letters to Harold Hedges urged him to write a letter to the editor of the *Arkansas Gazette* recounting the Hedges' experience on the Buffalo, describing its beauty and pleading for its preservation. That he did and promptly received a letter in protest from Gib Walsh of Marshall, the secretary of the BRIA.

Walsh's letter was a gentlemanly statement compared to some of the verbal explosives that were to follow. But it established a pattern of criticism of us by the big dam boosters that would persist until the last trump. Their sales pitch was that Searcy and its sister counties languished in poverty from which condition they would never recover unless they received a transfusion of federal funds into that free-flowing vein, the Buffalo River. Had not that remedy already worked wonders on the nearby White, Norfork, and Little Red rivers? Big dams were a panacea for every ill that the land and its inhabitants might be heir to. There was no alternative.

Anyone not actually living in the counties involved was an "outsider" and had no right to tell them what to do with *their* rivers; no matter that their program for the Buffalo would be underwritten by taxpayers all over the nation. The real Buffalo River landowners, the people who owned property in fee in its valley and on its banks, were totally ignored by the businessmen, newspaper people, and real-estate developers in towns like Marshall and Yellville. The bona fide landowners certainly wanted no dams, with their best bottom land permanently covered by the Buffalo.

Walsh and his coworkers decried the taking of land for a wilderness or park and ignored the fact that the big reservoirs they so avidly sought would require as much or more acreage than the proposed park.

The argument of the BRIA would become more devious, labored, and unreal with the passage of time. They would wind up tightly wedged in the corner of negativism, forced to demean the undeniably beautiful stream in every way that they could think of.

Meanwhile, very positive events were in the making for us nature lovers.

THE NATIONAL PARK SERVICE IN THE FIELD

Bill Apple was something of a conniver. He liked nothing better than to journey to Camelot, and there, to ferret out the choicest bits of news and to try to put ideas into the heads of those seated at the round table.

In a message to Evangeline Archer dated September 13, 1961, he revealed that he had "a very lengthy and satisfactory meeting with your President!" (Did he mean JFK?) They were, he said, very interested in establishing a recreational area on the Buffalo River and that the park service was sending a full crew to survey the Buffalo River in October. Then on September 19 came a letter to me from Bill Apple stating that the survey team would arrive on the twenty-eighth

Prof. Sam Dellinger, University of
Arkansas Zoology Department.
*(Courtesy of the University of Arkansas
Library, Special Collections. Picture
Collection)*

with Gene Rush and Harold Alexander of the Game and Fish Commission acting as guides.

In that communication the rest of us were pleased to note the U.S. Park
Service disapproval of Bill Apple's tributary dam idea. In his desire to mollify all
parties Bill wished to throw the Corps of Engineers a bone. He was *in* with them
on a number of projects, but on the Buffalo the rest of us didn't want them to
have even a nibble.

The next message in regard to the inspection team was from Gene Rush to
Sam Dellinger with some final instructions. He stated that Raymond Gregg,
superintendent of Hot Springs National Park, had worked out a schedule for the
survey team and that I had been invited to accompany them for part of it.

Gene Rush, chief of the Game Division of the Arkansas Game and Fish
Commission, was an old acquaintance of mine and both of us former students
of Sam Dellinger's.

Gene, a tall, robust, and intelligent fellow, had been born in Newton County
and was the one who knew the Buffalo country best. In the 1960s he was in
charge of the commission's effort to restock the black bear in Arkansas. It was a
successful undertaking, and Gene is the man who put the bear back in the "Bear
State."

But with all of these favorable events and more yet about to transpire, the
plot thickened.

On September 28 the *Marshall Mountain Wave* displayed this prominent
headline: "Funds for Study of the Buffalo River Project Approved."

The public works bill carrying funds for the study of Lone Rock and Gilbert
dams had passed the House and Senate.

Gene Rush, chief of the Game
Division, Arkansas Game and Fish
Commission. *(Courtesy of Mrs. Gene
[Jean] Rush)*

Congressman Trimble stated that a bill authorizing approval of the projects would be introduced early next year and that work would begin soon after approval.

The *Wave* also reported that the National Park Service was making a study of the river preparatory to the establishment of a park along it. Congressman Trimble said that the lakes created by the Lone Rock and Gilbert dams would only add to the attractiveness of the park, if one was authorized.

The *Wave* had been assured by Senator Fulbright's secretary that the senator would lend full support to the Buffalo dam projects when the time came.

How precarious had nature's realm become, that its fate depended upon the ever-fluctuating opinions floating through the halls of Congress. Our opponents were to learn, as well as we, the unpredictability of that political tide.

But important decisions are made at home as well as in Washington. Any citizen wishing to apply his own evaluation to a problem, and wishing to act accordingly, can influence the course of events. Tom Dearmore was such a citizen, and in a letter dated September 8, 1961, to Bill Apple, explained his quandary; his opinion in transition to a better decision for the Buffalo:

> Thanks for your letter of the fifth regarding the proposed National Park Service study of the Buffalo River area this fall. We carried an article August 17 quoting an assistant Secretary of the Interior, announcing that this survey was forthcoming, but I did not know just when the survey team was slated to arrive. ...
>
> I would like to sit down with some group, as you suggested, and analyze the Park

Service's report after it is submitted. Our informant in the article says this should be by the end of this calendar year.

I feel fairly certain now that the Department of the Interior wants to recommend some type of development for the Buffalo valley, and I believe folks in this region will consider whatever is suggested with open minds. However, I am personally not ready to abandon the U.S. Engineers if they want to make a serious study of the valley and if they get the study funds at this session of Congress. I think leaders in this region will also want to confer with the Engineers on the matter.

If the Engineers think the dams are feasible, there will be considerable support for the two dams and the issue can be very divisive. It is my opinion at this stage that the dams would have a more stimulating effect on the area economy than the park, but I'm willing to consider any evidence to the contrary with an open mind.

October 1, 1961, A Day to Remember

Senator Fulbright had indeed obtained an appropriation to fund the National Park Service inspection team expenses to the Buffalo River. It was for $7,000 in 1960 dollars, a ridiculous sum when compared to monies assigned to the Corps of Engineers: $135,000 just to study that stream. But the park service would make do with what they got. As it turned out, it was as decisive an expenditure as was to be made during the prolonged contest now beginning.

I was more than pleased to be invited to attend that inspection tour, although I could only be present for the first three days.

Greetings were exchanged at the Riverview Motel in Jasper before the first day's tour on October 1, 1961. Present from the National Park Service in the nation's capital were Carl Schreiber, in charge of the inspection team; John Russell, architect; Ernie Shultz, geologist; Jim Howell, historian; Hodge Hansen, planner; Raymond Gregg, superintendent of Hot Springs National Park in Hot Springs.

On this occasion the Arkansas Game and Fish Commission provided invaluable assistance in the form of personnel and equipment. From their ranks were Harold Alexander, wildlife biologist and naturalist; Gene Rush, principal guide of the party; Fred Bell, game warden and Jeep driver from Jasper; Jack Atkins, commission photographer and Scout driver; two guests, Dr. Dwight Moore, eminent botanist from Fayetteville, and I had been invited as a national park partisan and photographer, with special attention to 8mm movie equipment.

This was the last time that the Game and Fish Commission would take such an open and active part in the contest over the Buffalo River, not because they didn't want to but because of pressure brought to bear by the BRIA, by the legislature, and by state officials under the influence of the BRIA.

The first day was to be a tour of the upper Buffalo, unlike any other before or since. We were going to ride up that deep and narrow gorge in two International Scouts and one Jeep, over an ancient wagon road once negotiable by horses and mules but abandoned after the coming of the jalopy. We would need all four wheels and the winch as well on that route.

From up on Mount Sherman we took the long, steep road down to Camp Orr (the Boy Scout camp), a descent of thirteen hundred feet. When we dropped over the rim of the gorge into the blue infinity of that abyss there were gasps of surprise and admiration from our visitors. When we arrived at the bottom on the banks of the Buffalo, they all declared that they had no idea that there was any such grand scenery anywhere in this part of the country. "It's better than anything we've seen," they said.

From their enthusiasm it was obvious that this was a place that the National Park Service was going to want in their system and that we were going to have a real monkey wrench to throw into the Corps of Engineers' machinery.

From Camp Orr we took off upriver, fording it below four-hundred-foot-high Buzzard Bluff and exploring the hillside west of the bluff for a vertical shaft cave that Gene Rush knew about. It went straight down into the limestone for one hundred feet, and we wondered what might be at the bottom of it.

At the mouth of Hemmed-In Hollow we waded the Buffalo and walked up into the box canyon to see the two-hundred-foot waterfall, which was not much more than a trickle that day in October. We exited from Hemmed-In Hollow on the almost-impassable Center Point road, up the long hill, and then down again to Ponca and Lost Valley. The tree butchery previously discussed was still in evidence but was not a disqualifying factor. The earth features would remain and new growth would soften the injury, Raymond Gregg said, and that it has in the years since.

On our way to Cave Mountain for a final look at the upper Buffalo country we came, just south of the entrance of Lost Valley, to a newly opened rock quarry that was to furnish metal for a stretch of paved road then abuilding in Boxley Valley. Against the hill on the east side of the valley sat a line of giant yellow beetle-like machines, Le Torneau earth movers, to chew up and spit out dirt and gravel in the right places for a modern-grade, high-speed thoroughfare down this idyllic vale. Gone were the old stone fences, the sharp turns, and the forest patriarchs that once bordered that country lane. The inspection team could not know, but for me there was painful recall of what had been there, a recollection best borne in silence lest their enthusiasm be further dampened after the look at logged-out Lost Valley.

In Boxley Valley a considerate designer would have left that unobtrusive roadway as it was, to be black-topped *in situ*. But planners of modern thoroughfares go by the axiom that the best route between two points is a straight line, and that is what we were getting in the south end of Boxley Valley with its almost-nonexistent traffic count.

The Arkansas Highway Department would leave that completed stretch between Boxley and the Mossville hill as an example for seven years. Then they would come back in full force to perpetrate incredible alterations on all of the other roads into that bit of Arcadia still left in its Boston Mountain fastness.

But in 1961 under Gene Rush's leadership we ground our way up the nearly vertical Cave Mountain road to have a look at the enormous cavern where in 1862 the desperate Confederacy had established a facility to make saltpeter—an ingredient in gunpowder—from the large deposits of bat guano within. There the Confederates were surprised by a troop of Yankee cavalry who wrecked the operation. Parts of an old iron boiler still rest in the woods nearby for subsequent generations to ponder.

We had seen enough that day to stimulate lively conversation when we got back to the Riverview Motel. Hodge Hansen spread out the maps on the hood of the Scout for better scrutiny of where we had been and to better surmise possible boundaries for the proposed park.

He then said something to me that I cannot forget. He said, "Doctor, you are doing great things for the state of Arkansas."

That was a most kind compliment, and I accepted it for its intent, trying to conceal misgivings within. It was out of order at that point in time. Hodge Hansen was counting our chickens before they hatched. Could it be an omen for bad luck down the road? What I had done up until then wasn't very much: a lot of talk and a few inconclusive investigations.

Such optimism in the face of the Corps of Engineers' steamroller coming down the pike was disturbing. In spite of a now very likely National Park Service espousal of the proposed park along the river, we should, to be honest, not expect victory in any set-to with the Corps. Historically, their influence in Congress far surpassed all other federal agencies.

As soon as I arrived at home I sought out the aid of a new friend in Bentonville. Don Winfrey had retired from years of service as a geologist with the Pure Oil Company. He was an excellent draftsman, thoroughly experienced in the art of map making. He would be glad to draw off a professional-looking map of a national park along the Buffalo. The product of that undertaking was not inhibited in its extent. It contained over four hundred thousand acres, including a wide swath in Newton County, along with Marble Falls and the Boat Mountains in Boone County. It then narrowed down in the midsection of the Buffalo and expanded widely again near its junction with White River and the Sylamore Division of the Ozark National Forest.

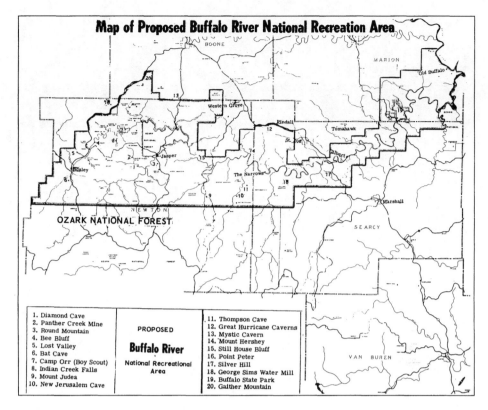

Map of Proposed Buffalo River National Recreation Area

1. Diamond Cave	PROPOSED	11. Thompson Cave
2. Panther Creek Mine		12. Great Hurricane Caverns
3. Round Mountain	**Buffalo River**	13. Mystic Cavern
4. Bee Bluff		14. Mount Hershey
5. Lost Valley	National Recreational	15. Still House Bluff
6. Bat Cave	Area	16. Point Peter
7. Camp Orr (Boy Scout)		17. Silver Hill
8. Indian Creek Falls		18. George Sims Water Mill
9. Mount Judea		19. Buffalo State Park
10. New Jerusalem Cave		20. Gaither Mountain

The first map of the proposed Buffalo River National Recreation Area, designed by Neil Compton and drawn by Don Winfrey. The final version was trimmed to one-fourth this size by the Department of the Interior.

Don and I were both proud of the result, and it was sent off to National Park Service headquarters in Washington, D.C. It was expected that it would be trimmed down, but we weren't prepared for the skinny product eventually brought forth by the park service.

More than one revision of the national river map was made over the next several years, and the park finally wound up with a little over ninety-five thousand acres within its boundaries. As many points of interest were left out as were included in it, but we had a viable creature nevertheless. At that size, the park would protect the immediate shoreline of the Buffalo River and would include that original bone of contention, Lost Valley.

Ken Smith, now at Berkeley, had been kept abreast of the goings-on in Arkansas and had become very interested in the role of the National Park Service as an active agency in conservation matters. He was interested in the possibility of a career in the service and had taken steps to join up. I encouraged him in that and furnished recommendations for his employment. Ken would accept employment in the U.S. Park Service, dropping his business course at

Berkeley, and for the next several years would, by quirks of fate, be at the right station at the right time to render us on the battlefront here in Arkansas invaluable service as an agent at large.

That December it was necessary to inform Paul Bruce Dowling of the expansion of the contest to include the entire Buffalo River and of Mrs. Primrose's increasing intransigence. The Lost Valley project was now moot and should be shelved. He agreed, suggesting that perhaps other Nature Conservancy projects could be activated in Arkansas. But from then on we became too deeply embroiled in the battle of the whole Buffalo to do much else. Sam Dellinger remained as president of the Arkansas Nature Conservancy for a few years, with the rest of us retaining our membership, but that was the best we could do.

THE FIRST HEARING AT MARSHALL

All of these shenanigans had not gone unnoticed by the Corps of Engineers. There was a fox in their chicken house, and it was time to load up the shotgun and fix the varmint good and proper.

First, they came up with a plethora of new, stunning, and irresistible projects for the Buffalo, in addition to Lone Rock and Gilbert dams.

Then they would do something that usually wasn't necessary in getting one of their big programs off the ground: they would hold a hearing somewhere (at Marshall, naturally) to give the American citizen a chance to have his say. Occasionally such hearings had been in Washington, D.C., but on many projects there had been no hearing at all.

Such hearings have no legal basis; they in no way constitute a court of law or an elective event. As they are conducted, we were to learn, they are a court after a fashion … a kangaroo court. But no matter; if you are for or against the project, you had better be there because the principal product of the affair is publicity. If there was to be no opposition to the big dams expressed at the hearing, their approval would be implicit in the press releases, and the federal agency and their backers in Congress would be home free.

The revved up motion in the Corps of Engineers' machinery is best related in an item printed in the *Arkansas Gazette* on Pearl Harbor day, December 7, 1961:

Engineers Begin New Study on Buffalo River

By Jerol Garrison of the *Gazette* Staff

The Army Engineers said yesterday that they had begun a study to determine the best method of developing the water resources of the Buffalo River and its tributaries.

The study will delve into all aspects of flood control, power generation and recreation, including the new "pump-storage" plan for producing hydroelectric power.

This latter method, which uses smaller reservoirs than the conventional method for producing power, may prove to be a compromise between persons who want to build dams on the river and those who want to preserve it in its natural state.

The Engineers said that one of the first steps in its study would be a public hearing to obtain the views of persons interested in the Buffalo, as well as governmental agencies. …

The present study was directed by the Public Works Committee of the United States Senate in 1958, but $30,000 to begin it was not appropriated until the 1961 session of Congress. In effect, it will be a completely new evaluation of the Buffalo River because of recent developments in the field of engineering.

The Little Rock District of the Army Corps of Engineers has been paying particular attention to the Taum Sauk hydroelectric power project being constructed by the Union Electric Company of St. Louis on the Black River in Southeast Missouri.

This project, not far upstream from the Little Rock District's Clearwater Reservoir, consists of two lakes, one of 40 acres and the other of 340 acres.

The 340-acre lake is on the Black River, and the 40-acre one is high up on the nearby Taum Sauk Mountain. Late at night, when electrical power is very cheap because of the low demand for it, a pumping station boosts the water from the lower lake up a 6,700-foot conduit into the higher one.

In the daytime, when electrical power commands a high price, the water flows from the high lake through turbines back into the lower one, producing a maximum of 350,000 kilowatts of electricity for residents of St. Louis 90 miles away.

At least that's the way Taum Sauk (named for an Indian chieftain) will work when it is completed in June 1963.

The Engineers' study will determine whether a project similar to Taum Sauk would be possible on the Buffalo River. They will also consider a different version—using two reservoirs on the river itself. Water would flow through the turbines of the upstream dam during the daytime and would be caught by the downstream dam. At night, pumps would return the water to the upstream reservoir.

Representative James Trimble believes the pump-storage system of power generation would make dams on the Buffalo economically feasible. It also is possible that the reservoirs could be built small enough to keep most of the Buffalo flowing in its natural state, except when the reservoirs were needed for flood control. …

The original Gilbert and Lone Rock proposals also will be reevaluated in the light of new information, and consideration will be given to only one large dam on the Buffalo.

"When you consider power generation, flood control and recreation there are dozens of possibilities," a spokesman said. "We will try to find the plan that is most beneficial and most desirable."

In this labored plan, total ruination of the Buffalo River (especially if we were to take into account some additions that surfaced later on) is recommended.

The pump-back projects deserve special comment. They are, to be honest, nothing more than Rube Goldberg systems to be set in motion with the waters of the river going around and around to come out here or there, like the music in the old jazz-age hit. Those later additions were to be pump-back jobs with, not pie, but lakes in the sky at Point Peter above Woolum and Compton above Hemmed-In Hollow.

With all these goodies held out to the BRIA and their true believers, an announcement of the hearing was awaited with anticipation by them and consternation by us anti-dam people.

Sixty-one went out quietly but sixty-two came in with a bang in the form of a public notice from the Corps of Engineers:

U.S. ARMY ENGINEER DISTRICT, LITTLE ROCK
CORPS OF ENGINEERS
Federal Office Building
700 West Capitol
Little Rock, Arkansas
3 January 1962

NOTICE OF PUBLIC HEARING

Pursuant to a resolution adopted June 20, 1958, by the Committee on Public Works, U.S. Senate, the District Engineer, Little Rock District, has been directed to make a survey (review) report on development of the water resources of the Buffalo River Basin, Arkansas.

The resolution reads as follows:

"RESOLVED BY THE COMMITTEE ON PUBLIC WORKS OF THE UNITED STATES SENATE, That the Board of Engineers for Rivers and Harbors, created under Section 3 of the River and Harbor Act, approved June 13, 1902, be, and is hereby, requested to review the reports of the Chief of Engineers on the White River and Tributaries, Missouri and Arkansas, published as House Document Numbered 499, Eighty-third Congress, Second Session, and other reports, with a view to determining whether the recommendations contained therein should be modified in any way at the present time, with particular reference to the feasibility of construction of a multiple-purpose reservoir at the Gilbert site, and provision of hydroelectric power facilities in the authorized Lone Rock Reservoir, on Buffalo River, Arkansas."

In order that the required report may fully cover the matter, a PUBLIC HEARING will be held in the National Guard Armory at Marshall, Arkansas, on 30 January 1962 at 10 A.M.

To liven up the contest, an unauthorized preliminary report from the National Park Service became known to us just before the hearing. It marked the position of the battle lines from which the struggle for the Buffalo River would be fought during the next ten years:

The Buffalo River is the last and the choicest of major free-flowing streams of the Arkansas Ozarks; it affords an outstanding opportunity for preservation of a unique segment of the outdoor scene of America. Here can be found an expansive climax of wildness, ruggedness, and elevation contrasts in two sections of the Ozark plateaus Province of the Interior Highlands Division of American Landscape Forms.

Geologically, the area is outstanding and typifies the Arkansas Ozarks with numerous caves, waterfalls, persistent springs, interesting side canyons and the impressive bluffs of the Buffalo River. The known flora of the study area is perhaps the most diverse of equal area extent in Arkansas and ranks noteworthy even on a national basis.

The varied wildlife of the area constitutes one of its chief attractions with practically every wild animal common to an oak-hickory type forest found in great numbers.

More than twenty archeological sites are known to be within the area and it is likely that many others are present.

The area, which is centrally located with respect to large urban populations, is well suited to provide an increasing volume of outdoor recreational opportunities for the nation.

Getting Our Act Together

These developments served as strong stimulants to cooperation and communication between the widely dispersed park supporters.

Harold and Margaret Hedges read of the hearing in the *Kansas City Star* and wrote to us asking for the proper procedure at such hearings. The O.W.W.C. had heretofore engaged in floating Ozark streams for the pure fun of it and had not attempted to influence public projects such as this. Some of their members would come, among them Margaret Hedges and Ray Lockard, her father. Harold would not be able to be present.

We sought the answers to questions and distributed them to as many compatriots as possible: Where the hearing would be held; the date and time of day; the length of statement permitted; the use of display material, maps, slides, and charts; disposition of written statements.

Those of us from the Fayetteville area didn't know what to expect any more than those in other quarters, but we were becoming a clearinghouse for all the questions about the Buffalo. Evangeline Archer with her typewriter and I with my home movies and my modest collection of color slides and sometimes grandiose ideas were kept increasingly busy as this storm on the Buffalo came closer. As an example, in the first week of the new year (1962), Doris Larimore, a businesswoman from Rogers who sometimes served as secretary pro tem for Mrs. Archer, relayed a request for a program on the Buffalo River question from Mrs. C. J. Craft of the Green Thumb Garden Club of Green Forest, Arkansas. On January 17, I traveled to Green Forest, showed the garden club ladies the film, to which a sound stripe had been added, and finished with a statement as to the desirability of the national park over dams on the Buffalo River.

As a result, Mrs. Craft and her garden clubbers drafted a resolution opposing the Engineers' program that they later submitted to the hearing officer at Marshall. These ladies remained as supporters of our conservation effort throughout the next ten years.

Our 8mm home movies were admittedly amateurish but they were not too long and contained points of sustained interest. We would later find that they were more effective than the too-smooth style of many professional films.

The appearance before the Green Thumb Garden Club was the first of three or four hundred similar programs that would be given over the next ten years to various women's clubs, service clubs, conservation groups, schools, and other organizations in Arkansas, Missouri, and Oklahoma.

It would mark the beginning of an accelerated tempo to my regular routine. Sometimes it was difficult to realize how much was being accomplished in a day's time. One simply turned to and did what had to be done then and there. To begin with, there were all of the tribulations of the practice of medicine, which took priority, but there would be those moments during the day when a letter could be put on tape for my personal secretary, for a phone call, or for an hour or so off for an appearance at a local service club. In the evening, whatever time could be found between emergencies at the hospital would be devoted to running prints in the darkroom, editing film, sorting slides, writing up reports, and reading the latest news about conservation in Arkansas. Then there were those meetings of our cohort with everyone loaded with their own special ideas trying to get a word in edgewise. Those sessions would be once or twice a week here in Bentonville or at Evangeline Archer's or Sam Dellinger's in Fayetteville, usually in the evenings, with general meetings once a month in available hospitality rooms in hotels, churches, or public service buildings.

An important responsibility was to keep abreast of every news item concerning the Buffalo in all the area newspapers. Thus it was that we were happy to see an excellent article pleading for preservation of the Buffalo in the *Arkansas Democrat* for January 12. It was by John Heuston, an outdoor writer for that paper. This devoted outdoorsman was to be one of our staunchest soldiers in the Buffalo River war and would suffer indignities for his unswerving stand against the big dams.

Searcy County Sounds Off

Headlines in the *Marshall Mountain Wave* grew larger and blacker as the month went on. The following are quotes from that newspaper:

Trimble Pleased with Budget for River Projects

Congressman Jim Trimble is well pleased by the request in President Kennedy's budget for work to be done in the Third Congressional District.

Kennedy asked for funds to continue construction in the Arkansas and White River basins. Beaver Dam would get $14 million, Dardanelle Dam, $16 million, Greer's Ferry, $4.9 million, and $3 million for bank stabilization on the Arkansas River.

$100,000 were asked for General Investigations which would allow for studies to be continued on the Gilbert and Lone Rock dams. ...

Searcy County Organization Backs
Engineers' Proposal for Buffalo

At a meeting of the Buffalo River Improvement Association Monday evening Association members voted unanimously to support the construction of Lone Rock and Gilbert dams on the Buffalo River, or *whatever projects the Corps of U.S. Army Engineers recommend for the improvement of the river.* ...

... Members of the Association expressed strong opposition to the National Park project ... attractions along the Buffalo were not sufficient to attract a large number of tourists or vacationers ... Park Service Personnel would prohibit and prevent private citizens from making business investments along the river ... this area would extend for several miles on each side of the river ...

No other attraction offers as many or such varied recreation sports for groups and families as large lakes ... The lakes on the Buffalo River is the only plan which would provide a higher standard of living for the citizens of the county to which they are entitled.

Engineers' Hearing on Buffalo
at Local Guard Armory Tuesday

A hearing to determine what can be done on the Buffalo River for the most benefit for a majority of the people will be held in the National Guard Armory Tuesday beginning at 10:00 A.M.

In the past several years many plans have been proposed for the Buffalo River by a group of individuals who opposed the creation of lakes in central and northern Arkansas. Their latest plan was to make the entire Buffalo River into a million-acre national park. ...

Mayor Ralph Treadwell announced Tuesday that all business houses in Marshall had agreed to close their doors from 10:00 A.M. to 12 noon to attend the meeting (hearing).

Ready or not, in the early morning hours of January 30, 1962, three of us took off for Marshall for a tilt with the Corps of Engineers. Along that day with me was Doris Larimore of Rogers and lawyer Clayton Little of Bentonville. Doris had a friend in Marshall, John Driver, a lawyer whom she had known at the University of Arkansas and who had confided his opposition to the big dam program to Doris on another occasion. We had hoped that he might initiate a movement in Searcy County to counter Jim Tudor's BRIA.

A short interview with him served to squelch any such possibility. There was so much hostility in that town to any big dam opposition that anyone taking such a stand would be in danger of social, financial, or even physical retribution.

The hearing, as we expected, was heavily tilted in favor of the agency calling it. In such circumstances, the Corps of Engineers participates as the author of the proposal and as the judge and jury to determine its fate as well.

That chilly January morning the new metal Armory building was jammed with dam-the-Buffalo enthusiasts. As the mayor had requested, the stores and offices were closed; school had been let out so that the high-school and junior-high students could attend the hearing, all wearing lapel cards with the caption: LET'S DAM THE BUFFALO.

Conducting the meeting was Colonel Dalrymple, district engineer, with headquarters in Little Rock; Major E. T. Williams; Mr. Paul Adams; and others, all from the district office in Little Rock.

Testimony was arranged in descending order of importance, with representatives of divisions of the federal government appearing first.

No officials of the executive or legislative branches of government were present; but Mr. Trimble was represented by Tom Tinnon, a lawyer from Mountain Home, with a predictable pro-dam speech.

Representatives from the Department of the Interior, including the Bureau of Sports Fisheries and Wildlife and the chief hydraulic engineer, pledged cooperation with the Corps of Engineers.

Two units of the REA furnished spokesmen recommending the completion of the dam because "people in this area have just entered into the electrical age."

Of state agencies, the Arkansas State Highway Department spokesmen pledged cooperation with the Corps if the project went through. The Arkansas Game and Fish Commission and the Arkansas Publicity and Parks Departments did not furnish spokesmen but did have observers present.

Chairman Tudor of the BRIA then introduced a series of local groups and individuals.

The first of these and principal spokesman was the Reverend L. R. Winners. From him we heard of the day of glory at hand when those dams were done, winding up with the exhortation: "Let's all pray for the dams."

Then the high-school students had their say and the chamber of commerce, the Rotary Club, and numerous individuals from surrounding towns and counties, all for the big dam plan, among them Tom Dearmore from Mountain Home.

The first and most conspicuous dam opponent made the podium late in the day, Bill Apple, representing the Arkansas Wildlife Federation, and a new outfit, GROW (the Great Rivers Outdoor Writers). His opposition to the big dams was firm and aggressive, but he had had to bring in his pet project, the small upstream impoundments, which neither side really wanted. He envisioned a wilderness area or a recreation area along the Buffalo to be administered by the Arkansas Game and Fish Commission, something the rest of us hadn't heard of. However, he endorsed the national park proposition along with the other ideas.

Also from Little Rock was Cleveland Cabler, the president of the Arkansas Audubon Society, giving a good address against the dams and endorsing the national park plan.

Then there was Ray Lockard, Margaret Hedges' father, representing the O.W.W.C., with a long, romantic description of the river, pleading for its preservation.

Fran James of Fayetteville, one of our special group, spoke for the Arkansas League of Women Voters, endorsing the park plan. She was followed by her husband, Dr. Doug James, ornithologist from the zoology department of the University of Arkansas and an original worker with our group but representing the Audubon Society.

Ben F. Payne of Fayetteville brought a petition containing 190 names favoring the national park proposal.

Maxine Clark (Mrs. Joe), destined to be an important personality in our organization, represented the Fayetteville Council of Garden Clubs with three hundred members, favoring the national park plan.

There were others: Dr. Horace Marvin, M.D., from the University of Arkansas Medical School, along with one of our old anatomy professors, "Daddy" Langston, both for saving the river, and various people of a similar mind milling about like lost souls in that sea of river dammers.

We weren't well organized and were put off until the last, with little or no opportunity to get a word in. I handed in a statement and let it go at that.

But this exercise in futility did serve one good purpose for those of our convictions. It did get us together in one place so as to become acquainted with important personalities in the contest that was to follow. After that, we knew who the Hedges, Ray Heady, Hubert and Mary Virginia Ferguson, Dean Norman, Chester Kelly, Nancy Jack, the McAllisters, John Heuston, Maxine and Joe Clark, and many others were and how to communicate with them. It was, after all, a worthwhile function, bringing us together and pointing up the urgent necessity of getting a hard-hitting anti-dam organization together to contest the preposterous fate now threatening this truly lovely Ozark waterway.

Our attitude was best expressed by the first sentence in the statement submitted by Ray Lockard of the Ozark Wilderness Waterways Club: "Five hundred years before Christ a Chinese philosopher wrote, Nature is already as good as it possibly can be. He who seeks to improve it will spoil it. He who tries to direct it will mislead it and become lost himself."

For reasons that we are not sure of, a quick decision was not forthcoming on this hearing. For one thing, the Corps of Engineers may have been surprised by the opposition from so many parts of the country. Criticism was not something that they had previously had to deal with, and they might possibly have thought that it would be best to revise their plan downward, to throw Lone Rock to us wolves and concentrate on the Gilbert dam.

Whatever the reason, we were going to have a little time to put together what we hoped would be an effective anti-dam organization.

ANTI-DAM SENTIMENT—A RISING TIDE

The first requirement was going to be the question of leadership. The anti-dam people urgently needed someone with name recognition, someone from the Ozark region, familiar with its environmental problems, someone willing to volunteer time and effort, preferably retired and energetic.

Sam Dellinger was a first choice, formerly head of the zoology department at the University, a former member of the Game and Fish Commission, and a well-known archeologist. But Sam had not been able to get out on investigative trips much. He said: "I am too old and fat for that sort of thing."

Bill Apple and Cleveland Cabler in Little Rock were the next candidates, and we approached them on the subject, but both declined for various reasons.

I wanted very much to generate the primary activity in Little Rock, the state capital and more or less the center of things. Bill Apple had agreed to assist in getting a group together down there.

He did arrange for a meeting between me and some of the conservation-minded citizens of Little Rock. One of these was H. Charles Johnston, Jr., a husky, ebullient fellow full of enough enthusiasm for a half dozen people.

Another was Everett Bowman, an engineer, who loved hiking and hunting; he was low key and dependable, a man to rely on when needed.

Following the Little Rock meeting, I was invited to meet with the Fort Smith Chamber of Commerce, which body afterward officially endorsed the national park program for the Buffalo.

That favorable action was reported by John Heuston in the *Arkansas Democrat* and begot a predictable reaction from the executive secretary of the BRIA, which epistle was sent on to us by Bill Barksdale.

In it Winners pleaded for a reversal of the chamber's resolution for these reasons: (1) It is not the duty of a chamber to decide on problems of other communities; (2) The Fort Smith chamber had not studied the problem and could not make an intelligent decision; (3) Fort Smith would not be affected by the Buffalo dams; (4) The Fort Smith resolution was an act of unneighborliness, and the BRIA would like to know the names of persons requesting such action.

The Fort Smith Chamber of Commerce did not respond, and there would be much more in the way of confrontation between the citizens of that city and the BRIA down the road.

The need for action to save the river was rendered all the more urgent by the publication of a document from the proceedings of the Eighty-Seventh Congress dated February 27, 1962: It was H.R. 10430, Congressman Trimble's bill for approval of Lone Rock and Gilbert dams on the Buffalo River, with amendments to allow generation of power in addition to flood control.

San Francisco and David Brower

In a letter to Bill Apple the reference to my trip to California involved two weeks' duty for training at the San Diego Naval Hospital. But there was more. Ken Smith was then at Berkeley and upon request furnished information on how to get in contact with David Brower, executive director of the Sierra Club, in nearby San Francisco. The Sierra Club had set the pace in opposition to ridiculous, environmentally damaging undertakings in both the public and private sectors. It seemed to me that the Sierra Club might be an ideal organization to take the Buffalo River under its wing and to do for us what they had done in California and elsewhere. Could we function as a unit of the Sierra Club and get anywhere with our program for the Buffalo?

The interview with David Brower was pleasant but inconsequential.

We would first have to organize a chapter in Arkansas and to continue the campaign already underway under the colors of the Sierra Club. Their experience and know-how would be most valuable to us neophytes. I left Mr. Brower with the understanding that we would notify him if we could get a chapter started.

But their dues were nine dollars per year, a sum that at that time was just too much for the people we expected to recruit. Brower had not volunteered to send an agent out to investigate or help us organize, something that might have made a difference.

Then our opponents were calling us "outsiders" in every other breath. There was not in the field any organization tailored to meet the specific problems of the Ozark region, to speak out and defend against Lone Rock and Gilbert; we all wanted to be Ozarkians defending the Ozark scene, and that is what we would be.

The BRIA Makes Ready

During this time the BRIA was also undergoing reorganization. New officers were elected with James R. Tudor staying on as president, as did Mrs. Okla Ruff as secretary-treasurer. Gibson Walsh, a local abstracter, was elected as a new vice-president.

The most important business was the hiring of Rev. L. R. Winners as a full-time executive secretary, with offices provided by the county judge in the county courthouse. From that source we would be the recipients of much sulphurous commentary before it was over.

One of BRIA's first acts at this time was to bring up a favorite bone of contention. It was the matter of who actually owns the river, its waters, and the bed on which they flow. The BRIA claimed that adjacent owners could produce warranty deeds proving that those waters, the bordering gravel and sand bars, and the riverbed itself were their property. The Buffalo was, according to them, not a navigable stream, which was a critical point of law. If it was not navigable, the adjacent property owner could claim the stream if he wished. According to the BRIA, some of the property owners had consulted a local lawyer who stated that they could legally fence off the river, if they desired, thus closing it to recreational use. They could close all access roads ending on the riverbank, also under the same stipulation.

This turned out to be a knotty problem that would only be settled as a result of a lawsuit on the Mulberry River years later, if it was indeed settled then.

JUSTICE DOUGLAS ON THE BUFFALO

As February 1962 drew to a close, Evangeline Archer and Margaret Hedges exchanged correspondence concerning something that would turn out to be a major media event on the Buffalo. Both of them were in contact with Justice William O. Douglas of the U.S. Supreme Court concerning his proposed trip to the Buffalo River. The O.W.W.C. would be the chief planners and expediters of that event.

It was an affair I wouldn't want to miss, and upon hearing of it I took steps to obtain an invitation by urging Evangeline to have Mrs. Hedges include me in to photograph the trip. In that I was successful, being assigned as a bowman to Dick Mosley in his new twenty-foot Grumman canoe. The O.W.W.C. would meet Justice Douglas and his wife, Mercedes, at the Harrison airport and take them to Ponca on the upper Buffalo.

The end of April arrived with all of us as ready as we could ever be for the big day on the Buffalo, April 30 and May 1, 1962.

There was a mass of humanity milling about at the Ponca low-water bridge, with canoes galore along the banks of the stream. I sought out Margaret Hedges, whom I had met at the hearing and was introduced to her husband, Harold. He looked every bit the part of the prime explorer of Ozark rivers. Tall and crewcut, he resembled pictures of Charles Lindberg, a proper fellow to show us the way.

Reporters were there from everywhere: the *Kansas City Star,* the *Arkansas Gazette,* the *Arkansas Democrat,* the *Harrison Daily Times,* the *Southwest American,* and more. They gathered around Justice Douglas with hundreds of questions, while the rest of us ranged about in the crowd renewing old acquaintances and making new ones.

Finally, the great float got underway with an unspeakable miscue on the part of our number-one guest being barely avoided. There was just enough air space under the low-water bridge to permit a canoe and its passengers to pass underneath. The proper routine was for the occupants to bend sharply forward so that nothing except perhaps the backside would make contact with the rough concrete. Harold, who was the paddler for Douglas, neglected to inform his charge of this, and the Justice bent backwards, a difficult twist at best. His nose cleared the concrete undersurface by a millimeter or less, and we all heaved a sigh of relief when he emerged unbloodied on the other side; but it all went well thereafter.

There was plenty of water in this upper reach of the river to carry us on at a good clip. The great palisades rose above the lush forested riverbanks on one side or the other. At that time no channelizing had been done by P. W.

The Justice William O. Douglas party at the Harrison, Arkansas, airport, May 1962.
Left to right: Ray Lockard, Mercedes Douglas, Harold Hedges, Justice Douglas, Margaret Hedges, Harold Alexander, and others unknown; Laird Archer, *far right. (Courtesy of Nancy Jack)*

Justice Douglas, *bow,* and Harold Hedges, *stern,* on the Buffalo River. *(Courtesy of Nancy Jack)*

Evangeline Archer, the Ozark Society secretary, greeting Justice William O. Douglas at the Harrison airport. *(Courtesy of Nancy Jack)*

Justice Douglas going under the Ponca low-water bridge, wrong side up, May 1, 1962. *(Courtesy of Nancy Jack)*

Yarborough, owner of the Valley-Y Horse Ranch below Ponca, and the great beech trees stood in splendor along its course.

With the expert navigators of the O.W.W.C. in charge we ran the course to Big Bluff and there beached our craft, pitched tents, built our campfires, and gathered around Mr. Douglas to hear his commentary on conservation problems in the U.S.A. and his opinion of what he had seen so far on the Buffalo River.

That evening as darkness descended on Big Bluff the scene that had been displayed across the nation in *Time* magazine was recreated. It was better than anything that could be recorded in color, on paper or film, and we knew that we had a cause not to be denied.

Later that evening, we became aware of faint lights coming down the precipitous west slope of Big Bluff across the river where the vertical face of the bluff shelved off. Who or what it might be we couldn't imagine, but soon three tried-and-true bushwhackers appeared on the west bank and waded across to join the activity. It was Gus Albright and Jack Atkins, employees of the Arkansas Game and Fish Commission, along with John Heuston of the *Arkansas Democrat,* come to interview the eminent jurist. They had missed us at Ponca and proved their devotion to the "save the Buffalo" idea by this impossible performance on the dark hillside. How they ever got back up again we never knew, but that they did.

Daybreak under Big Bluff was most impressive, with heavy valley fog rising up the face of the great precipice like a curtain. The Justice wanted to climb the five-hundred-foot cliff, and at the lower end of it we tied up our canoes and undertook the laborious ascent. Justice Douglas didn't miss a lick, and soon he and his wife, Mercedes, arrived among the stunted oaks and cedars on the very topmost ledge. From there the people left behind looked like an assembly of ants on an anthill down below.

Mercedes Douglas carried two Leicas around her neck and recorded the progress of the Judge all the way. On that topmost rim she held onto a tree with one hand and leaned far out to get a good shot of William O. up there on his perch. We all breathed a sigh of relief when she was safely through.

From Big Bluff we made it on down the river through the lively rapids without mishap to the mouth of Hemmed-In Hollow. Since I had been in there a few times I was honored to serve as a guide into the then-trailless box canyon to the two-hundred-foot-high falls. There the uninhibited Puck Acuff and a few others stood under the spray for a morning shower bath but not in the buff.

In the Eye of the Storm

For lunch we stopped on the shingle at the mouth of Hemmed-In Hollow under wild azaleas on the cliffs above. There I had opportunity to describe to Justice Douglas our plans for a regional organization to save the Buffalo. He agreed with everything but had no new or startling suggestions for us bold iconoclasts. At least he knew who we were and might possibly be able to lend a hand at the right moment later on.

At midday great masses of billowing cumulus clouds gathered overhead, and soon we heard the rumble of thunder as we drifted past the Buffalo's spectacular palisades. Then a couple of miles above Buzzard Bluff, a four-hundred-foot promontory, the rain came in torrents. We were in the midst of a watery world, both above and below, and through it all out of the curtain of rain and mist appeared Harold, our leader, and his charge, apparently enjoying this demonstration of nature. Most of us had brought rain gear and were fairly dry as we waited out the storm. It was then that a peculiar situation was noted. The Justice had on a serviceable poncho but was not wearing his hat. He had to be as wet inside as was everything outside. But it was not for us to wonder why.

An estimated one and one-half inches of rain fell in less than forty minutes and then was quickly over. We proceeded on down river in that freshly washed green world. Then as the lead canoes drew near to Buzzard Bluff, those of us in the rear echelon heard loud shouting from those on ahead. When we arrived under the sheer limestone face of Buzzard Bluff, we saw that no one was in trouble but that those were shouts of joy and amazement. They were stopped on a small gravel bar to witness a sight that none of us had seen before. Leaping waterfalls spouted away from the top to strike narrow ledges lower down and then on into the river below. It was as fitting a display as nature could contrive for the eminent, outspoken conservationist for whom all of this had been arranged.

From there it was on to Erbie where we camped the second night across from Goat Bluff. From Erbie the party broke up, with each going his separate way after this epochal voyage on the Buffalo. From that time on, this river of contention would figure in the news at local, state, and national levels, a fact that for us was most welcome.

Favorable and interesting write-ups appeared in all the news sources previously mentioned. Ray Heady gave a graphic description in the *Kansas City Star* of the spectacular rainstorm that we came through. He stressed the importance of our Supreme Court justices, saying that of our population of 180 million there are only nine selected to that high office. In comparison, he said, if we had 180 million rivers in the land, the Buffalo should be one of the best nine.

A few days later we were more than pleased to note the publication of an account of the Douglas visit in the *New York Times*, May 20, 1962:

Plan for Ozarks Dam Stirs Fight
Justice Douglas Aids Nature Lovers
Who Oppose the Project
Special to the *New York Times*

LITTLE ROCK, Ark., May 19—The Buffalo River, a spirited stream that flows a course lined by rocky bluffs, is a haven in the Ozarks for people grown tired of city living.

Justice Opposes Dams

Justice William O. Douglas of the Supreme Court recently made his opinion known. Arriving in the Buffalo country for a three-day canoe trip as the guest of a Kansas City sportsmen's club, he told reporters that the Buffalo was a "unique" stream. Arkansas should try at all costs to preserve it, he said.

Justice Douglas' remarks, which were reported in detail in Arkansas newspapers, also included criticism of the Army Corps of Engineers, the agency that is studying the proposed dams. "When you get a bureaucracy going, it's hard to stop them," Justice Douglas said. "They just keep on going."

Reaction to the Douglas visit by the BRIA was, as expected, verbally violent. Tudor's paper, the *Marshall Mountain Wave*, for May 3, devoted most of its front page to that event claiming that:

… The part of the river seen by Douglas would not be affected by the dams.

… He didn't hear both sides of the question.

… Fishing would be better above the proposed reservoirs.

… 72% of the homes in Searcy and Newton counties don't have running water or flush toilets.

… The annual per capita income in Searcy County is $468 and in Newton County $346.

Searcy County has lost 27% of its industrial payroll while reservoir communities have had a 34% increase.

… The tuberculosis rate in Searcy County is more than twice the national average.

… ARE THESE THE CONDITIONS TO BE PRESERVED? SHOULD THE PEOPLE OF SEARCY COUNTY BE FORCED TO FORGET THE DEVELOPMENT OF THEIR GREAT-EST RESOURCE, THE BUFFALO RIVER, FOR QUESTIONABLE AESTHETIC VALUES?

… All statements by opponents are based on false and exaggerated premises.

… Mr. Douglas is quoted as saying that there was a surplus of power from dams. This is in direct contradiction to known facts.

… Supporters for the dams have tried to "smoke out" the real opponents. It is well known that private power companies have opposed all federal projects that include power units.

... The BRIA has never been outspoken towards the few landowners who have opposed the dams.

... The recent hearing at Marshall was not unfair to the dam opponents. Col. Dalrymple would have stayed all night.

... The opponents of the dams have made public statements that they would spend one million dollars to see that the dams were not built.

... Who paid for Mr. Douglas's trip to the Ozarks?

... The Buffalo River Improvement Association is as poor as the people that support it. How can we hope to fight a million dollars?

With that the *Wave* printed the text of a telegram sent by the BRIA to President Kennedy, repeating much of what is quoted above.

The W. O. Douglas trip to the Buffalo was basically not a decision-making event in regard to politics or law. It in no way committed any department of government to a decision one way or the other. No law was exercised any more than was the case at the recent Marshall hearing.

The importance of Justice Douglas's presence and what he had to say cannot be overemphasized. It attracted nationwide attention to a truly beautiful waterway that was about to be destroyed by senseless ideologues.

Commentary by Harold Hedges in a letter to me emphasizes the above opinion:

May 4, 1962

We considered the trip down the Buffalo a tremendous success and hope that you did too. We are especially grateful to you for being so helpful with the side trips up the Bluff and into Hemmed-In Hollow and know that you contributed much to our pleasure and that of the Justice.

We have a note from Justice Douglas today and enclosed with it is a copy of a telegram from the BRIA that the justice found waiting for him upon his return to Washington. I quote: "If you render decisions in cases you hear in court with as little knowledge as you possess about the Buffalo River, and with no more consideration than you showed the areas along the other 150 miles of the stream, then may heaven help the poor souls who must depend on you to mete out justice."

We are speechless but we did apologize to Justice Douglas for this statement. What is your reaction?

Sincerely,
Harold C. Hedges

My reply to Harold on May 7, 1962, was more prophetic than I would now care to admit. The big dam enthusiasts were for a fact defenseless, increasingly committed to a position of negativism as they became evermore deeply mired in confusion and frustration, that despite the presence on their team of the unconquerable Corps of Engineers:

> We have made further progress in the organization of a regional conservation group for the Ozark area and the committee did approve shortening the name to Ozark Society which I think is much better. We have a constitution and bylaws in the process of being printed and will send copies to you.
>
> Neil Compton

In that message to Harold we see for the first time the name of the new organization that we had been laboring so hard to bring forth.

Along with the W. O. Douglas tirade in the *Marshall Mountain Wave* of May 3, 1962, there was another quarter heard from.

In the editor's column entitled "Here and There about Town," this information was submitted:

> Much has been said and written about the Buffalo River during the past week. Last Thursday evening and Friday morning the state daily newspapers of Little Rock carried a release purported to have been written by one Charley McRaven, who claimed to be a property owner of Searcy County, and opposing the construction of the proposed Lone Rock and Gilbert dams. The *Gazette* headed the article thusly "Buffalo Dams Are Opposed for the First Time." What a laugh. We thought that Bill Apple, the Outdoor Writers and Ray Heady claimed that distinction.

Later in the month I was to learn more about this "one Charley McRaven." On May 23, 1962, I received a letter from him which was an eye opener:

Dear Dr. Compton:

I was very pleased to learn from *Democrat* writer John Heuston that your group planned to form, and even more pleased to see in today's paper that this had become a fact.

I trust you have been following the issue of the Buffalo quite closely for the past few weeks especially, and you probably know that it was I who incurred the wrath of the Buffalo River Improvement Association by helping draw up and circulate a petition opposing the dams. Because the BRIA erroneously linked my company, AP&L, with my activity, I have found it prudent to bow out gracefully as far as any overt action as spokesman for the property owners is concerned.

The opponents of the dams who have the best reasons for being such (and who can command the most sympathetic ear, I believe) are those farmers who own property in the river basin who would lose this land. These people, I must warn you, are as violently opposed to a park or recreational area as they are a dam. They do not like the idea of restrictions being placed on the land they have been heretofore free to call their own. I have discovered this in talking to numbers of them and feel it may be useful to you if your group is to use the stand they have taken in your plans to oppose the dams. ...

Concerning the valley opposition petitions—these have several hundred signatures, with the amount of acreage of each. (this as of last week) One of the opponents in St. Joe, a man named Goggin, had called Rep. Jim Trimble, who advised them to photostat the petitions, send a copy to him, one each to Senators McClellan and Fulbright, and one to the Corps of Engineers. ...

Much of this I am telling you is probably old news, but perhaps not. For instance, on Friday, May 18, a meeting was held in St. Joe (unfortunately I was out of the state on business) to collect the petitions and elect a spokesman for the valley property owners. I will not know until this weekend the outcome of this meeting.

If your group deems it wise to fortify your activities with interviews of opponents of the dams who are bona fide river dwellers, I can direct you to some of the better informed and sincere men. Also, you will find the editorial staff of the *Democrat* disposed to aid in what they, too, believe is action motivated by a desire for dollars of a concerted minority. (They've met Rev. Winners, Tudor, and others of the BRIA *in delegation* following each article opposing the dams, you see.)

A word about the BRIA. Having had to confer with them at length for the purpose of extricating the Arkansas Power & Light Company from my own activities, I consequently learned much about their organization. It is run, literally, by Rev. Winners, whom you may know as a displaced Northerner who dabbles in real estate. He is a very clever man; *very* persuasive. (He has the BRIA paying him a full salary and raising funds to send him on a second trip to Washington to lobby for the dams.) His major political strategy is the wanton attacking of the personal character of any individual who opposes him. I suggest that you prepare yourself for all sorts of unpriestly epithets. He has caused a number of untrue claims and false figures (such as the per capita income of Searcy County, for instance, a matter of public information available through the Census Bureau) to be published. ...

Perhaps my endless patter will be of some value to your group. Perhaps not. If I may be of any real aid, do not hesitate to call on me. I am at our place on the river near St. Joe every weekend; my brother and his family live there.

I would be very glad to meet you, and glad for you to visit us on the river if possible.

Charles McRaven

Charley McRaven was a native of North Little Rock, a graduate from Ole Miss in journalism and a jack of all trades, lean and laconic with the proper dry wit. He had indeed organized the real landowners along the Buffalo in opposition to the proposed drowning of the valley. He was at the time in the employ of the

Arkansas Power and Light Company, and that fact was seized upon by the BRIA as proof that that company was trying to stop the dams and that it was the financial power behind the anti-dam movement.

Actually the AP&L could care less about the Corps and their dams. Few of those projects were in the realm of good hydroelectric power planning. They were projects that no profit-conscious company would undertake, necessitating those tacked-on, multipurpose features. The hydro power generated by the Corps represented no serious competition to the AP&L, who in reality assumed a stand-off position during the whole contest. But it would cost Charley McRaven his job nevertheless.

A short account of who and what the real landowners were was related in an item in the *Arkansas Democrat,* May 25, 1962:

Buffalo Dam Opponents Organize Association

MARSHALL—Property owners in the Buffalo River basin have organized the Searcy County Farmers' Association to oppose the proposed construction of two dams on the river.

W. L. Goggin, who was elected president of the association, said the group has obtained 360 signatures on a petition which opposes the dams.

The Buffalo River Improvement Association here supports the proposed dam projects.

"Until now," Goggin stated, "the public has been led to believe that we in the Buffalo valley want the dams." He added that "the activity in favor of the dams has been generated by a small group of merchants and land speculators in Marshall."

In stating the new organization's reasons for opposing the Buffalo dams, Goggin said:

"Most of us are farmers who own land on and around the river. We know that this river land is the best in the county, and we want to keep on farming it and have it for our children to farm."

The association has consulted an attorney, Goggin said. "We are being advised on every legal means open to use as free citizens to oppose the dams," he declared. The spokesman for the new organization said its members believe hydroelectric power on the Buffalo is impractical.

Other officers of the new association are Lunce Cash of St. Joe, vice president, and Mrs. Love Hensley of Leslie, secretary-treasurer. …

Goggin said copies of the association's petition are being sent to Arkansas congressmen "and others who may be able to help us."

We wasted no time in contacting Charley McRaven, who arranged an introduction with Will Goggin, a tall, elderly, distinguished-looking citizen of St. Joe. He and Charley went to extra trouble to acquaint us with the middle section of the Buffalo: Red Bluff, Blue Bluff, the Barns, the Narrows, Margaret White Spring, Peter Cave Bluff, Buckin' Shoals, and more.

Then about that time, while prowling around up on top of Peter Cave Bluff, Charley McRaven fell off, becoming a near fatality. He was rescued by passing float fishermen and spent a month in the Harrison hospital with an eventual good recovery. Thus we were deprived of an important worker during this formative period.

Eventually he was employed as an announcer on radio station KHOZ in Harrison where he occasionally dropped an anti-dam message, to the exasperation of the BRIA.

Charley and his brother Dan had bought the old Lynn Coursey place down on the Buffalo River and were bona fide riverbank landowners. Charley lived there in a cliff-top log house with his new bride but was not destined to stay. He took up blacksmithing at Hollister, Missouri, catering to the construction trade. He then published a book on how to build a log home and eventually wound up in Virginia as head of a construction firm specializing in old house restorations.

But with the departure of Charley McRaven there was no effective or aggressive leadership in the Landowners Association. The policy of intimidation by the BRIA scared off some; and as the battle proceeded, and as we gained a seeming advantage, many of the real landowners lessened their interest and the organization faded away. They wanted neither big dams on the river or a national park, but if forced to choose would prefer a park, since their land tenure within it would be possible. They were satisfied to sit back and watch the fur fly as the BRIA and the Ozark Society laid it on each other.

Their attitude was expressed by Lunce Cash, owner of sixteen thousand acres along the river, who said: "Doc, if they are not going to get those dams, why don't we just draw off and leave the Buffalo alone?"

That would have been fine with me (like it was when I first saw it in 1932) but knowing the momentum that had been generated and the methods of the Corps, it was obvious that that could not be. Jim Tudor had said on more than one occasion that in the end it would be one or the other, the dams or a national park, and that was the truth.

The Buffalo River Landowners' Association (the Searcy County Farmers' Association) had employed William S. Walker, an attorney from Harrison, to direct their activities. That he did to the best of his ability. I made his acquaintance and received good advice in regard to cooperation between the real landowners and our new anti-dam organization in Fayetteville. We understood the importance of that and managed to maintain contact with some of them throughout the contest, but their input was minimal after the departure of the McRavens.

MAY 24, 1962—BIRTH OF THE OZARK SOCIETY

As soon as we could regain our composure after the Justice Douglas visit, it was necessary to give full attention to the birth of our contemplated anti-dam organization. We kept in contact by phone and by mail with coworkers around the state. A communication dated May 19, 1962, to Gus Albright described some of the final activity:

> Dear Gus:
>
> Thanks for your letter of May 18th and for the comments offered. I am sorry you could not be present during some of the discussion that has taken place during the organization of the group in question. ...
>
> At a committee meeting last night a final decision was made as to the constitution, the aims and objectives, and as to the current officers. I have agreed to serve as chairman and Evangeline Archer will be the secretary. ...
>
> We will expect to have chapters or divisions and at the present time the Northwest Arkansas Chapter is the only one in existence. We have decided that groups of 12 or more may serve as chapters. These people must be sincere in their beliefs as to the objectives and willing to donate their time and effort. I am sure that we can find that many in and about Little Rock, and we will be glad to come down and attempt to establish a chapter there any time it appears that it would be successful.
>
> We are not yet ready to have a statewide meeting, but we are going to have an open meeting Thursday, May 24th, in the auditorium of Waterman Hall at the University. At this time formal announcement of the establishment of this society will be made. A movie lasting 15 or 20 minutes showing Justice Douglas' trip down the Buffalo will be presented. The aims and objectives will be read and discussed and questions answered. For those of you in Little Rock who haven't been able to attend organizational meetings, we will be most happy to suggest that you be present if at all possible.
>
> Neil Compton

On the evening of May 23 a preliminary meeting was held at the home of Harry McPherson, long-time president of the Northwest Arkansas Archeological Society, and an assistant in our conservation efforts for many months.

On that occasion we collected annual dues of one dollar after a symbolic gesture by McPherson, who held aloft a Caddo cooking pot in which to deposit that with which we hoped to defeat the mighty Corps of Engineers. He said: "And now we give our dollars to stop the dams on the Buffalo River."

Evangeline Archer, our acting secretary, provided a list of those contributors, the charter members of the Ozark Society: Mr. and Mrs. Laird Archer, David Burleson, Dr. and Mrs. McPherson, Mr. and Mrs. Roy Hayman, Dr. and Mrs. Meade, Dr. Compton, Dr. and Mrs. Whitcomb, Charles Stewart, Sam Dellinger,

Doris Larimore, Craig Rosborough, Marion Wasson, C. W. Lester, Barton Groom, Buck Fenno, George Kinter, Paine, Mr. and Mrs. Buran, Alice Letch, Tom Eads, and Mr. and Mrs. Ben Coffee, Jr.

Doug James had arranged for us to conduct the formal founding of the Ozark Society the next evening, May 24, in the auditorium of Waterman Hall, the Law School of the University of Arkansas. An announcement was distributed as widely as possible in the area and appeared in the local newspapers:

Arkansas Chapter
THE NATURE CONSERVANCY
Thursday, May 24, 7:30 P.M.

Showing of film taken of Justice William O. Douglas and his party on their recent float of the Buffalo River. Adjournment, followed by a meeting of the Ozark Society, Inc. TO SAVE THE BUFFALO RIVER. Now open to memberships.

Type of Membership	Annual Dues (tax deductible)
Individual	$ 1.00
Family	$ 1.50
Contributing	$ 5.00
Sustaining	$10.00 or more

If you cannot be present at this meeting and wish to mail your membership, please send to Mr. George R. Kinter, Treasurer, Ozark Society TO SAVE THE BUFFALO, Highway 45 East, Fayetteville.

The movie of the Douglas float was shown, the aims and objectives of the new society discussed, and officers elected.

Sam Dellinger had been increasingly unable to participate and would not be considered as the principal officer. But he did continue as head of the now all but inactive Nature Conservancy.

It was necessary that I accept the presidency. I did so hoping that an effective successor could be found before long.

For vice-president, Craig Rosborough, another Bentonville citizen, was selected. Craig was a soil conservation officer for Benton County, a man with long experience in greeting the public and in managing regional and community affairs. He was an acquaintance of several years and a man long disillusioned with the doings of the Corps, who often trod on the toes of the Soil Conservation Service. Craig, a quiet, observant man, spoke with a soft and easy southern drawl.

For treasurer we elected another quiet and unobtrusive fellow, George Kinter, from Fayetteville. He was an expert with figures but didn't have to exercise that talent too much in view of the minimal finances upon which we had to operate.

Evangeline Archer, a carryover from the Conservancy, was named the secretary.

When the meeting in Waterman Hall came to a close, I was presented with a telegram that bore a message of deep significance to us all. Upon reading it to the audience its words provided the most dramatic climax possible to the mission that we had undertaken:

DR. NEIL COMPTON, PRESIDENT DLR 7:30 P.M.
THE OZARK SOCIETY INC. WATERMAN HALL
UNIVERSITY OF ARKANSAS, FAYETTEVILLE, ARK.

CONGRATULATIONS TO YOUR NEW ORGANIZATION. THERE ARE MANY OF US IN MEMPHIS AND EASTERN ARKANSAS WHO TAKE THREE OR FOUR TRIPS A YEAR ON THE BUFFALO AND WE OFFER YOU OUR SUPPORT. WE HAVE RECENTLY ATTENDED THE EXTERMINATION AND FUNERAL SERVICES OF THE DECEASED AND NOW EXTINCT LITTLE RED RIVER AND ITS ONCE SPLENDID MOUNTAIN TRIBUTARIES, THE DEVIL'S FORK AND THE MIDDLE FORK. AS THEIR BEAUTIFUL SHOALS, THEIR SWIFT ROCKY CHUTES, THEIR LONG GRACEFUL WILLOW-STUDDED BARS OF SAND AND THEIR COOL DEEP HOLES WERE INUNDATED AND TAKEN FROM THEM TO BE SUBMERGED FOREVER THESE FINE RIVERS CRIED AND DIED. THEIR QUIET DEATH RATTLE OBLITERATED BY A PROFANE SOUND OF MONEY—MOTEL MONEY CAFE MONEY BOAT MONEY GOLF COURSE MONEY OUT-BOARD MONEY WATER SKI MONEY SUBDIVISION MONEY MONEY. WE HOPE THAT YOU ARE SUCCESSFUL IN YOUR EFFORTS TO PREVENT THE GUILLOTINING OF THE BUFFALO AT BOTH GILBERT AND LONE ROCK. PLEASE TELL US WHAT WE CAN DO TO HELP PRECLUDE THE MURDER OF THIS MAGNIFICENT RIVER.

JAMES G. OWEN, JR.—MEMPHIS, TENNESSEE
RUSSELL WOOD—MEMPHIS, TENNESSEE
JAN B. YOUNG—MCCRORY, ARKANSAS
TIM L. TREADWAY—MEMPHIS, TENNESSEE

Having adjourned the meeting, I went home and dispatched two letters to two very important individuals. The sentiment felt by all of us was stated as best I could in these messages. They also pointed out guidelines that we would attempt to follow in the forthcoming battle for the Buffalo River:

May 31, 1962

Hon. Jim Trimble
New House Office Bldg.
Washington, D.C.

Dear Jim:

On January 30th of this year some of us who have been concerned by the prospects of the loss of the Buffalo River to high dams attended the hearing at Marshall, Ark., before the Corps of Engineers. At this meeting we were definitely impressed by the thorough-going efforts of the people in the Buffalo River Improvement Association. They were so well organized that those of us who opposed the damming of the river had scarcely an opportunity to present our case.

We realized at that time that we would accomplish nothing without an organization and we have been working on this problem since and wish to announce that we have formed an organization to be known as The Ozark Society, Inc., whose current slogan is "To Save The Buffalo River." … We will be interested in all problems concerning the conservation of nature and outdoor recreation for the entire Ozark-Ouachita area. We intend to maintain a thoroughly scientific attitude toward all questions and we will foster the proper study of the region in regard to its geology, geography, ecology, its botanical and zoological resources as well as its human cultural attributes. In other words, we will try to encourage the proper investigation of archeological sites, preservation of historical structures, protection of its incomparable scenery and the restocking of desirable plant and animal species. We are not fanatics or extremists but are serious minded people who are concerned by the very rapid disappearance of many things that have in the past been a source of pleasure and sustenance to us all.

We think that the destruction of the Buffalo River would be the biggest single blow that could be levelled against these objectives. We are going to obtain all possible information about the Buffalo River and its surrounding area in regard to scenic values and its possibilities economically as a recreational area for future generations. We will disseminate this information through all channels now available and we expect to continue in these efforts regardless of what developments may arise. In other words, we will contest the destruction of this river at all stages whatever. We have laid the ground work for what I believe will be a permanent organization that will function for many years to come. In our organizational efforts we have encountered surprising and wide-spread support except for the communities of Yellville and Marshall. However, in these and many other communities we have made no effort so far, but we do have definite evidence of interest in our ideas in these very towns where the movement for the dams is so strong.

I would like to state here that we intend to work for the preservation of the entire river and not for any segment or part. This stream is one of the most beautiful in North America and in its entirety constitutes the reason for considering the location of a park or recreational area on its watershed. It does not truly reach the status of a river except in its lower portions where the first dams are to be erected. During dry weather this is the only portion of the river suitable for camping, canoeing, swimming and other water sports of this nature. The creation of reservoirs at Lone Rock and Gilbert will thus eliminate the main part of the recreational area during the summer months, except for perhaps water ski and power boat fans for whom too much has already been done by our government.

The Ozark Society has had its inception in northwest Arkansas and its headquarters are now at Fayetteville. For want of someone better known, the members have asked that I act as president for the time being, however, we expect soon to have chapters and more influential members in Little Rock, Fort Smith, Harrison and Hot Springs and also in many other cities outside of Arkansas.

We do not wish to become embroiled in quarrels and useless arguments, but will wish to present sane and sensible reasons for the objectives we have in mind, and if what we have to say may be contrary to what you may propose, we do not mean it disrespectfully. We, in turn, ask that you give us your respect and consideration for most of us will be your constituents as well as are the people in Marshall and Yellville.

The Ozark Society, Inc.,
Neil Compton, M.D., President

A similar letter announcing the formation of the Ozark Society was then dispatched to Senator Fulbright.

The Kickoff Event for Blanchard Cavern

The first general meeting of the Ozark Society after formal organization took place on June 22 with Hail Bryant and Hugh Schell showing Bryant's outstanding slides of Half-Mile Cavern, with an attendance of one hundred or more. What we didn't know was that a couple of them were from the Ozark National Forest Service office in Russellville. Half-Mile Cavern was on National Forest Service land and was so called because it was one-half mile up a draw from the big Blanchard Spring in the Sylamore recreation area.

The cavern had first been visited by a man with the most appropriate name of Roger Bottom, who went eighty feet down a rope to the bottom of the natural shaft entryway, hand over hand. He was a member of the CCC crew that built the recreation area at the big spring back in the 1930s. Bottom did not range far in that labyrinth, but he did find in one room the skeleton of an Indian who had somehow got down hundreds of years ago but couldn't make it back out again.

After that (before Bottom, Bryant, and Schell), the great cavern system remained unknown and undisturbed. What Hail Bryant had to show was a virgin fantasyland in stone draperies, tall columns, and an underground running river. Its waters were indeed the source of Blanchard Spring as the forest service soon proved by dye tracing.

The forest service attendants at the Fayetteville meeting reported the findings of Bryant and Schell to their superiors. The information went to headquarters in Washington, and the National Forest Service then made their own investigation. They had on their domain a rival to Carlsbad and wished to make it available to the public.

Congressman Wilbur Mills, in whose district it was, without delay secured an appropriation of thirteen million dollars to install a 250-foot elevator to the

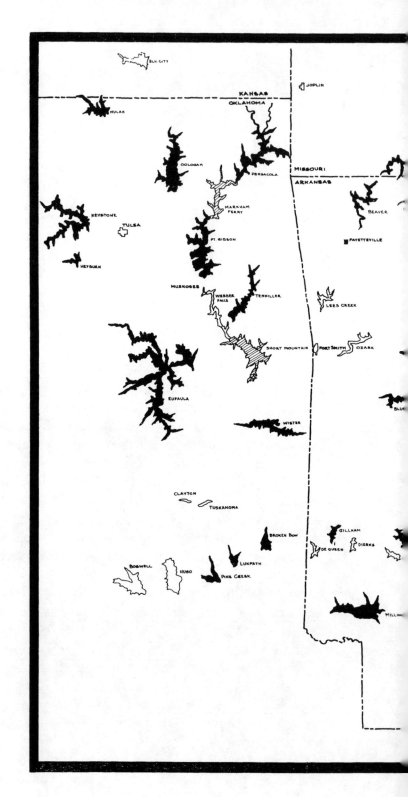

Reservoir map of the
Buffalo River basin.
(Courtesy of Don Winfrey)

114

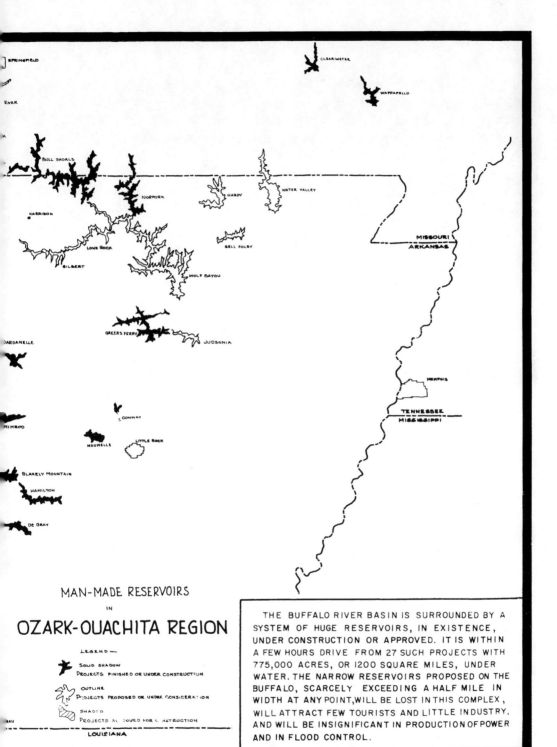

SPRINGFIELD

RIVER

CLEARWATER

WAPPAPELLO

BULL SHOALS

NORFORK

HARRISON

HARDY

WATER VALLEY

LONE ROCK

GILBERT

BELL FOLEY

MISSOURI
ARKANSAS

WOLF BAYOU

DARDANELLE

GREERS FERRY

JUDSONIA

MEMPHIS

CONWAY

TENNESSEE
MISSISSIPPI

NIMROD

MAUMELLE

LITTLE ROCK

BLAKELY MOUNTAIN

HAMILTON

DE GRAY

MAN-MADE RESERVOIRS
IN
OZARK-OUACHITA REGION

LEGEND —

Solid Shadow
Projects finished or under construction

Outline
Projects proposed or under consideration

Shaded
Projects approved for construction

LOUISIANA

THE BUFFALO RIVER BASIN IS SURROUNDED BY A SYSTEM OF HUGE RESERVOIRS, IN EXISTENCE, UNDER CONSTRUCTION OR APPROVED. IT IS WITHIN A FEW HOURS DRIVE FROM 27 SUCH PROJECTS WITH 775,000 ACRES, OR 1200 SQUARE MILES, UNDER WATER. THE NARROW RESERVOIRS PROPOSED ON THE BUFFALO, SCARCELY EXCEEDING A HALF MILE IN WIDTH AT ANY POINT, WILL BE LOST IN THIS COMPLEX, WILL ATTRACT FEW TOURISTS AND LITTLE INDUSTRY, AND WILL BE INSIGNIFICANT IN PRODUCTION OF POWER AND IN FLOOD CONTROL.

cathedral room, to build walkways, to install lights, and to erect a sumptuous visitors' center for the great cavern, now a national attraction.

All of this took place without fanfare or controversy. It was already on federal land, in the district of our then powerful congressman. All that was needed was a "kickoff event," as Hail Bryant said later, which demonstrates how the right word at the right time can be decisive in matters of conservation and outdoor recreation. Purists, of course, will decry these human activities in that underground world, but had it not happened as it did, it might well have ended up one day in the hands of garish commercial developers who might do untold damage to this national curiosity.

We played no further active part in the program for Blanchard Cavern, as it is now called, but did have Bryant and Schell back for other Ozark Society functions in the years following.

First Steps of the New Society

Just how touchy the situation was at this stage is illustrated by the words of Charley McRaven in a letter to me dated May 30, 1962:

> … Our biggest problem on the river, however, is getting started on what needs to be done. John Heuston, the *Democrat* outdoor writer, and I spent the weekend trying to get a letter-writing campaign started, with letters to go to the newspapers, congressmen, and even U.S. Senators Fulbright and McClellan. We must remember that farmers are notoriously busy folks and not given to the finer points of politics, etc.
>
> Last week the Buffalo River Improvement Association called the *Democrat* long distance, threatening a libel suit, of all things, for Heuston's innocuous column of the 22nd or 23rd, in which he lauded the formation of your group. The paper discounted this, of course, but has told Heuston to slow down in his writing. This is off the record, of course, as is the continued action both Heuston and I are taking. …
>
> Our limited activity has not taken us far outside my area, but we are reasonably certain any of the 360 names on our petition would be sound contacts. Mr. Goggin has these at St. Joe.
>
> Charles McRaven

That summer we labored mightily to get out a short information sheet on the Buffalo River situation. We needed something that would be inexpensive, suitable for mailing, and easy to hand out at various times and places. We haggled over every item in it at length. For instance, in the first draft I had learned from reasonably dependable sources that the drawdown on Lone Rock was to be 121 feet and 86 feet at Gilbert. When Evangeline Archer saw that in my report, she nearly had a seizure. If we published that, she said, she would "get out" of the

Ozark Society. So that there would be peace in our camp, the figures were reduced to 80 and 40 feet. Later, however, we found that the original statement was correct at that time. But as time went on and we learned more of the planning process in the Corps headquarters, it was evident that proposed drawdown figures are never firm.

The relationship of the Buffalo reservoirs to the plethora of others in the area was graphically shown by a map prepared for the Ozark Society by Don Winfrey. Our principal argument against the Buffalo dams was thus shown at a glance to any who might see it.

June 1962, A Trip with the Walton Boys

Realizing that I was sadly deficient in firsthand knowledge of the full length of the river, it was imperative at this stage of the game to rectify that. It had been thirty years since that long trip with the Haydens, and the memory was fading. I joined the O.W.W.C., those experts in the fast-growing outdoor sport of canoeing. They had a Tuesday trip planned in June, from Gilbert to the Buffalo River State Park, and I couldn't wait to get with them.

On that expedition I took along a good old boy who would soon wake up the business world of the nation, Sam Walton, and his three boys, Rob, John, and Jim, and my own son, Bill David. These sturdy youths were to serve as paddling crew for Sam and me.

At that time I was the possessor of a fourteen-foot aluminum johnboat, which I had used a few times to putter around on Lake Leatherwood, and a trailer to carry it. When we got to Gilbert and the Hedges and the rest of the Missouri River cruisers saw that rig, they laughed and laughed. But there was nothing else to do but put the thing in the river along with a beat-up, square-end canoe that Sam had brought along. We all had another laugh when I told them about what happened on the way over.

At that time Sam owned twenty-one dime stores scattered around the country, and we considered him a big businessman for sure. Facetiously, I sometimes referred to him as Samuel Woolworth Walton. Somewhere out of Berryville that day Sam realized that he had forgotten something important—his paddles. When we got to Harrison, there was nothing else for him to do but go into Woolworth's and buy a set.

But our bunch showed the Waterways Club a thing or two. Once underway, those boys took to the water like seals or otters. The canoe paddlers would look down in the clear water, and there would be a creature, with long arms and legs,

Sam Walton, *center,* and family at the Bat House Bluff, middle Buffalo, August 1965.

swimming along, Jim, Rob, or John, who would pop up from time to time to look us over.

We all enjoyed the outing immensely and learned a lot about canoe floating and gravel bar camping, but we would never be as efficient at it as the Hedges.

Sam and his family went out with us numerous times thereafter, but he never got into the Ozark Society. He was just too busy becoming the richest man in the United States of America.

As for me, when that float was over I had a session with Harold Hedges on how best to obtain a proper status in the canoeing clan. The first step would be to get rid of the dinky johnboat and get in touch with Mike Naughton, the O.W.W.C.'s canoe supplier, who could get special rates for members of the club and from whom I obtained two of them.

Park Economics

The summer of 1962 saw the appearance of economic reports favorable to the national park proposal.

According to Dr. Max Jordan, Cooperating Agent in Agricultural Economics with the Economic Research Service of the U.S. Department of Agriculture and the University of Arkansas Experiment Station, a national park on the Buffalo would within five years provide more than three and a half million tourist dollars to the local economy. That would be equivalent to the payroll of 108 industrial plants employing sixteen to nineteen people each.

That report was accompanied by much more detail, but the gist of its message was welcome material for our pro-park argument.

Progress in Missouri

At that time the question of the fate of the Current and Jacks Fork rivers in Missouri was still undecided. A hearing had been scheduled by the Corps at Big Spring State Park near Van Buren, Missouri, for June 23. It did not involve dams directly but concerned the question of whether or not those streams would be administered under the National Park Service or the National Forest Service, under the new National Scenic Rivers bill. In the event of a stalemate, the Corps would be prepared to step in and build dams on those streams, as originally proposed.

We wished to participate, but Big Spring was too far for most of us at the time. Therefore, Craig Rosborough was delegated to represent the Ozark Society at the first such function to be held after the founding date. Craig was able to make good contacts with the various partisans present, but the Ozark Society just wasn't ready to continue further in the problem of the Current and Jacks Fork. We had other fish to fry here in Arkansas. The Missouri proposition, with the full backing of their state and conservation agencies, would go on to a successful conclusion without further word from us. The Eleven Point River, in the same general area, was also under consideration. The Corps was eager to build a big dam on it down across the line in Arkansas at Water Valley. In that controversy we would become involved as a secondary group to one of the most determined anti-big dam citizens' organizations that ever came along, The Upper Eleven Point Association.

Good Words from a Few VIPs

Distribution of the first fact sheet was immediately productive. Its message was reprinted in many newspapers, from the *Tulsa World* to the *Memphis Commercial Appeal* and the *Memphis Press Scimitar*. Editors of those papers, Henry Renolds, John Spence, and E. J. Meeman, would continue to give favorable publicity to the free-flowing Buffalo from then on. In Arkansas, the only hostile press was in Marshall and Harrison.

Dr. Hugh H. Iltis had been an assistant professor of botany in the University of Arkansas in the 1950s and was the trip leader on Ken Smith's first visit to Lost Valley.

In a letter of congratulations to our newly organized society he said better than any of the rest of us could say what that country meant to us all:

> ... Botanically, biologically, this area is a gold mine of rare and interesting species. Endemics galore, plants such as grow nowhere else in the world, are sheltered on the hills and valleys of the Buffalo River region. I still remember when Dwight Moore, that venerable and indefatigable great explorer of the Arkansas flora, took me to a locality for Delphinium Newtonianum near Jasper. One of the wonders of evolution, this handsome species. And the quiet evenings along the Buffalo, camping near the gravel bars of chert, with their thickets of spring-blooming Witchhazel, a species whose ancient history leads us back to times when Arkansas was subtropical, and when the present flora was continuous from Canada to China. For in China grows its only close relation. All this, and much more would be lost if this senseless, short-sighted scheme is carried through. "Dam the Buffalo" is the same as saying "damn the Buffalo."

Unfortunately, Dr. Iltis was then at the University of Wisconsin and could not participate in the day-to-day give and take of the contest for the Buffalo River.

Other encouraging communiqués came in from various sources: from Bill Apple, news that the Omnibus Bill of 1962 did not contain money for the Buffalo dams; and word from the National Audubon Society that they back the program to save the Buffalo.

Greetings from the Marshallites

The BRIA did not learn until August that an in-state organization whose specific purpose was to stop the Buffalo River dams had come into being. That realization produced a fervid write-up that took up most of the front page of the *Marshall Mountain Wave* in more than one issue. Its headline proclaimed:

New Society Opposes Dams with Same Old Argument

… This week the Ozark Society, Inc., recently organized at Fayetteville and seemingly composed chiefly of brainwashed disciples of Bill Apple and Gus Albright, longtime foes of the development of the Buffalo River, distributed circulars in Marshall and presumably other areas of the county entitled "Facts You Should Know About the Buffalo River."

Instead of facts, the circular contained many half-truths, many untruths and much propaganda. The Buffalo River is already a wilderness area. Only a few productive farms will be covered by the Buffalo dams, and those studying the river report that only one modern house will have to be moved because of Lone Rock dam. …

All the beautiful scenery, the high bluffs, Hidden Valley and other historical spots they claim to want to preserve will be unaffected by any dam proposed to be constructed on the river. …

The Ozark Society circular states that "the Buffalo dams are unnecessary for flood control, for power, for water supply or for any other reason." The statement is not only false but was made only for the purpose of deceiving those gullible enough to swallow such poppycock. Today the Buffalo is being considered as a source of municipal water supply—the only source known to exist. The Buffalo was the main source of the May, 1961 flood at Batesville, and the lower White River valley. How absurd can the opposition get? How ridiculous can their arguments become? …

The BRIA did come by an important supporter in the summer of '62. The *Ozarks Mountaineer*, a small, regional, folksy-type magazine published at Springfield, Missouri, endorsed the dams on the Buffalo with a long article by Steele K. Kennedy, a freelance writer from Eureka Springs. This article containing some of his own pro-dam commentary actually consisted in large part of a verbatim rewrite of material that had appeared on the front page of the *Marshall Mountain Wave*. Craig Rosborough was acquainted with Steele Kennedy and tried to get him to change his tune but without success.

The *Ozarks Mountaineer* would remain in the BRIA camp for the duration but not as a major or wholly effective spokesman.

GROW

On October 6 the Great Rivers Outdoor Writers Association (GROW) held a meeting at Lakeview, Arkansas, to discuss the development of outdoor recreation programs in Arkansas, Kansas, Oklahoma, and Missouri areas.

I was pleased to be invited and to show our movie of the Buffalo River to the many important personages present.

One of these was Elbert Cox, director, Southeast Region, National Park Service, who gave a most informative address, principally about the Buffalo River.

He mentioned that park service interest was first focused on the Buffalo by Senator Fulbright, in March 1961, who had requested Secretary of the Interior, Stewart Udall, to investigate the possibility of placing the river in park service jurisdiction. Director Cox then read the preliminary report (already recorded here) which we knew of but hadn't been able to publicize. That made it an official statement of the National Park Service, and John Heuston who reported the meeting for the *Arkansas Democrat* included its full text in his article. The national park program for the Buffalo was now a thing in being and not just a wild idea of a lot of little old garden club ladies, birdwatchers, and dimwit nature boys.

Director Cox also informed us that two much more comprehensive studies were scheduled for the Buffalo. One of these was for an economic study for which a contract had been entered into with the University of Arkansas. The other would be an in-depth survey of the river and its surroundings with a final report anticipated in a year or two. The park service was not, by Cox's statement, committed to a final decision to recommend inclusion of the river into the park system, but it did place the park service on a par with the Corps of Engineers as a federal agency interested in its fate. At that time neither was ready to present a blueprint of what their product would look like. In spite of all the fanfare, H.R. 10430 (the bill including Lone Rock and Gilbert dams) had not come up in the last Congress, and the engineers were still on the starting line with the U.S. Park Service now on the track with them.

A Report from behind the Lines

Virginia McPherson, the wife of Harry, at whose home we had met several times, sometimes did secretarial work for us and was very interested in the program. She made an investigative trip behind the lines and furnished us with a most interesting log of her travels:

October 6, 1962

I am giving you the following detailed information that I picked up on a recent trip from Alpena to Little Rock (Highway 65) returning via Fort Smith (Highway 10) for what it may be worth or worthless: I distributed a small supply of Fact Sheets.

It is quite evident that the Marshall group got busy immediately after the hearing last spring. They may not be the best educated group in the state but it is plain that some of them DO KNOW HOW TO ORGANIZE and get the masses of the people to work for them. I contacted about 35 places of business and individuals. All of them had been contacted some time ago and signed petitions for the construction of dams on the Buffalo River. They had changed the minds of some of the farmers we had talked with 3 months ago. They were told that most of their farmland was practically worthless—that the government would "buy up most of it and pay them a good price for it and the dams would bring in much more money"—they would have power for new industry, etc. Some of them said that "even the birds would get more to eat from the dams than the dry Buffalo River."

Right or wrong ALL of them have signed petitions for the dams. I was told that "they didn't take time to learn about what the opposition had to offer besides saving the birds and the bees."

ALPENA—12 people were in coffee shop—all had signed petitions for dams. They ask for Fact Sheets. I left 6.

HARRISON—Left 4 Fact Sheets with customers in coffee shop. No one "cared" to express themselves for either side. ...

WESTERN GROVE—Talked with owner of department store. He said "*everybody* in Western Grove had signed up for the dams."

STATE HIGHWAY WORKER—Said he and some of his farmer friends signed the petitions because they were told that their farm land could be sold to the government for a good price, if the dams were built.

THE HURRICANE—Mr. Jones, owner, said they needed something that would help people over there—Buffalo State Park didn't help them any. "Nobody paid to go there." Political skullduggery is going on somewhere. That 60 percent of the people in Searcy County were supported by State Welfare.

F. L. STILL—Barber at St. Joe, said everybody he knew had signed the petition whether they knew anything about it or not. He also said if Arkansas had a Two-Party system the poor people would get a chance to learn something, also most all of the people were voting this year because others had paid their poll tax for them.

GIFT SHOPPE at Silver Hill—Owners, Mr. and Mrs. Fields, said the electric power there was so weak that light globes burnt out in a few days. They want dams for more power and new industry. They and all their many relations signed the petitions for the dams. (They now get their power from the Petit Jean Co-op at Clinton.) Mrs. Fields said Arkansas needs a Two-Party system—but when the few Republicans do try to do anything the Democrats steal their ballot boxes and destroy them. Also "if anybody wants to save anything do something to give those poor people over there an education so they can save themselves." "Let the birds look after themselves."

NORTH ARKANSAS WOOD PRODUCTS CO., just below Silver Hill—Owners Mr. and Mrs. Earl Martin. They know what it is all about and are against the building of dams on the Buffalo. BUT they are in business and have to go along with the others??? Mrs. Martin asked for several Fact Sheets. They are moving to Yellville soon.

I didn't even stop in Marshall or Leslie.

CLINTON—Earl Riddick, druggist, is very much against the building of dams on the Buffalo—But again, he has a good business in Clinton.

PETIT JEAN CO-OP, Clinton—Talked with office manager—he said all of them signed up for the dams. "The big boss had just left for Silver Hill to close deal on building new electric power house there."

Returning I made a few stops between Little Rock and Fort Smith. Had only two or three Fact Sheets left.

PERRY—Talked with owner of small store. He said "he guess he was for saving the Buffalo but if we wanted to get anything done about it he was sure Paul Van Dalsam would 'hep' on it. He is a smart boy."

OLA—Mr. and Mrs. Gene Cobb, own drug store. They would hate to see the Buffalo dammed up.

DANVILLE—Talked with R. L. Fisher, Editor and Publisher of the Yell County Record. He seemed to be very much interested in saving the Buffalo. He knows the Cobbs at Ola and will discuss it with them, et al.

Virginia L. McPherson

It is to be regretted that we made no further similar surveys in that territory. Here is demonstrated the importance of personal contact in trying to sell an idea or to make a point. The BRIA people had already been there and had come away with signatures. If we had had the right salesman there ahead of them, we might have been able to nip the big dams in the bud. It is to be noted that in those places where the big dam people hadn't been, the sentiment was in our favor, a situation that would prevail throughout the rest of the state before it was over.

SIX

THE DRY BUFFALO

The first important event of 1963 was the organization of a chapter of the Ozark Society in Little Rock. As has been mentioned, we had met on a get-acquainted basis in Bill Apple's office, and H. Charles Johnston, Jr., had taken the lead of the Little Rock people.

Evangeline Archer and I drove down to Little Rock on February 12, 1963, for the occasion, which was held in Johnston's place of business, the lobby of the First Federal Savings and Loan Association at 312 Louisiana Street. The evening meeting was well attended by the new members of the Society, recently recruited by Charley Johnston, and by a goodly number of folks who paid their dues that night. In addition to the program we had brought along the provisional map that we (not the National Park Service) had prepared. It was three by four feet and was displayed on a table in the center of the room. The proposed park area was over four hundred thousand acres. After the meeting was well under way, we had some unanticipated and not exactly welcome guests: Jim Tudor, L. R. Winners, and Gib Walsh. Exactly how they knew of the meeting we weren't sure. We, of course, knew who they were from our experience at the hearing. We kept our cool and proceeded with the business at hand, which was better than what they did after they got a good look at the map. After their initial outburst over actually seeing the size of that "land grab" they subsided enough to permit us to proceed. They left before the party was over, but it would not be the last time that they would crash one of our affairs.

Throughout the long, drawn-out contest over the river, they never responded to our reminder that the Corps of Engineers required land, lots of it, federally acquired, and totally committed to inundation, with no possibility for inholdings, as in national parks or forests. Before it was over, over five hundred thousand acres of Arkansas's good bottomland would be permanently under water, put there by the BRIA's (they-can-do-no-wrong) Corps of Engineers.

John Heuston reported the meeting in the *Arkansas Democrat,* February 13, 1963:

Buffalo Defenders Organize

Approximately 50 local citizens who oppose damming north Arkansas' Buffalo River met here last night to form a Little Rock chapter of the Ozark Society to Save the Buffalo, Inc.

H. Charles Johnston, Jr., vice president of the First Federal Savings and Loan Co., was elected president. H. K. Brewer is vice president and W. M. Apple, secretary. ...

Mrs. Sonya McRaven of St. Joe, secretary of the Searcy County Farmers' Association, which opposes the dams, was present to explain the attitude of the landowners along the river. ...

The local chapter plans to organize canoe trips, nature studies, hikes, cave explorations and other outdoor activities, Johnston said. The first canoe float on the Buffalo has been set for April 13-14. ...

The Ozark Society's First Float

By April of '63 we felt sufficiently well organized with enough momentum going to schedule a float of our own on the Buffalo River. We gathered at the Buffalo River State Park (now Buffalo Point) on a drizzly morning with a good turnout from both Fayetteville and Little Rock.

Evangeline Archer had come to participate in this, her first and only trip on the Buffalo or any other stream. She would be paddled down the river that day in an authentic, old-time, wooden johnboat by Wayne Dillard (of the Dillard's Ferry people). One of the passengers in that craft was a General Cort, a retired army engineer, generally sympathetic with our objectives, but who had this to say after arriving at Rush Creek and after passing through several narrow rocky chutes: "What you need to do with this river is to bring in a bulldozer and scrape out those narrow places."

Also present to use my spare canoe was Ken Smith, back for a visit from California. His partner, a random traveler, also from California, was Ron Guenther, who like Ken had never been in a canoe before. After getting the bow turned downstream, they went the course without a bobble.

Attending to the car shuttle and furnishing boats where needed, was Delos Dodd, manager of the Buffalo River State Park. Delos was a real native and a strong "save the Buffalo" man. We would see much more of him from then on. We didn't demonstrate that "relentless efficiency" that some have applied to the O.W.W.C., but we got the job done and knew what to do next. It was to be a hike down in the Boston Mountains with the Ozark National Forest people guiding us into the Hurricane Fork of the Big Piney to see the natural bridge up there and to explore the shut-ins in that remote region. That was accomplished under the general direction of Charley Johnston and Doug James, who were old friends already.

After that we felt that we could fulfill our goal as an outdoor conservation organization, but we needed writers and photographers to tell the world of our activities and accomplishments.

Good News and Bad

Nineteen sixty-three brought a series of upsets to the river improvers in Marshall.

The first piece of bad news was reported in the *Marshall Mountain Wave*, March 28, 1962, under this headline: "Report on Buffalo River May Take a Few Years, Engineers Say."

Because the people of Searcy County and other counties traversed by the Buffalo River would like to see immediate action and construction of proposed Lone Rock and Gilbert dams, the report in both the state papers that the engineers report would not be ready for several years yet caused great concern among our people.

The article appearing in the state papers follows:

Buffalo Report Long Way Off, Engineers Say

... Col. Charles D. Maynard, district Army Engineer, said the Buffalo River investigation had been incorporated in a comprehensive study of the White River basin which will require several years to complete. ... He said the study would cover flood control, navigation, hydroelectric power, water supply and recreation throughout the entire basin.

In talking to Congressman Trimble, he stated that after Congress passed the bill authorizing study money for the Buffalo, the bill was amended in the Senate to include the whole White River Valley without his knowledge. Mr. Trimble stated that he was in the process of contacting the Engineers, members and chairmen of the various committees, in an effort to get quick action on the Buffalo only. ... John Newman, editor of the *Harrison Daily Times,* said it may be the obituary of the Buffalo River project. ... What's good for Boone, Marion, Baxter, Carroll, and Benton counties would have been good for Searcy County. The triumph of the propagandists is to be regretted.

Meanwhile, we too had been in contact with Mr. Trimble. The Rogers Chamber of Commerce had adopted a resolution endorsing the park proposal for the Buffalo; and Doris Larimore, our secretary pro-tem and a member of the

Rogers chamber, in notifying Congressman Trimble of that endorsement, received a most revealing reply in which he emphasized his commitment to Lone Rock and Gilbert dams. According to him the Buffalo was no more sacred than the White, which he had been damming for years to protect its area and to provide people with recreation. To him the Buffalo was no different.

A Conversion in Mountain Home

The struggle for the Buffalo would not necessarily become a contest for regional chambers of commerce approval, but when an opportunity arose we tried to exploit it. One of these was in Mountain Home, which was in the big dam camp then.

Sam Thorn, an old friend of mine and of Craig Rosborough, had moved over there a few years before. We contacted him to see if he could help, and that he would. He arranged for Ozark Society programs in that town with the Wildwood Garden Club, the Kiwanis Club, the Rotary Club, Presbyterian Supper Club, and the Lion's Club. Craig, being retired, had the time to go to Mountain Home with our Buffalo River movie and fill those appointments.

From the interest generated there, Pete Shiras and Tom Dearmore, both of whom were editing the *Baxter Bulletin* and both of whom had serious misgivings about their town's endorsement of the big dam plans, requested our program for the Mountain Home Chamber of Commerce.

Craig and I drove over there one day and presented our program to a packed meeting room. As a result, the Mountain Home Chamber of Commerce and the *Baxter Bulletin* reversed their stand and endorsed the national park program for the Buffalo to the everlasting wrath of the river dammers in Marshall.

The effect that had on Congressman Trimble is explained by a note I received from Tom Dearmore long after the event.

> After Pete Shiras and I changed the *Bulletin's* position and came out against damming the Buffalo, I went to Washington and visited Rep. Jim Trimble in his office to explain the reasons. He would not even discuss it, and plainly felt let down in the most painful way by our move. He was kindly as always, though—a good and gentle man—and it grieved me to offend him. But the rescue of the Buffalo was far more important; times and realities change and we have to change with them, in our view of a new order of importance.

By now our amateur movies of the Buffalo were paying good dividends and would continue to do so. We had come by two extra Kodak sound-8 projectors, which we had been able to purchase at a big discount. With my personal machine, that made three, and we had stout wooden cases made for them so that they could be shipped, usually by bus, to wherever necessary. If not the projectors, the small 8mm reels could be mailed with ease and little expense to those where 8mm sound projectors were available or rentable. As an example, Park Naturalist Robert W. Carpenter in Hot Springs sent a letter to the Ozark Society requesting the loan of the "Save the Buffalo" film.

That particular film made the rounds in Richmond and in park service offices in Washington. Before it was over, I was able to put together eight or nine movies on conservation subjects but with the Buffalo River as the central theme for most of them.

These were particularly effective because the construction of Beaver reservoir on White River nearby was going on at the time. It was not a lot of trouble to run over there and to film a particular gruesome sequence of the giant "cutters" clearing away the timber on the slopes and the riverbank, the burning of the downed trees, and the destruction and burning of Monte Ne. A small tape recorder was carried along and the live sound later dubbed in along with ominous orchestral passages and appropriate commentary. It was amateurish but effective and cost no more than seventy-five dollars for each film, including the added sound stripe.

Upon learning of this nefarious activity, the BRIA swore by the gods on high that the Arkansas Power and Light Company (or if not them, the Arkansas Game and Fish Commission) had funded each of these movie productions to the tune of thirty thousand dollars each. What a compliment!

The footage shot of the clearing of Beaver reservoir is the only such recording of a Corps of Engineers' project in progress, made from the conservationist point of view, that I know of. What there is left of it should be of historic interest some day.

THE NATIONAL PARK SERVICE FIELD REPORT

Memorial Day, June 4, 1963, was a significant day in the contest to determine the fate of the Buffalo River. We had known for several months that the park service had a survey team in the field preparing an in-depth report on the desirability of including the Buffalo in the national park system. We would have liked to have assisted in this, but they did not want such avowed partisans as us involved. Their decision was awaited expectantly.

The principal Memorial Day event was the dedication of the newly created Pea Ridge National Military Park at the Pea Ridge Battlefield, and Conrad Wirth, director of the National Park Service, came to make the dedication. He then also announced the decision on the Buffalo River study. It was contained in a lengthy illustrated report entitled "Field Report on the Buffalo National River," and it put the National Park Service unequivocally in favor of including the Buffalo in its jurisdiction as a "national river," a new designation in the system.

That was news! It was reported in all area newspapers. Then the *Arkansas Gazette* published a news release that illustrates the anxiety that we and our opponents were having to endure. Our public officials were sending upsetting signals that they might be willing to compromise away the basic objectives of either side:

Trimble Backs Buffalo Dam but Fulbright Favors Park

WASHINGTON, June 3, 1963—Two key Arkansas Congressmen are in disagreement over which comes first for the Buffalo River—the park or the dams.

Representative James W. Trimble of Berryville reiterated today that he placed first priority on two dams proposed for the Buffalo. He added that he would like to see the remaining section of the River after the dams were built transformed into a "national river" by the National Parks Service. About two-thirds of the River's length is above the reservoir sites.

Senator J. W. Fulbright, on the other hand, attaches priority to park development, he disclosed today, and also is for the dams if "technical details" can be resolved.

Senator John L. McClellan, who also has taken part in discussions on the future development of the River with National Parks Service officials, said today he had not altered his neutral position. "I haven't made up my mind," he said. ...

Although both Fulbright and Trimble would like to see the park development and the dams, there seems to be confusing engineering reports on whether it is possible to have both. The present attitude of Park Service engineers is that both are not possible. ...

At a recent secret meeting among Fulbright, McClellan, Trimble and top Park Service officials, Fulbright said he had been authorized to say Governor Faubus felt that the "national river" proposal should be favored, although he was not opposed to the dams.

All of this illustrates the importance of sustained gentlemanly suggestion by any group attempting to obtain a legislative decision. We had no hope of altering Congressman Trimble's stance, but we would subject him to friendly persuasion anyhow. Senator Fulbright, known to favor our side, needed ongoing support. Were we to become indifferent or to cease activity, he might well succumb to compromise. Governor Faubus needed a firm base, something to stand on, before he would make a decision favorable to the preservation of the river. It was our obligation to provide such a base. Senator McClellan represented opportunity, a legislator undecided but willing to listen, a big Corps of Engineers supporter though he was. We would be there later to take advantage of that willingness.

Our position was now better than we could have expected so early in the game. Overconfidence raised its head in some quarters. In a report to members the good news was announced and some of the field report was summarized, but it also contained an admonition against a letdown in our efforts to finish the job.

The announcement by the National Park Service of its approval of national river status for the Buffalo found us short on information of the mechanics of its accomplishment.

Other Agencies Heard From

The distress of the Corps of Engineers protagonists had been compounded by the almost simultaneous announcement by the National Forest Service of a proposition to widely extend the boundaries of the Ozark National Forest (not a bad idea in our book) but not a "cause" that we were looking for.

The plan was described in a letter to Charley Johnston dated May 6, 1963:

> ... Since I arrived home, I did get down to Fayetteville and interviewed Mr. Arnold, who is in the Forestry Department in the University of Arkansas and who is preparing the report for the National Forest Service regarding the forthcoming expansion. I can assure you that this is a very serious move on the part of the National Forest Service. They intend to expand almost all of the way to Batesville and all the way to the Missouri Line in Marion County. It will include all of the area that we have in mind for a park or recreational area for the Park Service. However, I understand that the National Forest Service is not an aggressive organization and that they would be perfectly happy to see any part of this new territory in the hands of the Park Service if they so desired. I have been assured that there will be no conflict what-soever. At the present time this report is confidential. I do have a map of the area but have been, more or less, instructed to keep it under cover until they announce it. This move has been brought about through some of our delegation in Washington. Nobody seems to know for sure which one of the representatives or senators it might have been. It was not initiated by the National Forest Service itself. ...

That idea was received in classical fashion by the *Marshall Mountain Wave*, July 4, 1963:

Government Would Take
All of Searcy County for Forests

A plan to take ALL the lands of SEARCY COUNTY and Incorporate it into the boundaries of the Ozark National Forests by the National Government was disclosed last week by State Forester, Fred Lang, who was quoted in an article in the June 20, issue of the *Arkansas Democrat*.

The government would acquire the western and southern part of the county in their first grab—which could be any time—and the second portion of the county would be taken in the second grab. More than two million acres including all of Stone County, except a few acres along the east boundary, all of Izard County, all of Newton County, and portions of Van Buren, Madison, Washington and Crawford counties would also be included in the area.

Most citizens of the county, who have been harassed during the past several years while trying to develop the Buffalo River, and this area, are now facing the prospect of having it made into a National Park, which will take several thousand acres of privately owned land along the river course, have now become disgusted since the disclosure of the proposed grab for the Forestry Department and some of them have been heard to say, "The time has come for us to load up our shotguns and be heard on this proposition." ...

Already there is strong talk of county consolidation, and once Searcy County is taken into the Forestry system, it will only be a short time until the county government will be consolidated with Boone or one of the neighboring counties, and all of Searcy county will be left to Bobcats, wolves (with rabies), hoot owls and soaring vultures, not to mention the ticks and chiggers. ...

We soon had contact with forest service people who, though interested in the idea, were not hard pushers for it. Its accomplishment would have required full-scale, sustained support of a number of citizens' organizations. There was none to undertake it, and the proposal died away within a year or two. Who the instigator in our congressional delegation was we never learned.

The initial reaction by the BRIA to Conrad Wirth's public recommendation of a national park on the Buffalo was remarkably subdued. His praise of the river was printed in part in the *Marshall Mountain Wave* on June 3, more or less as it was delivered. The very optimistic report of the University of Arkansas's study as to the number of jobs the park project would bring in and the amount of money expected from the tourist trade in the park was quoted as stated. No undue animus was detected in comment concerning the seemingly defeated dam program.

But on June 13 there was more fire and brimstone on the front page. In "Here and There about Town" we learn that:

All of our people have heard that a national park has been proposed for the Buffalo River. True to characteristics of the dam opposition, the Department of the Interior has continued the propaganda of untruths and half truths in their arguments in an effort to justify the taking of the area for a national park. ...

The fight over the Buffalo River is not a fight between residents of the area and sportsmen of the state. To sum it all up, the fight is just another of the age-old fights between private and public power advocates. ...

The Park Service has never made a complete study of the river ... Now they want to grab 103,000 acres along the river, an area that not only would take millions of dollars to acquire but would also require several hundreds of millions additional to improve to the extent that it would ever be of any benefit to the area.

The originators of such wild statements were discredited in most quarters by their obvious emotional context. Any statement by opponents in such a set-to needs to be carefully researched and couched in reasonable terms. That is what the Ozark Society tried to do, knowing all the while that we too could be guilty of overstatement when the going gets tough.

The bugaboo of hundreds of millions of dollars for development of a national park already in situ along the Buffalo was a figure that should by rights be applied to the financially high-flying Corps of Engineers, whose estimates in 1960 dollars for the Lone Rock and Gilbert projects ran over fifty million dollars each.

Land values in the counties involved in those days averaged fifteen dollars per acre. It was still the time of exodus, and much of it could be had for taxes. The National Park Service estimated purchase price of the proposed park area was then nine million dollars. Its development over a period of years would not exceed that, a much less burden on the national treasury than the series of big dams. Unfortunately, serious negotiation for the park area was fifteen or more years in the future, following the great real-estate tidal wave that hit the Ozarks, as well as the rest of the country. Rough hill land by then was going for four hundred dollars per acre and some of the better bottomland for two thousand dollars per acre. The advantage of having been a landowner in the proposed park area in 1963 should be obvious today.

THE LOSING STREAM

In May 1963 the Ozark Society had noted its first anniversary. Two chapters were in existence, one in northwest Arkansas and one in Little Rock. We had about four hundred members, that including adults in families. A general meeting was in order to promote cohesion of the group, to get acquainted, to review what had been accomplished and to generate publicity for the Buffalo National River.

Fort Smith was the natural place for it. The Chamber of Commerce there had stuck its neck out for the Buffalo. Bill Barksdale, the editor of the *Southwest Times Record,* the most important western Arkansas newspaper, which was published in Fort Smith, was now a long-time friend and supporter and would give us good coverage.

The Goldman Hotel was selected and the date set for Saturday and Sunday,

Udall's "Gem" Is Big Mudhole. *Arkansas Gazette,* July 13, 1963.

July 13 and 14. Displays of maps, photographs, newspaper clippings, and other documents were set up in the lobby.

Altogether forty or fifty were present to hear Robert C. Squier from the Southeastern Regional Office of the National Park Service and Courtland T. Reid, acting superintendent of the Hot Springs National Park, describe the national park program for the Buffalo.

Charley Johnston and other Little Rockians were there, and we all took advantage of the opportunity to meet and greet.

The *Southwest Times Record* gave us good coverage, both in the editorial section and on the front page. The headline, however, was a bit much considering what was in store. It said: "Buffalo River One Step away from National Status."

How many more steps would be necessary were highlighted by an item brought in that Saturday by Clayton Little, who laid it out for us to see, saying: "This is going to hurt." It was the current copy of the *Arkansas Gazette,* and there on its front page was a headline: UDALL'S "GEM" IS BIG MUDHOLE. Under that was a photograph of the Buffalo River, taken a few days before just below

Woolum. On its bone-dry gravel bottom reposed a johnboat occupied by three smirking gentlemen: Jim Tudor, Gib Walsh, and an unidentified farmer. The text stated that:

> The above picture was taken on the Buffalo River (or where the river once was) about a mile below the old Woolum post office. Opponents to the dams, who are trying to get the river area converted into a national park have said that "virtually the entire river is floatable the year around." This picture and others in the process of developing will prove beyond a doubt that the lower two thirds of the river is wholly unfit to be included in the national park. ...

That photograph on the front page of the prestigious *Arkansas Gazette* did hurt. The legend of the "dry Buffalo" was born with it and persisted for twenty years, losing some supporters for us and dampening the enthusiasm of many others. The Buffalo didn't run full in droughty times like the North Fork once did. We knew that but couldn't make some doubting Thomases understand.

Sometime after Jim Tudor's counter coup, close observers studying the photograph noted the presence of canoes in the distance descending a narrow passage in the river along a bank and under the branches of trees. We then learned that there were floaters descending the river that day, in cramped quarters to be sure, and unnoticed by Tudor and party. The editor and photographer had arranged the exposure so as to show the wide gravel bar and as little water as possible. But no matter, it was a scoop for them.

But anyone giving it a second thought should by all rights pose a question. If the river is that near to nothing, how could any engineer design a power plant that could generate electricity from it?

We did not and do not deny that at that particular place the Buffalo River is a "losing stream." There is water there flowing toward its union with the White River, but it is underground, especially so during long dry spells. The land surface there is technically known as karst topography, a place where slightly acidic surface water has dissolved out channels down below in the soluble limestone and dolomite. In very severe drought the bed of the Buffalo really is dry for three miles below Woolum. The stream ducks into the underground passageway and actually flows under one of the high hills below, not following the valley at all, to come boiling up again in its bed, cooled by the underground passage at White Spring, a short distance above Margaret White Bluff.

The Richland Creek coming in from the south does the very same thing for a few miles above its junction at Woolum. A similar situation exists for about a quarter of a mile on the Buffalo again above the mouth of Hemmed-In Hollow. These are facts of nature that a true conservationist must accept without emotion or condemnation. It is simply the result of one of the mechanisms by which our Ozark streams have carved and molded our interesting hills out of the old coastal plain.

It is a process meaningful to dam-building engineers as well as to fishermen, birdwatchers, and nature buffs. Shortly before this, boatmen on Bull Shoals reservoir noticed a strange whirlpool and upon investigating found a hole in the bottom of the lake. The Corps checked it out and found that water was exiting through an underground channel, coming up again in the riverbed above Cotter. It was water that they could spare, and no effort was made to plug it up. But during the construction of the Bull Shoals dam, they did hit a similar conduit in the local limestone. It took twenty-two trainloads of cement to seal it off. There are many more of these solution channels around the artificial lakes. They are the means by which nature will some day open avenues around the big dams so that the rivers may reclaim their valleys.

This fact gained minor notice in regional newspapers twenty-five years after the completion of Beaver Dam:

Northwest Arkansas Morning News, September 8, 1989

Funds Allocated for Dam Repair

WASHINGTON—A House-Senate conference committee Thursday approved legislation allocating $9.3 million for work next year repairing leaks in the Beaver Lake dam.

The money tucked away in an $18.5 billion appropriations bill for federal energy and water programs for the fiscal year that begins Oct. 1. The conference agreement must be ratified by the full House and Senate, but that usually is a formality. ...

At Beaver Lake, the earthen dam and dike north of the lake spillway has leaked for several years. The Corps of Engineers plans to hire a contractor to build a concrete seepage cutoff wall measuring two feet wide and 1,500 feet long.

Estimates for the project cost range as high as $20 million.

There had been leakage in that structure from the beginning, but it had been ignored until the loss threatened the lake level and the integrity of the withholding earth and concrete barrier.

Nineteen sixty-three was the driest year on record in this part of America. At its end the total rainfall was twenty-two inches, about half the normal expected. It was an isolated drought limited to that year only. Meteorological observations now indicate that droughts in this country occur in eleven-year cycles, lasting three or four years with each episode. Stream flow rates in the Ozarks are thus dependent on such vicissitudes of nature. Nineteen sixty-three did not fit into that pattern, but it did provide the BRIA with effective ammunition, negative though it was, to bombard the proposed Buffalo National River. It came to be almost their only argument.

The *Marshall Mountain Wave* for July 25 quoted some of the landowners, most of them latecomers and some even absentees:

> Homer Blythe, owner of 723 acres: ... I moved on the river many years ago and the river has never failed to dry up each summer ... We have no attraction worthy of a park ... I would join with other property owners along the river and kill all of the timber on my land adjoining the river to help keep out the park. ...
>
> Larry A. Potter, owner of 1,500 acres on the lower Buffalo: ... I have supported the project for the proposed dams 100 percent. I own six miles of river frontage. Every year the river gets so low that to actually float it is impossible. ... all the big holes are filled with moss ... willows and cockleburs are smothering the gravel bars. The large holes are filled up with silt from cultivated and cut over lands of the mountain tops.
>
> C. F. Stuart of Muskogee, Oklahoma, owner of 600 acres on the lower Buffalo: ... The national park will be worth nothing to the government as an attraction. ... I will spray and kill all of the timber along the river on my land. ... we can make the river and its adjoining lands so unattractive that the Department of the Interior will not want it.

The Spotters

With such commentary rampant, some of it even present on the front page of the *Arkansas Gazette,* Charley Johnston and I decided that something had to be done about the "dry" Buffalo. We would float it if we could. We loaded our canoes and camping gear and got in touch with always dependable Delos Dodd at the Buffalo River State Park and arranged for him to move our cars to Buffalo City on White River, a mile upstream from the mouth of the Buffalo. I took a friend, Jay Hoback, along and Charley, and his son, Warren Mallory. Two of the rangers from the Hot Springs National Park also came along bringing a large, heavy, square-stern Old Town canoe, which we knew would never make it. In late July we put on the river at the state park and, as expected, it was low. River-level readings at Gilbert were published daily in the *Arkansas Gazette.* The readings on that date were the lowest ever recorded, and we expected the worst, but the ease of our passage was surprising. In the big holes one couldn't tell that there was a drought on the land; the rocky chutes were narrow, but with any degree of expertise we went through without contact. The riffles were shallow but floated us over, even the heavy Old Town with the park service rangers.

The water was clear and warm, just right for swimming and in it we saw the moss. It was no loathsome vegetation but was the yellow water stargrass, a beautiful aquatic plant which flourishes in many of our Ozark streams, especially in those spring-fed rivers of Missouri. It is a marker for pure and wholesome running water. It had filled some of the holes like the one at Rush where the spring-fed creek comes in. It also liked to grow on the shoals in swift water where long streamers of it would wave back and forth in the current, a pleasure to drift over and admire. At that season it was in bloom with its small yellow flowers opened even below the surface of the stream. This plant possesses the ability to grow out on the riverbank as well. There it makes a low ground cover looking exactly like a lawn. That day it provided a most pleasing border to much of the riverbank along our way, especially with its sprinkling of yellow blossoms open in the air as well as in the water.

Since then we have been able to observe that it does not make such a show every year. It does better in low-flow years when flood waters do not scour the river bottom and uproot its colonies. It was just one factor to add to the numerous delights of floating the Buffalo, hot dry summer or no.

But something not exactly delightful awaited us at the foot of Clabber Creek Shoals. That rocky concourse that day was floatable, even to the heavy Old Town canoe, but we had to choose and pick the route. No zooming over with a flourish through the lively haystacks. Charley and Warren Mallory were out ahead when I rounded the sharp bend below Rush to enter the rapids. There I saw Charley walking up the shingle on the righthand bank with a worried look on his face. There was a note of anxiety in his voice: "Neil," he said, "Jim Tudor is sitting down there on a big rock."

It was a matter to give us pause. For the two principal antagonists to come to a physical or even verbal confrontation here on the banks of the Buffalo would be a bit much, especially with us so unready for such a thing. What headlines might be generated in all those newspapers that had been following the increasingly turgid standoff between the big dam and free-flowing river people? How did the Marshallites ever find out that Charley and I had planned this trip to test the dried-up river? How did they know exactly when and where to intercept us? The answers to that we would never know, but henceforth we would know that we were under surveillance wherever we might be.

There was nothing else to do. Mentally calculating what possible unpleasantries might await, I turned my craft toward the ledge on which Jim Tudor sat, like doom itself, watching my uncertain maneuvering as I got out and ascended to his perch. I was relieved to observe nothing more lethal than a Speed Graphic camera about his person. In my best Rotary Club style I identified myself and exchanged greetings, initiating whatever small talk that came to mind. He was interested in none of that but wanted to know how much trouble we had had in getting to Clabber Creek. When I disclaimed any trouble at all, I could tell that he "knew" that I was a liar, but he was kind enough not to call me one. He wanted to know where we thought we would get to that evening, and when I said to below the mouth of Big Creek he was incredulous. We were out of our minds to undertake such a thing. It would take four or five days of hard dragging to get to Buffalo City.

That riverside summit didn't last long, but before it was over the subject of a meeting between our opposing factions to see if any possible accord might be arrived at came up. Nothing definite was arranged, and we embarked for the rest of our journey hoping that no further or more unpleasant surprises awaited us. There were none except for the puzzling fact that a few hours later we were buzzed by a small aircraft.

Much later it became evident that it was Tudor's purpose to get a series of pictures of us stranded on the rocks of the dry riverbed, laboriously toting our canoes and camping gear from one pool to another. He got nothing like that and used pictures of me in the *Mountain Wave* getting out of my canoe in order to greet him as an example of the necessity of a canoeist to walk instead of ride down the Buffalo in July. What sort of aerial photographs were obtained must have been the worst possible, otherwise they would have made the front page of the *Wave* and possibly the *Gazette*.

The dry Buffalo had at this stage become the forte of the BRIA, an argument that, in the long run, would not hold water but would itself go dry.

A Conflab in Harrison

By now we were having misgivings as to the possibility of establishing a national park in an area where local citizens were so increasingly opposed. Even those in Searcy County who didn't get on with the Tudors were influenced by the inflammatory front page of the *Marshall Mountain Wave*. It was time to pursue any opportunity of reaching some sort of accord with the BRIA if at all possible. Those local people who favored the park over the dams were now afraid to comment because of strong hints of reprisal. Their opinion was important and sorely needed.

Thus when I arrived home from the dry Buffalo trip, opportunity presented. It was a telegram from Jim Tudor requesting that some of us principals in the Ozark Society come to Marshall for a get-together with the BRIA. They would take us on a grand tour of the river and its environs to show us just how ugly it all was and how wonderful it would all be if dammed and reservoired. That we didn't care to do, but a conference to talk it all over was suggested.

In my reply I had the satisfaction of letting Jim know how quickly we arrived at Buffalo City:

July 29, 1963

Dear Mr. Tudor:

This is to acknowledge the receipt of your telegram which arrived during my absence. As you already know this was because of the weekend outing that we were on on the lower end of the Buffalo River.

After leaving Clabber Creek Shoals we experienced little or no difficulty and arrived at Buffalo City at about 12:45 on Sunday. ...

It might be possible for some of us to come over there on a weekend for a conference if you would consider attending such a meeting.

I believe that there are a good many points that have been misunderstood on both sides of this argument and it might do something to clear up the situation. The time, place and conditions for such a meeting would have to be arranged later on. If you are interested in such a possibility, please advise.

Neil Compton, M.D.

Tudor's response sounded sensible, but the final decision was that it would be the best if just the two of us could get together for a planning session and then to have a meeting of a panel from each side.

Trying to figure out what might be a possible compromise without sacrificing the mainstream of the Buffalo, I noted on an old geological survey map the broad valley of Bear Creek only two or three miles from Marshall. It narrowed down sharply at its lower end, a dandy place for a dam creating a large lake, but nothing like Lone Rock and Gilbert. I drew it in in blue and took the map along hopefully but with also a modicum of shame for it would cover up the best farmland in Searcy County, except for that around Woolum.

On Saturday, August 18, 1963, I drove to Harrison to meet Jim Tudor at the Seville Hotel. He took one look at my map and would have no part of the Bear Creek lake. It would put those good farms under water, it was no good for flood control and would produce little, if any, electricity. All of that was true, and I was forced then to listen to an hour-long monologue about the nasty, moss-filled Buffalo, all of the conditions already recorded here and much more. It wound up with a recitation of the prolonged cases of sore ears that his boys got from swimming in the unsanitary waters of the river when they were growing up. His commitment for Lone Rock and Gilbert was rigid, without any hint of compromise in that hour-long lecture. My position on the national park was best left undiscussed, and we parted with the understanding that we would try to get our people together sometime; but we both knew it would be useless, and it never came about.

The battle for the Buffalo would be fought hammer and tongs down to the last dog.

From Harrison I drove down to Russellville for a visit with Alvis Owen, then the supervisor of the Ozark National Forest. The U.S. Forest Service, of course, would as soon not be embroiled in the Buffalo River fuss, but they would cooperate on boundary lines if the national park should become a reality.

Investigation of Blanchard Spring Cavern, under Alvis Owen's jurisdiction, was then beginning, and from that visit we obtained an invitation to see it while that work was going on, before it had its walkways, lights, and elevators.

The Boxley Grange

On August 24, Jim Tudor and I were destined to meet again.

Craig Rosborough's good friend, Carroll Kemp of Rogers, was a worker in the Grange and had made arrangements with Orphea Duty (the civic leader of the Boxley community) for us to give a program to the Boxley Grange. The meeting was held in the picturesque Boxley Community Building with forty or fifty local citizens present. While engaged in describing the policy of the National Park Service and how it would affect the residents of the area and the reasons why it should be established, in came unexpectedly Jim Tudor, his wife, Gib Walsh, and L. R. Winners. They were not hesitant in interrupting my discourse with loaded questions and by injecting statements of their own concerning the benefits of the big dams, which would be built fifty miles and more downstream. The result was surprising and gratifying. Many of those people were for preserving the river and had hot words for Mr. Tudor. Since he was doing his cause more damage than good, I elected to sit down and let him go on with his tirade.

When that was over I had opportunity to exchange the time of day with Mrs. Tudor to show that there were no hard feelings on my part.

Knowing a little about medieval history, I asked her if Jim was a descendant of that famous Tudor, King Henry VIII. "Yes, indeed he was," she said, and for a fact could we have clad Jim in royal raiment he would have looked remarkably like that famous king of England.

"Well," I said, "did you know that Will Compton was King Henry's best friend?" She didn't know. "But things have surely changed," I said. "I am sure glad that Jim doesn't have the authority over people that cross his path that King Henry did—you know what would happen to me," I said jokingly. That she declared her Jim would never do; but all the same I was glad that times had changed and that they had no headsman in Marshall.

As that affair came to a close, Mrs. T paid me a compliment that I shall not forget. She said, "Oh, I wish we had you on our side."

I went home feeling just a little sad. It was a pity that we couldn't be on the same side. The people in Marshall were no different than Americans elsewhere, most of them good and a few with shortcomings. The Tudors were an important family and leaders in the community. In the contest that we were in I didn't want to think of any of them as enemies (a sentiment not always returned). They were opponents, such as contestants in an athletic affair. We were out to outdo them, and they us, but we would all try to abide by the rules of the game, the law, the regulations, and the decision of the courts and of Congress. When it was over and the score on the board, I hoped that we could be friends then.

In the long dry summer of 1963 tempers went up along with the temperature. Anyone in the area who might voice an opinion favorable to the national park proposal might be set upon, if not physically, then electromagnetically; they were subjected to telephone harassment, sometimes even during all hours of the night.

Paul Buchanan, editor of the *Batesville Guard*, found this out and reported it all in his column, "Two Cents Worth," March 13, 1963:

Long-Range Firing (At Me)—

The Buffalo River controversy over whether to "dam it or park it" has developed into a bitter squabble, a temperamental tug of war. This I learned the hard way.

Several weeks ago this column came out in favor of a national park in the Buffalo River area instead of dams. It was the only time I have ever written anything about Buffalo River, and I had almost forgotten about it.

In its Arkansas Press column yesterday, the *Arkansas Gazette* published excerpts from this particular column, touching off a long-range artillery barrage.

Yours Truly received two long-distance telephone calls—harsh reminders that it is always open season on columnists. Both callers were impolite and highly disrespectful, to put it mildly.

In so many words of insinuation, innuendo and flat accusations, they:

—Charged that a power company had paid me to oppose the dams (One of them kept asking me how much money I received to write the article).

—Charged that this newspaper and many other newspapers in Arkansas took editorial stands in accordance with pressure from big advertisers.

—Accused me of doing an about face on the Buffalo River debate.

—Claimed that I was not qualified to express an opinion on the Buffalo River issue.

—Insinuated that I was a traitor to my own community because of the "terrible flood damage in the Batesville area" in 1961, caused by the Buffalo River.

—Inferred that Yours Truly and everyone else who favors a national park in the Buffalo area are off our collective rockers.

—Wanted to know who held a gun at my head while I was writing the article.

In summation, the callers made me feel like two cents.

The Negative Refrain

Having been able to generate such a controversy, it was now imperative that the one-year-old Ozark Society do something more than to inspire an occasional news clip about the Buffalo in order to dampen the dreary, monotonous, dry Buffalo refrain in the *Marshall Mountain Wave*. Some examples of that:

Float Fishermen Shun Buffalo

August 8, 1963. Two weeks ago a family drove to the Buffalo on Highway 65. They noted the sign at the end of the bridge on which was printed the words, "Buffalo River." They then walked out on the bridge and took a look at the river. After several moments of looking at the moss-clogged waters, the slow-moving current and the small amount of water flowing in the stream bed, they climbed back in their car and drove to Mom and Pop's Fish Shanty and casually asked, "Where is the Buffalo River we have read so much about?" After Mr. and Mrs. Curtsinger had informed them that they were looking at the Buffalo River, they drove away in disgust.

Numerous carloads of sportsmen and fishermen have visited the area in the past several weeks only to find that they have been disillusioned by the glowing reports published and circulated by the Ozark Society to save the Buffalo and other anti-dam groups who have been trying to pawn the river off on the national government for a national park.

Camping and Recreational Area Proves Unattractive on Buffalo River in Summer Months

September 5, 1963. A pilot camp and recreational area, financed by the federal government and property owner Lawrence A. Potter, so far has proven unattractive to the tourists and recreation seekers who have visited the Buffalo River to date, Lawrence A. Potter stated this week. ...

... the Park Service will have a white elephant which they will never fully develop and which will not compare to other national parks over the country.

Here and There about Town

September 26, 1963. During the past three months we have written much about the Buffalo River. We realize that all we have had to say during the past three months when the river has been dried up in several places—from the head of the river in Newton County to almost the Marion County line, has not been pleasant to those who have been trying to pawn the river off to the national government as a national park.

However, we feel that the federal authorities should at least know what they are getting—they have not been told about its habit of going dry each summer, the condition of the water during the hot months and other features which make it unfit and unworthy of a national park name.

For us proponents of the free-flowing Buffalo the situation was, as editor Tudor implied, "unpleasant," not so much because of low flow on the river as because of the flood of verbiage issuing from the headquarters of the BRIA. It seemed for a time that our argument for the Buffalo might well be drowned in that torrent of words.

As a result, we were beginning to hear from some of our own converts snatches of the heresy of the "Dry Buffalo," just as it was revealed in the *Marshall Mountain Wave.*

SEVEN

THE STATE FAIR FRACAS

Something needed to be done without delay to present the better side of the Buffalo River to the public. In that case pictures are better than words. We had looked for a good photographer for our team without success, and it fell my lot to exercise my hit-and-miss talent in that field as best I could. Having dabbled in black-and-white prints, color slides, and home movies, I had by this time a limited inventory of those formats, and we would put them to use wherever we could.

The end of summer and early fall was upon us and fair season was coming to Arkansas.

After the usual lengthy discussion session, the Ozark Society board decided to operate booths in some of these with photographic displays for the public to see. At the same time we would solicit signatures for a petition to save the Buffalo River.

Charley Johnston and his Little Rock members would participate and forthwith they came up with a fine banner for the front of the booth. I had color prints made for a large display board. We would also operate a slide projector and run the movies from time to time.

After a brief warm-up exhibit at the Prairie Grove Clothesline Fair, we rented a booth at the annual Benton County Fair in Bentonville. It was an unqualified success under the direction of Craig Rosborough, assisted by such members as had time to man the booth.

We elected then to take a big jump to the Arkansas State Fair and Livestock Show in Little Rock a few days later. Some of the planning for that event is best related in a letter to Charley Johnston dated September 23, 1963:

> You might like to know the results of the local fair here were satisfactory in our opinion. We got about thirteen hundred and fifty names on the petition. The last two of these were a couple of Army Engineers who don't agree with their organization any too much. We had out a donation box and got the sum total of 25 cents in it. ...
>
> I am going to make a few suggestions for items for you to have on hand by

Saturday. First of these will be two tables. These tables should be high enough to accommodate the projector so that it will throw an image at a satisfactory level against the back wall of your booth. ...

We used a couple of tarps for the sides of our booth being as how that it wasn't enclosed. If yours isn't enclosed we will bring these tarps along for you to use. I will also bring along your signs which are still in good shape. We hung the two large cardboard signs on the tarps and stretched the roll up sign across the front of the booth and it looked very nice. I will bring along about 180 slides which should be enough to fill up two of the round magazines that go on the carousel projector. ...

Charley Johnston, having good connections in Little Rock, knew the right people to see about a booth at the state fairgrounds, a Mr. Magness in charge and Senator Byrd (state senator from Pulaski County). The Ozark Society was assigned a place in one of the permanent buildings, on state property, to our later discomfiture. We went down to Little Rock, looked the booth over, and were pleased with the attractiveness of our display. The beauty of the river and its environs was obvious, so much so as to counter much of the dried-up Buffalo talk from the BRIA. That they could not endure. They accorded us a fine demonstration of their influence with our state agencies. We didn't know just exactly who they made telephone calls to, but Magness and Byrd soon confronted Charley Johnston with the news that he was inhabiting a state facility under false pretenses and would have to vacate. It turned out that there was tent space at the fair not in state ownership, and the Ozark Society was forced, with tail between legs, to gather up all of those pretty pictures, the hand-out literature, and petition sheets and camp out so to speak.

The satisfaction of the BRIA people was reflected in the issue of the *Marshall Mountain Wave* for October 3, 1963:

Here and There about Town

Members of the Ozark Society to Save the Buffalo, who depend on the outer edges of the state for their membership, purchased space for a booth at the Arkansas State Livestock Show, and in the "This is Arkansas" building recently erected on the State Fair Grounds.

In their booth they portrayed pictures of the Buffalo river country made in the early spring months depicting the Buffalo full of free flowing water, the scenic attractions, some of them great distances from the river, and were hawking their reasons for opposing the construction of the proposed dams on the Buffalo, and having people sign their petition in opposition to the dams.

After a few telephone calls to Little Rock Tuesday evening, the booth was closed down and the material moved out of the building.

The people of the Buffalo river area become indignant when state agencies are used to oppose a proposition which means so much to any section of the state as the construction of the dam will mean to the Buffalo river area. We believe, and stoutly maintain that whatever is done to the Buffalo river should be decided by the people

Craig Rosborough (in glasses), the
Ozark Society's first vice-president,
attending the booth at the Benton
County Fair, October 1963.

of the area only. They are the ones which will be affected most by the construction of
the dams, a National park or whatever manner is used.

One of the pieces of literature the Society was circulating from the booth at the
Livestock Show said: "Actually, the upper Buffalo is not floatable for small boats during the moderate droughts that may occur every three or four years. Usually during
the summer. During more severe drought, short sections of the middle Buffalo likewise are not floatable. But the lower Buffalo from Gilbert to White river has been
floatable by small boats even during the worst drought on record."

As of this day, October 2nd, the Buffalo river is too low for float fishing anywhere
in Searcy county.

Above Highway 65 bridge there is still a three or four mile stretch of the river that
has been dry since the last of June.

Meanwhile, our man in Little Rock was suffering compounded indignities as
our mail dated October 2, 1963, reveals:

Dear Neil:

By the time you receive this, I will have either been victorious or suffered total
defeat at the hands of the "Marshall Hornets Nest"!

Last night I received a telephone call which was gratifying in that it must have
depleted the BRIA exchequer by at least $15. One Gib Walsh honored me for at least

45 minutes with a most interesting filibuster, the gist of which was: (1) The Washington Private Electrical Power is paying us off. (2) The East Arkansas Duck Hunters are paying us off. (3) Don't Laugh—We are being paid off by the Hot Springs Gambling interest. ...

Mr. Walsh says he is going to close us down and is going to march on Little Rock some 50 strong today. My only thought is that I'll have every newspaper and TV Station in town on the spot and we will at least get some good publicity out of the bloodletting.

We filled up about 8 sheets of petitions yesterday and will continue our brave and courageous effort today.

Yours truly,
H. Charles Johnston

And then on October 7 more yet from H. Charles:

... I will package up the petitions we got and mail them to you. I would estimate 2,000 signatures.

I ran into lots of teenagers from the Marshall area and most of them have a lot more sense than their elders. One in particular, or two, live in the Snowball area and they stated that the engineers are presently surveying for the upstream impoundments on Bear Creek and, he thought, the Richland Creek, and I did not even get his name. The gist of it is that these boys stated that the engineers are saying that they are doing this as part of the National Park Service Proposal. My only thought at this point is that we had really better get to work and state our position on the upstream impoundments, and do it in a hurry, because, from what I can grasp from this, the corps is already at work trying to outflank us.

... I talked to Senator Byrd before we rented the small booth and explained to him what we were doing, although I do not remember whether I specifically mentioned petitions, but, as you also remember, we discussed the matter with Mr. Magness that Saturday, and as far as I am concerned everybody knew our position. ...

Fortunately, the state fair episode terminated without physical injury to anyone and without the application of adversary law. Our hurt feelings were compensated by over two thousand new signatures on our petition and some invaluable publicity around Little Rock.

After that we made a short appearance at the annual fall festival in Eureka Springs. A few months later we were able to send in to the Secretary of the Interior a petition favoring the Buffalo National River with 6,090 names on it. By that time we had been advised that such petitions do not carry much weight in the political arena in Washington because of the ease with which they can be manipulated and falsified. It was suggested that we could spend our energies better elsewhere, and we did not try to make the public fair circuit after that.

However, we did learn much in regard to the undeniably favorable attitude of most of our citizenry to the national park proposal. We learned that any publicity regardless of its slant is good for the cause (people will know who you are and your friends will be identified). Whether or not it cut any ice in D.C. we were proud of the fact that we came up with over six thousand names in a short time and that they were every one authentic.

A Few Words from Camelot

The BRIA's satisfaction in giving us the boot in Little Rock was more than complemented by another event in Arkansas at almost the same time. It is best related by the editor of the *Marshall Mountain Wave* on the front page for October 10, 1963:

Here and There about Town

The editors of your county newspaper attended the dedication of Greer's Ferry Dam at Heber Springs last week, and listened while President Kennedy enumerated the blessings and growth which the Dam and the Lake would bring to the state and nation.

President Kennedy said: "It will provide recreation to hundreds of thousands of visitors to this Beautiful Lake and shore line. ... In short, this project will benefit Arkansas, the South and the Nation. It will make the Maximum use of our resources. It will increase employment opportunities. It will expand the growth of our economy. IT WILL RETURN TO THE NATION AND TO THE FEDERAL TREASURY FAR MORE THAN ITS ORIGINAL COST.

"Last week I saw similar projects in the Western part of our country. And when I review the dividends to be gained from this kind of investment—When I consider the benefits ALL Americans receive from this kind of project—I CANNOT UNDERSTAND THOSE WHO WOULD DENOUNCE THIS KIND OF WORK with such phrases as 'Pork Barrel and Boondoggle.'

"Which is more wasteful: The loss of life and property caused by recurrent floods—or the cost of a multipurpose project which will ultimately pay for itself?

"Which is more wasteful: The failure to tap the energies of our streams and rivers when power is needed for industry—or construction of hydro-electric projects to serve the homes and farms and factories of this area?

"Which is more wasteful: To let land lie arid and unproductive and resources lie untapped, while rivers flow unused—or to transform those rivers into natural arteries of transportation, reclamation, power and commerce with billion dollar benefits as in the case of the Arkansas and White river Basins?

"These projects protect and create wealth—new industries, new income, new incentive and interests. And the wealth they assure to one region becomes a market for another. ...

"There will always be opposition to these projects. TVA was called a "White Elephant"—Grand Coulee was considered part of the "Pork Barrel"—Hoover Dam was thought to be a "Blunder." But they continue to operate for the good of the people and the country ... and so will the projects of this state long after their critics are gone."

The people of Searcy and other Counties in the Buffalo river area can take heart from the address of President Kennedy. His remarks bore out the contention that the members of the Buffalo River Improvement Association have argued for the past several years. Anything other than the two proposed lakes on the Buffalo river will not develop the river to its full potential value for the area, the state or the nation.

Several of the most ardent opponents of the dams, who are now arguing for a National Park have admitted in private conversation with the writer, that the dams and lakes on the Buffalo would attract far more visitors and tourists than a National park. However, they want the river left in its present state for the personal use of a few selfish individuals who like to float the river once or twice a year to get away from the routine of everyday life and "let their hair down."

Editor John Newman of the *Harrison Daily Times* of Harrison said in his Times Topic column Tuesday of this week:

"The big Hurrah raised at the Greer's Ferry Dam dedication should give Searcy Countians new strength in their campaign for a dam instead of a park on Buffalo River.

"They can quote President Kennedy directly that the Multi-purpose dams are the best thing that can happen to a community, state, and nation. I'M READY TO BACK THEM UP IN A NEW DRIVE TO CANCEL DELAY, AND PROCEED WITH CONSTRUCTION."

At this late date (1992) it is useless to try to assess President Kennedy's words at Greer's Ferry, especially in view of what was to happen six weeks later. But it would be interesting to know the source of his ideas. They were identical in context with what we had been reading in the Marshall paper and in Corps of Engineers' reports for many months. Kennedy's words served to inspire great confidence in the minds of the big dam faction. But it would all come to naught. His Secretary of the Interior, Stewart Udall, was at the moment one of the strongest supporters of the Buffalo National River proposal in Washington and would remain so throughout the next several critical years.

That fall the BRIA, attempting to follow our lead, circulated a petition that failed to provide them with any advantage due to a dearth of signees outside of Marshall.

At the same time the Marshall Chamber of Commerce issued a newsletter to all chambers of commerce in the state of Arkansas pleading for endorsement of the Gilbert and Lone Rock dams. They had just heard of the endorsement of the national park proposal by the Conway, Arkansas, Chamber of Commerce and were apprehensive that others would follow suit.

A strong appeal was then made to the various chamber presidents to endorse the dams by Harold Reid, president of the Marshall chamber, but if Reid obtained any supporters we didn't hear of it.

But if the Marshallites were caught on the horns of a dilemma they were no more so than us would-be dam busters.

On September 17, Evangeline Archer received a letter from Raymond Gregg, formerly superintendent of the Hot Springs National Park. He wanted us to initiate a move to get Senator Fulbright to introduce a bill in the Senate for a national park on the Buffalo because the administration policy was becoming increasingly opposed to such new programs. He was asking that we fly in the face of opposition from Mr. Trimble and the rest of the Arkansas delegation in the House. Such a move he said would be a "foot in the door" for a more positive effort later on.

It therefore fell my lot to dispatch a letter to Senator Fulbright emphasizing the favorable impression that we had recently obtained from the majority of those in attendance at the various fairs. It was with misgivings that he was asked to submit a bill for the Buffalo National River in that session of the Senate.

The senator's reply left us somewhat shook up, to say the least. In it the Dry Buffalo had come home to roost along with all those little dams in all those hollers:

November 4, 1963

Dear Dr. Compton:

I wish to acknowledge your letter of October 28, 1963 concerning the proposal to dam the tributaries of the Buffalo.

As you probably know, one of the most serious objections raised against the park plan is that during several months of the year certain stretches of the river are not floatable because of low water. It was suggested several years ago by some of the original proponents of the park that this problem be solved, at least partially, by construction of small dams on the tributaries so that water could be released when needed in order to maintain a fairly stable flow in the river. This suggestion has been considered by the National Park Service and I quote from a letter I received from them earlier this month.

"In a discussion the National Park Service had with representatives of the Corps of Engineers on June 5 in Little Rock, the effect of such dams on the national river proposal was discussed. The Service took the position that any such dams constructed outside the suggested boundaries of the national river would be beyond their jurisdiction. Further, it was pointed out that there might be an overall benefit to recreation by utilizing these impoundments.

"The representatives of the Corps said they have stepped up their studies in the Buffalo River Basin (part of the White River Basin) and expect to have a preliminary report completed within two years. Currently, they have eight or more possible damsites tentatively located on tributaries of the Buffalo River.

"Our view then, is that development of impoundments as presently proposed on tributaries of the Buffalo River outside the proposed national river boundaries would not be incompatible with our plans, and in some instances recreation might even be benefited."

You can be sure that I will continue to do all I can to insure that the park is authorized by the Congress at the earliest practicable date. It is my understanding that the National Park Service is in the process of preparing a bill for submission to the Congress, but it is obvious that even if the bill is ready in time for introduction this session that no action would be taken on it this year in view of the overloaded schedule of business.

With best regards, I am

> Sincerely yours,
> J. W. Fulbright

Just how solid a brick wall we were up against is described in a communication from Tom Kimball, executive director of the National Wildlife Federation, to Bill Apple on November 5, 1963. In it he said:

> ... Congressman Trimble has his hand on the lid for this project and will not let it out. Even though the Park Service has approved, until the Secretary of the Interior also adds his endorsement, no request for legislation can be prepared. ...
>
> There are no promises, but it is unlikely that the Secretary will send the plan to the Congress with a request for legislation unless there is more in the way of local interest expressed. We understand that Senator Fulbright also is interested but he, too, is awaiting a ground swell of public opinion. ...
>
> This information was received from the executive department concerned, not the legislative side.

We could plainly see that 1963 was not going to be the year of the national park, but then it wasn't going to be the year of the dams either.

The year wound down with a minor spat as reported in a letter from me to Charley Johnston on December 10, 1963:

> ... Still nothing in the *Marshall Mountain Wave* (It just occurred to me that in tectonic terms that's an earthquake, isn't it?). Not a word about the Buffalo in over two months but never fear, Tudor is still there. ... He called me on the phone the other day to correct me on the 121 foot drawdown at Lone Rock, says it is only 40 feet, says the Engineers have changed their plans. I said yes they have, they are going to build ten dams not two if they can. He didn't understand that. I said we would have a park over there someday and that he just as well come along with us. He said they would dynamite the bluffs first and I said that was the sorriest thing that I ever heard of. Then he got on the dried up state of the river and finally on what a tough time we had going down it last August and how Charlie Johnston knocked a hole in the bottom of his boat, and so on and on. Strangely enough through this harangue he called me Neil. When it was over I felt like I'd been on the hot line to Moscow. When that bill goes into the Senate we are really going to see some histrionics in Searcy County. ...

Sheriff Siezes Untaxed Whisl
Beer From Group On Buffalo R

Sheriff Beal Sutterfield raided afternoon from citiz- a group of from 20 to 35 fisher- area regarding the men camped close to the river drove to the river bank at the old Maumee Ferry and it was reported t

EIGHT

THE MARSHALL ARMORY— SECOND TIME AROUND

The contesting factions entered the new year on low key after the state fair standoff. But on January 23 from the BRIA there came a rumble. The *Mountain Wave* declared that 1964 was to be the year for the Buffalo River dams and that something must be done THIS YEAR!

BERNIE CAMPBELL ON BOARD

Ozark Society people who had been attending to their homework were pleased in late February to learn of a new and important personality on the scene. His presence was announced in a letter from Charley Johnston to Evangeline Archer, February 19, 1964. His name was Bernard Campbell ("Bernie" we would come to call him):

> Mr. Campbell, the new superintendent at the Hot Springs Park, was in to see me last Friday. He is on the Buffalo this week. ... On the humorous side, he tells me (he was not humorous about it necessarily) that Courtland Reid who was acting chief at Hot Springs had asked that he not be sent up to the Buffalo to make the survey this week because he was afraid he would get shot. I say this in all seriousness, the man was absolutely scared to death to come up to the Buffalo. I only hope that Mr. Campbell's faith in his fellow man or his courage, or both, are not badly misplaced.
>
> ... This man seems to be a good man—we just don't want him shot. I gave him your name and he will probably contact you (he had your name already).

Bernard Campbell, a tall, balding, suave gentleman had been born in New York State. He was a career park service man, and years before had been a field man for the service during the acquisition of lands for the Blue Ridge Parkway. Perhaps that is why the park service had sent him to Hot Springs. He knew how

153

Bernard Campbell, *bow,* floating the Buffalo with John Heuston, *stern;* middle man unidentified. *(Photo by Harold Phelps. Courtesy of Arkansas Publicity and Parks Commission)*

to cope with irate citizens who might want no part of a park in their territory. I don't know about his faith in his fellow man, but courage he did have, and he was going to need it. As superintendent of the Hot Springs National Park, he was that agency's top official in Arkansas and was in charge of all park service policy and development in the state. That included what was going on up on the Buffalo River.

LEARNING THE ROPES IN D.C.

The time and place for two weeks' duty for training sessions in the Naval Reserve was optional, and in 1964 I selected the Naval Hospital in Bethesda, Maryland. Bernie Campbell, upon learning of that, was pleased to furnish a list of people to see while there. It was near the center of everything, politics, legislation, and government. There the fate of the Buffalo River hung in balance and now would be the time to check the framework that suspended it.

In March 1964 Tillman Morgan, the commanding officer of our reserve unit at Eureka Springs, and I made the long drive to D.C.

The Warriors of the Eleven Point

On our way to Washington we stopped in Corning, Arkansas, for a visit and dinner at the home of Charley Black, an old Razorback from my days at the U of A. He was the forceful and outspoken leader of another anti-dam organization, the Upper Eleven Point Association. Charley and his people had had fifteen years of experience in contesting the Corps' relentless determination to build another monster dam on an eastern tributary of the White River at Water Valley on the Eleven Point. We had heard of them from time to time; and our investigative secretary, Evangeline Archer, had made contact with Helen Bly, their secretary, and the Ozark Society had elected to help them out in any way we could. Mutual assistance in these two concurrent struggles was of obvious benefit. The Eleven Point lacked the tall bluffs and the wide gravel bars of the Buffalo, but it was one of those steady-flow, spring-fed rivers of the eastern Ozarks, originating in the grand upwelling of Greer Springs and Morgan Springs across the line in Missouri. There was no earthly reason to put a dam on it—any more than on the Buffalo.

As a result of Evangeline Archer's contact with Mrs. Bly, we had been advised of the latest of several hearings on the proposed Water Valley dam to be held at Pocahontas, Arkansas, on January 29, 1964. Some of us from the Ozark Society drove to Pocahontas to make statements against the project and to see how the veteran Upper Eleven Point Association behaved at a Corps of Engineers' hearing. It was not unlike the one at Marshall two years before, the armory building full of locals, but in this case there were no droves of high-school students wearing pro-dam tags and no unending string of speakers for the project from the local citizenry.

Charley Black, the big, bald-headed, old Razorback tackle highlighted the event.

Colonel Charles Maynard, the officer in charge, had instructed those testifying that they should "not impugn the integrity of the Corps of Engineers" in their comment.

Charley Black started off with some plain and simple adverse remarks about the "Mudhole Plan" for the Eleven Point and then went on to say, "For years local residents have been forced to attend hearings to save their homes from a ruthless, arrogant, federal bureaucracy that calls itself the U.S. Corps of Army Engineers."

At that Colonel Maynard rapped the table with his gavel and in a loud voice declared, "Mr. Black, you are out of order. You have impugned the integrity of the Corps of Engineers. I must ask you to desist."

Charley and Barbara Black on the Eleven Point River.

Charley looked down at the Colonel over his reading glasses, like he was being pestered by a small, bothersome boy, and said, "Colonel Maynard, I hope what I am going to say from here on won't cause you to have to interrupt me again."

And impugn them he did while the Colonel sat it out in glum silence. That produced loud guffaws from just about all of those present.

I went home with every word and gesture inscribed in memory and was most happy to recount that episode to those present at the fiftieth-year reunion of the class of '35 at the U of A in 1985 with Charley Black in attendance.

Our alliance with the Upper Eleven Point Association in 1964 would last until the cessation of hostilities on both threatened rivers. We would come to the Eleven Point for several floats from B. B. Morgan's aquatic farm down to Dalton, in order to gain firsthand knowledge of the fast-moving Eleven Point and to lend assistance to its defenders.

They, in turn, would come to subsequent hearings on the Buffalo, adding authentic citizen support to that cause.

The Upper Eleven Point Association would eventually be victorious in their protracted war with the Corps of Engineers. That came after many seemingly hopeless confrontations without any alternative, such as the proposal of a recreation area along its course, as we had along the Buffalo. It was the result of grim, relentless determination by the citizens of the area who did not intend to lose their farms, ranches, and homes to "flood control." They are entitled to all the more honor for that.

We departed from Corning that day in 1964 determined to learn whatever we could of the probable fate in the halls of Congress of the Eleven Point as well as of the Buffalo.

No new or startling facts were learned during that sojourn in the nation's capital. But having made an appearance in some of those high offices was important. Important legislators had, at least, heard of the Buffalo River and had met a citizen proponent of the national park plan for it.

Some of these encounters are best related in words written at the time. On March 30, 1964, I got off an epistle to H. Charles summing up the situation in D.C.:

> I have been trying to get wound up to write you a letter for several days since my return from Washington but haven't found an opportunity until now. ...
>
> We did have a short session in Fayetteville Thursday evening and did talk over the results of what I learned while I was up there. Evangeline went into sort of a flap when she found that it is likely that a bill for the Park is not immediately forthcoming. I hope she hasn't made it appear more dismal to you than it actually is. I might say here that the people in the National Park Service headquarters feel that we are in fairly good position for the establishment of a national river on the Buffalo at sometime in the not too far distant future. ...
>
> First I visited Jim Trimble one Saturday morning. He was most gracious and hospitable, and we spent an hour or so talking about things in general. We spent another hour talking about the Buffalo river situation. He took us to dinner in the House restaurant. ... He makes no bones about being committed for the construction of dams, as you might readily know, but he did seem a bit regretful that he was unable to support the Park Service proposal. ... I do believe that he may not labor too hard to thwart us, however. There are reasons for this which I will explain to you when I see you. One must always remember that he is a politician and that anything he may have said to me may have been for some ulterior purpose. ... I was genuinely disappointed to find that he really does intend to run for another term. Actually it is obvious that his health is not such that he should do this. There are other complicated reasons, though, for him having made this decision. We will most likely have to put up with him for another session or for another term.
>
> My visit with Senator McClellan lasted for about thirty or forty minutes and was most interesting. Contrary to how he sometimes appears on the television, he is a very energetic and robust person and asked many pertinent questions in regards to the National Park proposal. It seemed to me that he was uncommonly ignorant of the general situation although he did possess the field report by the National Park Service and also the economic survey by the University. He admitted that he had never read these. ... We have reason to believe that he is favorable toward the Park Service proposal even before this interview, and during the conversation he stated definitely that

he does favor the preservation of some of our choicest scenery in Arkansas. He also stated that he believes we have enough dams and reservoirs in the state by this time. I believe if we continue to approach him in a reasonable manner that he will not be against it and may even work for this project as time goes on. He simply needs to be given more information on the subject by every method that we can think of.

I made several efforts to see Senator Fulbright and went to his office on three separate occasions and was never successful. He has been busy with a great many other problems, and this probably accounts for the inability to arrange an interview. I did get to talk to Pat Fleming, one of his assistants, for about half an hour. During this talk it became evident that the Senator is not considering introducing a bill in the Senate for the National River anytime soon. Their attitude is that it is impossible to get anywhere at all with a project which is opposed by the Congressman from the district in which it is located. This I knew long before and had to, of course, agree. I do think that we need to contact Senator Fulbright personally at the first opportunity to determine just exactly what can be done about introducing a bill. I think as you do that the introduction of a bill, even though it might not get very far, would be helpful. If it is not done, however, it is not a fatal omission.

I had two visits with the people in the National Park Service headquarters, and this consisted mainly of a general discussion of the situation; nothing really new came out. ...

I also had a visit with Anthony Wayne Smith who publishes the *National Park Association Magazine.* ... Mr. Smith will publish an article on the Buffalo river as soon as we can get one ready, and I am writing to John Heuston to see if he will do that. ...

The American Motors Award

What I didn't know then was that another trip to Washington was in store within a few weeks.

An announcement of the reason for that appeared in the Ozark Society report to its members for May:

> The president of Ozark Society, *Dr. Neil Compton,* has received an American Motors Conservation Award, one given in Washington each year to ten non-professionals selected by a committee of distinguished conservationists. It is indeed gratifying that Dr. Compton's efforts for the Buffalo River have received national recognition.

Doug James, who kept up with such things, had asked me some months before if it was okay for him to turn in my name for a nonprofessional award. That was agreed to, and I more or less forgot about it, thinking that there was little chance of receiving such recognition. Again, this would be a premature honor, and I was, to be honest, not deserving of such, certainly not yet, not until some positive protective measure for the Buffalo River was actually realized.

The award winners were quartered at the Statler Hilton where the award dinner took place, conducted by Ed Zern, director of American Motors' Conservation Award Program, with welcoming remarks and presentation by Roy Abernathy, president of American Motors Corporation.

For me the most satisfaction from all this was derived from making the acquaintance of Dr. Edgar Wayburn, M.D., internist from San Francisco, and another nonprofessional award winner. Dr. Wayburn was the president of the twenty-three-thousand-member Sierra Club, which group we were striving to emulate. While I had received very little critical comment from my patients and acquaintances for being so involved in such obvious extracurricular activity as trying to save the Buffalo, there had existed in my own breast misgivings. Was it really proper for a dedicated physician to become so embroiled in such a non-medical controversy? The presence there of Dr. Wayburn, now head of John Muir's Sierra Club, dispelled that feeling for all practical purposes. I was aware, of course, that down through history many physicians great and small had been important preservationists, leaving the natural world better off for their pres-ence. Soaring home across the long ridges of the Alleghenies and over the Mississippi and the Arch, I felt finally comfortable with the situation. I could now return to the contest without qualms as to the propriety of my actions out-side my office and the hospital. Those entrusting their physical well-being to me would receive first priority, as they always had, but as time permitted I would redouble my efforts to salvage the last best river in our Ozark homeland.

THE GREAT RAID AT MAUMEE

Meanwhile, Dr. Doug James had been busy with something else. He wanted to know if it would be proper to organize a chapter of Ozark Society on the campus at the University of Arkansas. The board could think of no reason why not, and the April 8 issue of the *Arkansas Traveler*, the campus newspaper, carried a front-page item stating: OZARK SOCIETY FORMED—PLAN FLOAT TRIPS AND HIKES. According to that report, over fifty students had paid dues. That included men and women (boys and girls if you will). Hiking trips had been planned and a float trip on the Buffalo at the state park scheduled for April 18 and 19. That news item was picked up by Bill Simmons of the Associated Press and was run in the *Arkansas Gazette* where it, no doubt, came to the atten-tion of the BRIA think tank in Marshall. For them the bell of opportunity had rung. Here was a chance to catch the Ozark Society with its pants down—

camped out on the Buffalo—boys and girls carousing on the gravel bar. To apprehend them in the midst of all that hanky panky would make a better story on the front page of the *Gazette* than did the Dry Buffalo Bombshell the year before.

On April 24 a longtime acquaintance of mine, then living in Mississippi, came into my office with a bemused look on his face, bearing news from Macedonia, which he thought I should know about.

Charles Ivy was an original Bentonville citizen, a veteran of the Second World War, and a graduate of the U of A Law School in Fayetteville. Curiously enough, he had run for Congress against Jim Trimble in 1950 and although lacking in name recognition had come within one thousand votes of defeating Trimble. That was because of the regard that the public then had for veterans and not because Trimble had been an unpopular representative.

Charley Ivy was an oil company agent, one who attends to the legal aspects of leasing or purchasing land prior to drilling operations. As such he was a member of the Landsmen's Association and its allied group, the *Micropterous Dolomieui* Society (Smallmouth Bass Society). That constellation was based in Fort Smith and claimed members in all the states around, all men engaged in that aspect of the oil industry. They were not a bunch of bums. For several years it had been their custom in spring or summer to hie away to some Ozark river, there to set up a well laid-out camp for a week of fishing and relaxation in the out of doors.

In 1964 they selected the big gravel bar at Maumee Mines on the free-flowing Buffalo. On the evening of the second or third day the first echelon had the camp arranged in good order. Steaks were on the griddle in the commissary tent; a few were still casting lures along the river, and some were just relaxing with a snort of Old Crow.

At that point the lights of two cars were noted coming down the hill, and when they drove up and stopped a passel of men got out. One of these was the Searcy County sheriff, Beal Sutterfield, and another, the principal spokesman and interrogator, an abstractor (Gib Walsh, surely). They approached their first prospective victims wanting to know a number of things:

First—"Where are the girls?"

Then—"Are you for or against the Buffalo dams?"

And—"Do you know Dr. Compton? Is he here?"

And—"Aren't you the Ozark Society bunch from the University?"

Finally—"Do you have permission to camp here? You are trespassing, aren't you?"

Being an outfit loaded with lawyers, there was no dearth of people to talk back to that posse:

No—"There aren't any women in this camp!"

Yes—"Some are against the dam on the Buffalo!"

Yes—"One or two know Dr. Compton, but he is at home."

No—"We aren't the Ozark Society!"

Yes and no—"We have permission from the landowner to camp here! We are not trespassers."

By now the Marshallites had begun to realize that they had caught a different bunch of birds. They were just too mature for U of A students, and there really were no women present. The inquisitor was especially infuriated to find that some were against the dams on the Buffalo, and loud argument broke out between the lawyers and the abstractor and his buddies. But the sharp-eyed sheriff had noted something. There was not just a little booze to be seen here and there. What was more, some of it didn't appear to have Arkansas tax stamps on it. In spite of belligerent protestation from the Landsmen's lawyers, the sheriff gave orders to search the camp. Thereupon, those that could grabbed up their untaxed refreshment and slunk off out of sight to bury it in the gravel and sand where some of it may remain to this day to fortify weary canoe paddlers if they just knew where to dig.

In the midst of all the hubbub in drove a car from Louisiana with the back end full of beer untaxed by the state of Arkansas. The *Micropterous Dolomieui* Society had been had.

In trying to prove the trespassing charge, an abortive effort was made to locate the property owner who, according to the Landsmen's group, had given them permission to camp at Maumee. But he could not be found that evening, and no warrant was to be had from him. Regardless of that, the Maumee crossing had been used for over fifty years as a camping, fishing, and swimming area by all visitors.

After two hours of harassment, the Marshall raiders left with a carload of confiscated spirits after issuing a summons for the campers to appear in court in the county seat. The sheriff's parting words were: "If you boys want a drink, there is plenty of moonshiners around here."

After that the BRIA people were in disgrace with some of the local merchants who had sold several hundred dollars worth of supplies to the campers.

Boat line operators who used Maumee as an access point were ready to go to war.

The Smallmouth Bass people who were apprehended that evening were obliged to appear in Justice of the Peace court in Marshall where they were fined $550, with only those who had said they were opposed to the Buffalo dams being assessed. Charley Ivy was permitted to send in his part of the fine in absentia.

OH, MERCY!

Now Progressive Searcy Countians, Americans, Buffalo River Landowners, and the Corps of Engineers have been . . .

GORED

. . . by some erstwhile "conservationists" and nature lovers from Fort Smith, who ran an ad in the March 14 issue of the Southwest Times Record. They intimated we are about to destroy a thing of joy and beauty forever when a much needed impoundment on the Buffalo River is created for the State of Arkansas.

CHUG A LUG

They didn't mention to you good people that: If the dam is constructed it might inundate their "Favorite Drinking Preserve."

HERE IS THEIR RECORD

. . . as published in the Marshall Mountain Wave weekly newspaper, April 30th, 1964.

Of those arrested (named in the following news story) the following persons also sponsored the misleading diatribe against the dam last Sunday: Dorsey Ryan, Tom Mueller, Robert W. Vater, Jim Chesser, and Gerald Delung. They were defended by the great defender, Heartsill Ragon.

Sheriff Siezes Untaxed Whiskey, Beer From Group On Buffalo River

Sheriff Beal Sutterfield raided a group of from 20 to 35 fishermen camped close to the river bank at the old Maumee Ferry on Buffalo river Wednesday evening of last week and confiscated 16 Quarts of Untaxed whiskey and approximately 10 cases of beer.

Sheriff Sutterfield said that he received reports Wednesday afternoon from citizens of the area regarding the group that drove to the river Wednesday, and it was reported that some of them were carrying a considerable amount of intoxicants.

The Sheriff, accompanied by Justice of the Peace L. A. Potter, and a number of deputies arrived at the river about 9 p. m., and found a number of the

men drinking and intoxicated. Upon searching the cars a total of 16 quarts of untaxed liquor, bearing Louisiana tax stamps, was found and many cases of beer. All the beer bore the Arkansas tax stamp but many of the cars contained more than the three gallons which is permitted under the law.

Warrants were issued for Dick Mueller, Max Hare, Dan J. Gaither, Henry Scheilder, Charlie McRae, George Farmer, Jim Chesser, Charles Ivy, Cal Remmel, Thomas C. Mueller, Gerald DeLung, for possession of untaxed liquor and additional warrants were issued for Richard O. Miles, Clyde Gains, D. M. Ryan, Gerald DeLung, John P. Shields, and Robert Vater for over possession of intoxicating liquor in a dry territory.

Hearing on the charges were heard Saturday afternoon by J. P. L. A. Potter in the courtroom of the county courthouse.

The group was represented by Heartsill Ragon, an attorney of Fort Smith, who joined the group Thursday.

At the hearing John P. Shields, who stated that his company had organized the group for a fishing trip, plead guilty to owning all the untaxed liquor and was fined $100.00 and costs. Richard O. Miles, Clyde Goins, D. M. Ryan, Gerald LeLung and John P. Shields, also plead guilty to the charge of over possession and were fined $50.00 and costs each on the charge. Total fines and costs amounted to $515.50.

The group at the river Wednesday evening stated that they were the early arrivers of a group which would gather at the river from four states, and that more than 100 people were expected before the end of the week.

Although they denied they had any connection with the Ozark Society and had no opinion as to whether or not the river should be dammed—some of them were well acquainted with Dr. Neal Compton of Bentonville, head of the Ozark Society and leader of the opposition to the proposed dams on the Buffalo.

Some complaints of the group driving on the roads between Marshall and the river were made by the citizens of the Morning Star community and by some of the residents along the ridge road to the river. Some of the party that came to Marshall after ice and were reported to be in an intoxicated condition when they arrived in town.

The group left the river late Saturday afternoon or early Sunday morning.

TRUTH

1.—

Ozark Society defenders forgot to tell you good people that more than 45 miles of the river above the lake would remain free-flowing— Enough for a 3 or 4 day float trip. Below the dam there will be more than 50 miles of river remaining in natural flow and condition.

2.—

Hemmed-in-Hollow, Goats Bluff, Diamond Cave, Fitton's Cave, and other attractions WILL NOT be disturbed.

3.—

Where some scenic bluff might be partially inundated the water will back up around the hill, create a long finger into another hollow, and Lo—another bluff has been Unlocked that was not accessible before. WHO KNOWS, the dam may unlock MORE natural beauty than it now has.

4.—

LOVE THE BUFFALO . . they said! Who has a right to love it more—The Landowners and Nearby residents who claim it in heritage, or the gamehogs and drinkers who come to sink beer cans in it while being paddled around by convicts from the Arkansas State Penitentiary??? National Treasure indeed!!!! Please

help us keep this for you and I— not the political kings and their slaves.

5.—

This entire area desperately needs that water resource for city water, industrial development, recreation etc. The proposed lake is in a naturally economically depressed area. Construction of a lake would result in more tax dollars to the state— less relief and welfare money— greater attraction to tourists and residents alike. A national park would result in taking land OFF the tax rolls, a disenfranchised citizenry, government ownership of more land . . . . . and what?? The government already owns thousands of acres in National forests and parks in Searcy, Newton, Stone and other surrounding counties. Do we need more of the same. If this brings such prosperity — this section should indeed be prosperous already!!!

6.—

YES, PLEASE DO WRITE YOUR CONGRESSMEN, as the Ozark Society advised. Only, tell them to vote FOR PROGRESS, and For the Buffalo Dam. Tell them that you are a loyal Arkansas citizen who wants to see ALL of Arkansas prosper and Grow More Beautiful. Tell them Selfishness and Greed should not prevent the development of this lake for all Arkansas. Please do it today!!!

The people of Searcy county only ask the right to develop their own property to the best interest of all our people. 75% of the landowners in the lake site favor construction of the proposed dam. And they have expressed their position by a petition to Congress and the Board of Rivers and Harbors. We solicit your help. Please do not hurt us.

Buffalo River Improvement Association

James R. Tudor, President   Gibson Walsh, V-President   Kate Ruff, Secretary-Treas.

Jim Tudor broadside.

162

At first the BRIA people must have felt that they had accomplished great things for the dam program and that more was in order. We heard that a few weeks later a similar fishing camp consisting of people in the plumbing business and their attendants, all from Little Rock, were apprehended on that part of the Buffalo and were likewise fined five hundred dollars. Even their hired cooks were fined in spite of loud protestations that they had nothing to do with the wine mess.

The Landsmen's Association people stayed their week out but went home madder than hornets. Their first act was to run a full-page ad in the Fort Smith *Southwest American* strongly endorsing the national park proposal and condemning the Gilbert and Lone Rock dams.

When the editor of the *Marshall Mountain Wave* saw that, he blew his top and printed a broadside in supposed rebuttal, but which only served to mire him and his cohort deeper in the mud of that classical example of how not to get a dam built on anybody's river. One of our snoopers obtained a copy of that broadside from which we had printed several hundred extras for distribution in the right place at the right time.

On April 28, one of the Landsmen who had been through it all sent a communication to the Ozark Society voicing continued support of our stand on the Buffalo:

April 28, 1964

The Ozark Society

Gentlemen:

A group of my friends and I have supported your campaign to save the Buffalo River for two or three years. This support has consisted of small contributions and small efforts at winning the support of others.

Last week our group of float fishing enthusiasts were molested by some citizens of Marshall, Arkansas, who had their Sheriff (Mr. Sutterfield) and a "Judge" (Justice of the Peace) as their accomplices. Our camp, consisting of about twelve members at the time it was raided, was illegally searched without warrant, without complaint, without any cause. Quantities of beer and whiskey were seized because of overpossession or because it was from out of state. As a matter of fact, the beer and whiskey were community property and, as such, were not of sufficient quantity (on a per man basis) to constitute a violation. Eventually we were fined several hundred dollars.

We were told by some of the "raiders" that their only reason for bothering us was to prevent a recurrence of anything like the Justice Douglas trip. They thought we were there to persuade our fellow campers (of which there were about fifty by week's end) to oppose the dam.

Actually we were there to fish and enjoy floating and camping beside the beautiful Buffalo. There was no rowdy behavior. The landowner consented to our using his property and resisted (we are told) the efforts of others to have us thrown out. The place we camped in is cleaner now than when we found it. This has long been an annual event sponsored by four of us in Fort Smith. We have never done anything to incur the disapproval of local residents, and this year was no exception. We were victims of circumstances.

I believe that the incident did far more to defeat the dam than our adversaries realized at the time. My resolve is strengthened by this episode. If you will send me some literature and petitions, I will work as much as I can toward the realization of the Buffalo National River. Several others share my interest.

Dorsey Ryan

After some time the Maumee raiders began to realize that they had pulled off a big boo-boo that had done their cause more harm than good. No one except Sheriff Sutterfield would admit that they were present on the scene. A year later the *Pine Bluff Commercial,* which had a special reporter out to cover the Buffalo River, printed an interview with the sheriff in which he absolved Tudor and Walsh of having any part of the raid. The sheriff stated that he had acted on the complaint of a "landowner and other individuals" in the area, and that was all.

He did admit that Tudor and Walsh "appeared on the scene—just looking on like a lot of people would when there is something going on."

That bit of action served as a warning to us in the Ozark Society. We knew that if we were going to defend the river we needed to visit the area and to study its every aspect. But after the Maumee affair, we could be certain that we would be set upon if there was ever any opportunity. We should be sure that put-in and take-out camping sites were cleared for our use and that we had landowners' permission of access.

The prime irony of all this was that the University of Arkansas chapter of Ozark Society had set their trip to the Buffalo up a week ahead of the date reported in the AP release. It was already over (properly chaperoned by Doug James—we hope—), when the Marshall vigilantes went looking for the women that weren't there.

SUMMER DOLDRUMS 1964

During those weeks and months while the wheels in Washington ground slowly and small, the effort to stamp out illegal cohabitation lurched along in Searcy County, and the Ozark Society struggled to establish an honest image as a gung-ho outdoor outfit.

To prove that, we went on our first official hike up the Hurricane Fork of the Big Piney to see the natural bridge and the shut-ins. We were spellbound by the delightful scenery, and in our enthusiasm to take in everything many became lost. Husbands were lost from wives, parents from children, and officials of the Ozark Society from one another, but all were reunited before it was over. Some said that we should adopt a version of close order drill before we tried it again.

If we were revealed as a bunch of greenhorns by our discipline on the trail, our navigational performance on the river was even more suggestive.

That June, full of enthusiasm to obtain some knock-out photographs of the lower Buffalo, we set out for Big Creek with a small crew: John Heuston and his wife; my wife, Laurene; Dave McDonald, and a few others.

The year before Charley Johnston had secured some good footage with his 8mm movie camera of John Heuston's canoe wreck on Clabber Creek Shoals. That qualified him for membership in the Over and Under Club, but on the Big Creek trip, I gained even greater distinction by becoming a member of the Out on a Limb Club, as related in a letter to H. Charles afterward:

> We came to a long riffle curving to the left and about one fourth of the way down it a big willow leaned out at a 45-degree angle from the water line. I swung the bow to the left, clearing my inexperienced, overweight and anxious lady passenger in the bow, of the tree. This put me you know where, in next to the bank and under that tree. It was just too late to go over the side so I gave a great leap forward and landed astraddle that willow like the Lone Ranger on a hoss. Down below the canoe without its navigator was picking up headway for the mouth of Big Creek. So in true cowboy style I reached down and latched onto the stern line and after some wrangling got it back up to where I could drop down into the stern where I belonged.
>
> While this was going on two canoes came up behind at the head of the riffle, one of them with my wife, Laurene, aboard. Dave McDonald, noticing the strange goings-on, said, "What in the world is Doctor Compton doing in that tree?" From Laurene he received this happy explanation: "Didn't you know Neil is the world's worst ham, he is up there hoping that someone will come along and take his picture."
>
> We sure needed you on this one, but I wouldn't mind doing a re-run someday if you want to shoot it.

The Buffalo River Country

But it was not all fun and frolic that season. We were hard up for something more than just hand-out information sheets pleading for a comprehensive protection plan for the Buffalo. We needed a good reporter to keep an ongoing account of developments in the news from day to day. While we enjoyed the favor of the press, except for in Marshall and Harrison, we had no committed spokesman for our side, as was, for instance, the editor of the *Mountain Wave.* Various other editors were pleased to print items sent in by us, but it was just too much to attend to everything else and journalistic chores as well. Up in Missouri they had, along with Ray Heady on the *Kansas City Star*, Leonard Hall, a first-class professional writer, reporting for the *St. Louis Globe Democrat* and other papers, keeping a play-by-play account of developments on the threatened Current and Jacks Fork rivers before the eyes of the public. Leonard Hall had also written a book, *Stars Upstream,* a moving plea for the preservation of those two outstanding rivers then threatened by high dams.

We also lacked a journal or bulletin, an official organ to broadcast our aims and objectives and accomplishments, something like the Sierra Club bulletin perhaps. But where were we going to find people to undertake such tasks? John Heuston, Charley McRaven, Gus Albright, and Bill Barksdale were all trained journalists, but they all had commitments that would prohibit such intensive participation.

We kicked that problem around constantly, and it inevitably came to the attention of Ken Smith, by then well embarked on a career in the National Park Service and stationed at Yosemite. His education as an engineer was being fitted to that course. But back in the fifties Ken had done those interesting reports on Lost Valley, Big Bluff, and Hemmed-In Hollow that were published in the Little Rock papers.

After listening to our agonizing for a time, Ken Smith decided to do something. In March 1964 in a message to Neal Butterfield, assistant regional director for the National Park Service in Richmond, Virginia, he had this to say:

> I am thinking of taking about six months' leave of absence from the Park Service to explore and photograph the Buffalo River country, and to try a little writing and publishing to promote preservation of the Buffalo. ... The biggest thing I want to try is a book, a sort of combination of guidebook and preservation appeal ... as it looks now, the best time to begin a leave of absence would be about November this year, so that six months would run until May 1965.

Ken had mentioned this to us in the Ozark Society, and we were nonplussed at first. Would it really be possible for him, a full-time park service employee, to undertake such a venture? He was not without experience as a writer, and we had confidence in his ability, but how much would it cost? He had an answer for that—thirty thousand dollars. That was more than all our income since the first day and was more than we would be able to come by in the next ten years.

When told of that hard fact, Ken suggested that we generate a loan that he would pay back from royalties on the book. That, too, was beyond our ability as was brought forth by Charley Johnston at our annual meeting in August. The board regretfully had to vote no. Ken Smith then said: "I'll do it myself!"

Ken with his own money would write the book and pay for its printing (with expenses to be repaid from proceeds of book sales), and the Ozark Society would be named as its publisher.

While it was obviously an enormous undertaking for one man, it was not entirely implausible. Ken had no family. He drove a Volkswagen bug and lived frugally on his modest park service salary. He had been able to set aside something as he went along to eventually pay the printer.

The Ozark Society was exhorted to help Ken in any way possible. I loaned him one of my two canoes, and we arranged our outing schedule so as to accommodate Ken's program for seeing and photographing all of the river in the allotted six months. Evangeline Archer had a good friend, Rabie Rhodes, a banker in Harrison who sympathized with the plan to save the Buffalo. At his home in Harrison he had a comfortable guest cottage which he offered for thirty dollars a month as quarters for Ken. A better central location could not have been asked for.

In June 1964 I dispatched a letter to Carl Stratton, associate director of the National Park Service, requesting that Ken be granted leave to do research for the book to run from December 1, 1964, to June 1, 1965. Response to that and to Ken's personal request was slow, during which time he became restless, even intimating that he might resign from the park service to perhaps work full time for the Ozark Society. I tried to downplay that (we certainly didn't have funds to pay a full-time employee). I encouraged him to work up in the service and to one day put in for the job of superintendent of the Buffalo National River if that should ever happen to become a reality.

However, these speculations were terminated shortly thereafter by approval of Ken's six-month leave, and soon after Christmas he was in residence at Harrison in Rabie Rhodes' cottage. There he quickly became associated with important area personalities who shared our viewpoint: Gene Waters; Charley McRaven and Sonya, his sister-in-law; Jimmie Dale White; Jim Schermerhorn; Jack McCutcheon; and others. They would serve as guides and informants during Ken's sojourn in the Buffalo River Country.

On August 29 and 30 the second Ozark Society general meeting was held in Mather Lodge, Petit Jean State Park.

The Upper Eleven Point Association was on the agenda, giving valuable information on their twenty years of harassment by the Corps of Engineers. William Bryant of the Ozark National Forest gave an account of activity in regard to the development of Blanchard Cavern, and Jim Schermerhorn and David Taylor reported on other cave systems in the Ozark area.

Members were brought up to date on the Buffalo River battle, and at the annual election of officers the incumbents were all reelected.

During that summer tremors from the *Marshall Mountain Wave* increased in intensity, reaching 7.5 on the Richter scale by June 11 when hints of violence were suggested …

Here and There about Town

A great majority of the people in our county are consecrated to OPPOSE making the Buffalo into a National Park, especially that portion of it flowing through Searcy County. Many property owners along the river have affirmed unrevokingly that they intend to spray and deaden the timber over all their lands lying along the river as soon as the bill is introduced to make it a park. Others have forcefully stated that bluffs along the river would be dynamited into the river bed once the park project is approved.

… a proper course of action for soreheads and poor losers that we would hear more of as time went on.

A Day of Tragedy

Then in the waning days of the summer of '64 tragedy befell one of us. Doug James appeared at one of our sessions at Evangeline Archer's with news too sad to believe. One Sunday morning in Little Rock Charley Johnston's wife, Mimi, and his son, little Warren Mallory, lost their lives in an unspeakable tragedy. There are no words to describe the remorse that we all felt upon learning of that appalling fact.

For Charley there would be nothing that we could do to ease the pain of his shattered existence. We thought that we had seen the last of him, that he could not now face up to the continuing controversy that we were engaged in. He did depart Arkansas for a sojourn into the peace and solitude of the Okefenokee in an effort to restore his exhausted spirit. From that communion with a world in genesis must have come some good, for upon his return and finding us preparing for the second hearing at Marshall (one of the major engagements of the Buffalo River war), he took up his sword and waded in with us.

But there was another problem. We had been aware in the summer of '64 of something undesirable going on across the river from the Buffalo River State Park. Bulldozers could be heard grinding away over there, and tall light poles and power lines were making their appearance on the wooded slopes. We learned that a fellow from Chicago who had been with us on one of our trips on the river had elected to establish a retirement and tourist center across from the state park. He bought the land and started operations, to the dismay of our State Parks Commission and the rest of us as well. I had written to Senator Fulbright, the national park authorities, Secretary of the Interior Stewart Udall,

Warren Mallory Johnston at Buffalo Point, Buffalo River State Park, August 1963.

and others to see if there was anything that could be done to stop such depredation. There was nothing. Until a public law came into being to protect it, the
Buffalo and its environs were up for grabs.

Then the state parks department decided that they would try to buy the
developer out. Their limited budget could not quite cover the amount. By early
1965 the state parks' effort to buy the land was stalled. Construction was about
to begin again when Charley Johnston stepped in and saved the day. There was
sufficient money from his son's life insurance that he would donate as a memorial. The *Arkansas Gazette*, June 19, 1965, took note of that:

$5,000 Given to Help Buy More Land for Buffalo Park

A gift of $5,000 toward the purchase of 205 acres of scenic land along the
Buffalo River has been made to the state Publicity and Parks Commission by H.
Charles Johnston of Little Rock in memory of his son, Warren Mallory Johnston. The
land has become part of the Buffalo River State Park.

Johnston, vice president of First Federal Savings and Loan Association, long has
been identified through the Ozark Society with efforts to preserve the Buffalo River
in its natural state. He is chairman of the Little Rock Chapter of the Society.

The state took title to the land for a total of $30,000 about the first of this year. It is

The memorial plaque to Warren Mallory Johnston, near the old restaurant.

on the east side of the river and includes part of the famous towering bluffs that rise above the stream.

State Parks Director Bryan Stearns said a marker is being prepared as a memorial to Johnston's 8-year-old son, who died last year. The marker will be set on the west side of the river near the park's restaurant, overlooking the newly purchased land.

The Arkansas State Legislature also took note:

S.R. NO. 3
(Brandon)

Expression of Appreciation by the Senate of the 65th General Assembly for the Donation Made by Mr. Charles Johnston to the Buffalo State Park in Memory of His Son, Warren Mallory Johnston, a Former Legislative Page.

WHEREAS, Warren Mallory Johnston was a Page during the 64th General Assembly of the State of Arkansas; and

WHEREAS, Warren Mallory Johnston departed this life not many months thereafter, and his death is mourned by all those who knew him; and

WHEREAS, Warren Mallory Johnston's father, Charles Johnston, has donated one-fifth of the cost of adding 205 acres of beautiful land to the Buffalo River State Park as a memorial to his son; and

WHEREAS, a memorial plaque is being placed at the most scenic point overlooking this new addition to the Buffalo River State Park, with the following inscription thereon:

There Are Little Corners of This Earth Put Aside By Nature to Be Discovered by and to Bring Joy to Little Boys. The Lands Over Which You Look Here, Across This Beautiful River, Are Such a Corner; and the Arrowheads to Be Found There, the Tiny Box Canyon With Its Waterfall and the Springs Above, Are Set Aside Forever for All Little Boys, In Memory of Another Little Boy Who Did Discover Freedom and Joy Here—Warren Mallory Johnston.

NOW THEREFORE,
Be It Resolved by the Senate of the 65th General Assembly of the State of Arkansas:
S.R. No. 3—1

1965 Regular Session.

The marker stands today by the old State Park Cafe, now on national park land for all to see. Many think that it is to commemorate a young boy who lost his life in an accident on the river. But it is instead dedicated to a charming little boy from Little Rock who was not permitted by fate to attain his manhood. Today, we who knew him shed a tear when we think of that.

BACK TO THE ARMORY

In March 1964 came the sound of a distant drum. The Corps of Engineers were rallying their troops for a final campaign to put us river defenders to rout. They, like the BRIA, intended to make 1964 the Year of the Dams (or dam). When we heard those first sounds of activity our people were alerted.

INTERIM REPORT TO MEMBERS
March, 1964

PUBLIC HEARING ON THE BUFFALO

The Corps of Engineers, which has been making extensive further studies on the Buffalo, states that another public hearing will be held "sometime next fall." Also:

"Comprehensive studies are now being made ... for development of water resources in the Buffalo River Basin, including studies of large dams on the main stem, small dams on tributaries ... and various other combinations of projects."

"Notices of the hearing will be mailed at least 30 days prior to the hearing. A description of the plan or plans under consideration will accompany the notice."

This bears out our statement in the November, 1963, report to members, that the Buffalo River is in a perpetual state of crisis. When time and place for this hearing are announced you will be notified. (Or you can ask the Corps of Engineers, 700 West Capitol, Little Rock, Arkansas, to be placed on their mailing list to receive announcement directly.)

What did they mean by "further extensive studies?" How comprehensive would their campaign be? Had they not already drawn detailed plans for big dams at Lone Rock and Gilbert? If we were going to have to testify at another hearing, we sorely needed specifics as to what the new plans were. For the first few months of 1964 we floundered about trying to find out, only to be stymied by their evasiveness.

Evangeline Archer did manage to ferret out the cost of the ongoing "studies" of the Buffalo River from the U.S. Army Engineer District, Little Rock. They stated that:

> … Available records in this office indicate that $256,553 has been charged to studies on the Buffalo River through 31 May 1964. Of this total, $125,900 is in the "preauthorization" category and $130,653 is in the "postauthorization" category. …

From then on, those of us in northwest Arkansas came up with very little specific information concerning the probable revisions in the big dam plan. We knew that the Corps had ready a three-volume *Interim Report on the Buffalo River Basin, Arkansas,* but for us it was not yet available. A state of uncertainty existed in our camp as to how to present our case at the oncoming hearing. A date had not been set, and we awaited its announcement with misgivings.

During that time there did come word from Washington that was welcome. Bill Apple sent a copy of a message from Senator Fulbright, stating that the Interior Department had submitted to him a draft bill to authorize the Buffalo National River but that its introduction would be delayed as long as there was opposition in the House.

From that it was obvious that we were playing for time. If the pro-dam slanted hearing could be stalled, and if we could somehow come up with a favorable congressman some day, the rest would be easy.

Then came a news release from the Chief of Engineers himself, couched in typical engineering double-talk describing the compatibility of it all and throwing to us coyotes the rear end of the river, Lone Rock dam and all:

For immediate release
U.S. Army Engineer District
Little Rock, Ark.

CHIEF OF ENGINEERS PROPOSES
CONSTRUCTION OF GILBERT DAM

LITTLE ROCK, Ark., July 30—A plan for development of the Buffalo River in north Arkansas that would preserve scenic and wild river areas and yet permit reaches of the river to be shared for flood control, hydroelectric power, water conservation, and public recreational uses essential to regional growth has been proposed

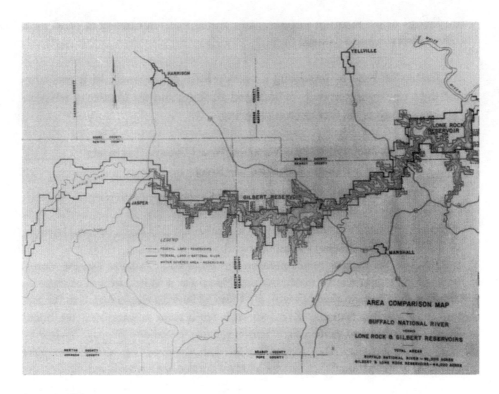

Area comparison map.

by the U.S. Army Corps of Engineers. Information on the proposal was received today by the Little Rock District of the Corps.

The proposals were presented in a draft report sent to the State of Arkansas and Federal agencies concerned by Lt. Gen. William F. Cassidy, Chief of Engineers, for comment in advance of preparing his final recommendations for transmittal to the Congress. The plan calls for construction of a dam at the Gilbert site. A dam already authorized for the Lone Rock site is part of the flood control plan for the White river basin and is recommended for deauthorization because it would back water up into the reach of the river considered suitable for preservation of the wild river. Corps of Engineers field offices and the Board of Engineers for Rivers and Harbors had noted that only through construction of both could full economic development of the water resources of the Buffalo river be secured. However, it was concluded and General Cassidy concurred that development of the Gilbert site alone would provide a balanced plan in which the intangible benefits of the national river proposal would be preserved along with the tangible benefits of flood control, power, and recreation. ...

Also, it would prevent destruction of camp grounds, boat ramps, and similar facilities by large floods. On the basis of experience of existing Corps reservoirs in the north Arkansas area alone, I am convinced that the Gilbert project would be an attractive and intensely used recreation facility. I am in full agreement with the conclusions of the Board of Engineers concerning the compatibility of plans for a Gilbert Reservoir and a national river.

We wasted no time in providing a press release of our own. In it was contained the basic argument that we had used all along and the essence of what we would have to say at the forthcoming hearing.

Ozark Society Presents Views on New Corps of Engineers' Proposal for Dam on the Buffalo River

Dr. Neil Compton, president of Ozark Society, stated following the Society's board meeting at Russellville on October 28 that "The plan as proposed by the Corps of Engineers is entirely incompatible with any other use that might be considered for the Buffalo River for outdoor recreational purposes."

Officers of the Society commented on the implication contained in the Corps' notice of the public hearing in Marshall on November 18 that some sort of compromise may have been worked out with the National Park Service in regard to the recommended Buffalo National River. No such implication is warranted, the board states. The words used in the Corps' announcement are strictly language devised by the Corps of Engineers. The proposal for the dam was drawn up with the complete ignorance of the National Park Service and that Service has not been consulted at any time in regard to it. The suggestion in the notice that the Gilbert Dam would be compatible with the National Park Service plan for a recreational area on the Buffalo is completely false.

In describing the proposed high dam at Gilbert, Ozark Society states that it would flood 45 miles of the middle section of the Buffalo and would completely destroy the river as a free flowing, undisturbed mountain stream. Water would back up all the way to Pruitt on Highway 7 on the Big Buffalo and almost to Jasper on the Little Buffalo. The drawdown of about 53 feet would produce a fluctuating shoreline of the reservoir during drouth and result in a most unsightly and undesirable situation, seriously affecting the use of this part of the river for stabilized recreation purposes.

Below the dam the water would remain throughout the year at a temperature of 58 to 60 degrees from the release of cold water from the bottom of the reservoir. This would eliminate practically all of the native fish such as the smallmouth bass for which the Buffalo is famous, along with perch, goggle eye, and catfish.

Above the impoundment the river would be invaded by rough fish from the reservoir. These fish are mainly shad and suckers and would displace the game fish in this section of the river.

Lowered water temperature would prevent the use of the river for swimming, and the daily fluctuation along the stream from release of water from the dam would render it dangerous. An analogous situation exists at the present time on the White River below the large dams on that stream, where a bole of water descends the river each day and poses a threat for fishermen and for persons camped along the stream.

Ozark Society challenges the statement made by the Corps that the Gilbert project is economically justified, and insists that further large impoundments of this nature cannot possibly be economically justified in the face of the multitude of others in close proximity. It is known that many of the developments already existing along these reservoirs are in economic distress in one way or another.

Ozark Society disputes the claims made for the proposed impoundment for purposes of flood control and power production. The Buffalo River has not been the source of any consequential flood in decades, the Society stated, and the destruction of such a beautiful stream as the Buffalo is unwarranted for either flood control or power production. Generation of electricity is much more economical by steam plants now being constructed by the REA. Power generation at the Gilbert site would be infinitesimal in comparison with already existing facilities.

Stabilization of flow is another claim of the Corps for the proposed dam. This is a completely unnecessary consideration, Ozark Society asserts. The Society points out that all streams throughout the Ozarks fall to low levels during drouth and any program to stabilize flow in any stream, whether it be the Buffalo, the War Eagle, the Little Red, the Mulberry, or the Big Piney is an unjustified expenditure of taxpayers' money. The Buffalo below Gilbert is a relatively large stream and during the worst drouths has never gone dry. There is a small section of the river within the reservoir area which runs underground during severe drouths, but this has not disqualified the Buffalo as a National River in the judgment of the National Park Service.

Ozark Society takes issue strongly with the claim of the Corps that the proposed reservoir would "enhance" the river at Buffalo State Park. It points out that the Buffalo at the State Park possesses unexcelled beauty as it now is, and that the location was selected because of that fact. On the contrary, they say the cold temperature of the water would prevent swimming, and the stabilization of stream flow would result over the years in the growth of weeds and brush on all the gravel bars, as has already occurred on White River. One of the chief attractions of the Buffalo at the State Park and elsewhere is the clean white gravel bars and sand bars available for campers, a situation maintained now by occasional rises in the river which scour and clean them.

Proponents of the dam will undoubtedly say that trout fishing will be available below the dam. Such fishing is strictly an artificial put-and-take activity, resulting from federal expenditures of over a hundred thousand dollars a year in hatching fingerlings and stocking. The Buffalo in its natural state needs no such operation. Interviews with commercial boat line operators indicate that a large percentage of their customers prefer the Buffalo and its bass fishing.

"All people interested in the salvation of the Buffalo River should act now," said Dr. Compton. "The time is short in which to act for the preservation of this beautiful stream. Arkansas has a wonderful opportunity to take its place among other progressive states in outdoor recreation if only a balanced and diversified program can be established. To build another large dam and to destroy our last remaining free flowing river can only condemn us to mediocrity. The Buffalo River is one of the loveliest streams left in North America. It has been acknowledged as such by the National Park Service, whose business it is to know."

On October 16, the Corps of Engineers finally announced the date of the hearing, November 18, again in the Marshall Armory—as one might know. We had a month to get ready for it, not time enough for us to work up a good game plan. We would have to make do with such skimpy data as we could come by and rely principally on the emotional impact of the thing.

But down in Little Rock H. Charles Johnston had the gall to demand from the district engineer facts and figures that we sorely needed in a letter to them dated October 22, 1964:

> … we herewith request that we be furnished with the following documents and studies which we understand have been necessary in your determination to this point:
>
> 1. According to paragraph 2, page 2, of your notice, a determination has been made that regulation of the flow of the Buffalo River is necessary. Please give us complete data concerning the methods of arriving at such conclusion.
> 2. The statement is made concerning all proposed "plans" that "it is economically justified." Please furnish us with complete data, and all studies related thereto, concerning your methods and legal requirements for arriving at such conclusions.
> 3. Please furnish this organization with all maps, and other documents necessary to show exact water levels of impoundments created on all plans, and delineating thereon the level of each designated "pool." It is our understanding that such maps are in existence and are prepared for each plan for use in your determinations.
> 4. All studies furnished by other governmental bodies used in preparation of your determinations.
> 5. Inform this organization of type of release of impounded waters, i.e., warm or cold release.

From the district office of the Corps in Little Rock, our man Johnston received notice that the Corps was not required to furnish information on any of the questions posed by him. Charley was told that after they had decided the case that he could come to their Little Rock office to study the three-volume report if he wanted to. The mechanics of the handling of their decision through the various agencies concerned was outlined.

That was short shrift for one so impertinent, but they had not disposed of pestiferous Charley Johnston. A front-page item in the *Arkansas Gazette* for November 13, 1964, let them know what kind of a gadfly they were up against:

Society Files Suit to Block Dam Meet
By Jerol Garrison
Of the *Gazette* Staff

The Ozark Society, an organization devoted to the preservation of the Buffalo River as a free-flowing stream, went to court Thursday in an effort to block the plans of the Army Engineers to hold a public hearing next Wednesday on proposals to dam the river.

The Society filed suit in Federal District Court at Little Rock, charging that it had been denied pertinent information that it needs to comment intelligently on the proposals. It asked for a temporary restraining order to cancel Wednesday's hearing and for an injunction to keep any hearing from being held until it gets the information it wants.

Judge Gordon E. Young scheduled a hearing for 1:30 P.M. Monday on the motion for a temporary restraining order. There was no immediate comment from the Engineers on the lawsuit. ...

The defendant is Col. Charles D. Maynard, district Army engineer. ...

Johnston said at a press conference Thursday afternoon that he felt the Engineers were being unfair by denying information to the Society that was available to the Engineers. ...

He said the Society members would have no way to tell whether the figures presented by the Engineers were correct. ...

Then four days later on November 17, 1964 (one day before the hearing), the predictable outcome was related in the same paper:

Dam Hearing Foes Lose Suit

U.S. Judge Refuses to Block Meeting

Federal District Judge Gordon E. Young said Monday that he didn't have the authority to stop the Army Engineers from holding a public hearing at 10 A.M. Wednesday on proposals to dam the Buffalo River. The hearing will proceed as scheduled at the National Guard Armory at Marshall. ...

The ruling came at the end of a one-hour court hearing on the motion for a temporary restraining order. James E. Youngdahl, attorney for the Ozark Society, said he would continue to press the lawsuit but would amend it to ask that the Army Engineers be ordered to hold another hearing in addition to the one Wednesday. ...

Assistant United States Attorney Walter G. Riddick argued that in order for the plaintiffs to obtain a court order blocking the hearing they would have to show that they were threatened with the loss of some property right. He said no property would be taken as the result of Wednesday's public hearing.

Says Study Made at Congress' Request

Youngdahl accused the Engineers of being a "bureaucracy," but United States Attorney Robert D. Smith denied this. "I have never seen the Engineers in any bureaucratic attitude," Smith said.

It would appear from all this that Charley Johnston and the Ozark Society had gotten their comeuppance, but it was, in fact, a publicity event advantageous to the anti-dam campaign. That he was a first-class schemer anticipating just that is borne out in a message that Charley had relayed to Evangeline on October 23:

> ... I am quite confident that the simple filing of a lawsuit for an injunction to stop the hearing would be worth $5,000 to us in publicity. I am absolutely certain that we could hit the front page of one of the newspapers here, as well as all the TV stations. I must point out to both you and Neil that we should not act as if we are going to protest the hearing until the very moment that we make our final move. All our efforts until that point should appear as honest efforts to get all the pertinent facts.
>
> Of course, all this reasoning is blown completely out of the tub should they actually offer us the information. And, one other thing—should they offer us a glimpse of the records up at their office, I should think that our course would be to go up there and take one look at the voluminous material and then write to them and tell them that we have to have the material in our possession because most of us live out of town and can't afford to stay up there to read it.

During those few weeks and days before the second hearing, the idea of a compromise or agreement between the Corps of Engineers and the National Park Service reverberated ever more loudly throughout the land. Officially the Corps did not claim that a "compatible" arrangement had been arrived at.

But language in the press generated by the Corps or their cohorts led the casual reader to believe otherwise. That is best exemplified by a report in the November issue of the *Ozarks Mountaineer*.

> The Engineers' plan is for a high multi-purpose project at Gilbert for flood control, hydro-electric power and recreation. ...
>
> The plan is economically justified and is "considered the most compatible with the Buffalo National Park Service," the Engineers said. Only about 45 of the 128 miles of the River that would be in the National River would be inundated by the Gilbert Reservoir, the Engineers said. ...

Such commentary gave rise to doubts and misgivings, even in our own ranks, and caused many of our sympathizers to say, "Well, why not? Why continue such a hopeless struggle when it can be settled so easily in this fashion?"

Our quandary was stated by Charley Johnston (also in his letter to Evangeline Archer on October 23, 1964):

> ... Oh, yes! I am enclosing a copy of a letter from Charley McRaven. ... He expresses fear of collusion between the Corps and the Park Service. This is inconceivable, and should be inconceivable, but—I *can* conceive of it. This is something for Neil and all of us to talk about next Wednesday. We must figure some way how to

make certain that the Park Service stays "hitched" to their previous stated position of not having any interest in the River unless it remains in its natural state. Once again, this should be inconceivable, but who knows? ...

And by another communication from me to Charley Johnston on November 2, 1964:

... I just had a telephone call from Elbert Cox [regional director for the National Park Service] in Richmond. It appeared to me that they were not too desirous of sending a man to the hearing; however, I urged that they do this because of the fact that the Corps of Engineers and now Tudor and the others over at Marshall are trying to make it appear that the Park Service has agreed to the compatibility of Plan D. A refutation from an authoritative member of the Park Service in regards to that is necessary in order to squelch once and for all this talk about a compromise or compatibility. It is a positive fact that there has been no agreement of any kind whatsoever and that the National Park Service was caught by complete surprise by this development. I learned this in a talk the other day with Raymond Nelson, superintendent of the Pea Ridge National Military Park, and verified it in my discussion today with Elbert Cox. Cox has agreed to have Bernard Campbell come by to see you and perhaps me before the hearing takes place and to draw up a good statement for them to present. ...

The official position of the National Park Service was then nailed down by a letter from Director Cox:

Dear Dr. Compton:

We, too, were just recently apprised of the Corps of Engineers' latest plan for reservoir development on the Buffalo River in Arkansas, which you describe in your letter of October 31.

Our position on the Buffalo remains unchanged; that is, we have determined that this resource possesses national significance in its natural, free-flowing condition. Secretary Udall endorsed our position enthusiastically in a speech before the Arkansas River Valley Development Association in Little Rock, Arkansas on July 9, 1963. Further endorsements have been made by the Advisory Board on National Parks, Historic Sites, Buildings, and Monuments; numerous conservation organizations across the land; and by literally thousands of interested citizens such as yourself.

Lest one be misguided, we would like to state emphatically that the Corps' announced plan is *not* "... compatible with the Buffalo National River. ..." We are certain that the proposed reservoir would destroy the river's national significance as a free-flowing stream by inundating its highly scenic midsection.

Elbert Cox
Regional Director

The question of the fate of the Buffalo, then, was building up to a watershed decision. We and our big dam adversaries were aware of that. The press and the public sensed the tension and awaited the decision with the deepest interest. If the Ozark Society and their supporters could really stop the mighty engineers, then there might not be another Greer's Ferry, Beaver, or DeGray to glorify the landscape.

At this precise point in time came another important event from which we drew no comfort—the election of 1964. Although Congressman Trimble was of advanced age, no effective successor had been arranged for by the Democrat party power structure. There was no new personality in that category for us to approach with the national river idea. The incumbent was obviously in for life.

The prospect of a Republican replacement was even more bleak. In 1962, Cy Carney, a lawyer from Rogers, had run against Mr. Trimble. Carney was a strong supporter of the national park program and had even made a public speech in Marshall endorsing the idea, a thing that would have caused him physical and emotional trauma had he lingered in the town. He was soundly defeated in the general election that fall. But in 1964 the GOP would try again. Winthrop Rockefeller down on Petit Jean was trying to improve the fortunes of the party in Arkansas. Jerry Hinshaw, a Springdale businessman and cattle rancher, was persuaded to run against Trimble. That summer Hinshaw had a talk session over in Marshall with Jim Tudor (ironically, like me, another Republican) and came away a supporter of the great dam plan at Gilbert. When we learned of that I invited Jerry Hinshaw and his wife to our house for a debriefing. They were shown one of the better Buffalo River movies and left convinced that we had the best plan for the Buffalo. Hinshaw wisely did not make an issue of it in public, but that didn't matter in the long run. He was defeated in November by a surprisingly narrow margin. There was cold comfort in that, in addition to the fact that I wouldn't have to listen to Evangeline talking about "that skunk Goldwater" anymore, and that Charley Johnston could now quit trying to save my political soul.

The handwriting was on the wall for the Buffalo River. The upcoming hearing would produce a favorable decision for the Gilbert dam. Plan D would proceed through legislative and departmental channels to reality unless we could come up with a significant champion in public affairs. There was one, and to him we would appeal. His name was Orval Faubus.

Finally the day arrived. On November 17, Evangeline Archer, Craig Rosborough and I, ready or not, repaired to one of the new cabins at Buffalo River State Park in order to be in easy driving distance of the Marshall cockpit the next morning. At the cabin we would meet H. Charles Johnston, Jr., and a few others.

In addition to my prepared statement I brought along what was believed to be some important display material. During the summer I had made in my darkroom a collection of 11" x 14" black-and-white prints, one series showing the scenic beauty of the Buffalo and another showing the contrasting ugliness of parts of a nearby reservoir (Table Rock) during a thirty-foot drawdown. There was the long avenue of dead snags lining the course of the inundated Roaring River, the expansive mud flats, the stranded boat docks and the shrunken reservoir in Viney Creek. Then there was the current destruction going on above Beaver Dam, then abuilding, the big bulldozers and cutters felling the timber, piling it, and burning it. Though they were only photographs, they were almost as offensive to the eye as the real thing. These pictures were mounted in pairs on large cardboard backing leaving room for captions. Each pair was supported by a wooden lath attached to the back and cut to a length so that the picture stood about eye level from the floor when propped against the wall. They were easily transportable, and a goodly stock was brought along. They were the subject of an amusing episode the next day.

Arriving at the armory we confronted the expected mass of local citizenry glaring at us "outsiders" who had come to do them out of their birthright. But we went in, registered at the command post, arranged our photo display along the west wall of the armory, and tried to make ourselves comfortable.

The first item on the agenda was to ask for statements from representatives of the branches of the federal government starting at the top:

From the offices of the president of the United States and the House of Representatives there were none.

William Bryant, representing the National Forest Service, said that more information was needed before that agency could deliver an opinion.

James T. McBroom, representing the Bureau of Sports Fisheries and Wildlife, gave a convincing statement in favor of preserving the Buffalo River in its natural state.

Jackson E. Price, assistant director of the National Park Service, made a firm and resolute plea for the preservation of the Buffalo and clearly stated that no possible compromise existed between the National Park Service and the Corps of Engineers in regard to the Gilbert or any other dam that might be considered for the Buffalo.

Following that, several officials of the Rural Electrification Administration appeared, all speaking for the proposed dam.

Various persons representing the Buffalo River Improvement Association spoke on behalf of the dam, offering various preposterous reasons for its construction. A main argument was that the town of Marshall has been having difficulty with its water supply and that the Gilbert reservoir, although about ten miles distant, would supply the needed water. This implies that the expenditure of around sixty million dollars for a dam is needed, in part, to supply a town of 1,045 people. Other reasons were that somehow the project would decrease the tuberculosis rate in Searcy County, that the lack of indoor plumbing would be rectified by the building of the dam, that the wells and springs in Searcy County have fallen to low levels or have gone dry during the last three years of drouth and that the reservoir would in some manner cause the springs to flow and the water to rise in the wells, and that the dam would attract industry to the town of Marshall. The National Park Survey report on Buffalo National River was dismissed as worthless and the 173-page economic study as untrue. One of the BRIA members even gave an account of the raid on the fishing camp at Maumee Mines last May.

In sum, it was claimed by the BRIA that all of the problems of Marshall and the area would be solved by the building of the dam—whether related to schools, employment, industry, health, home facilities, or water supply.

Members of the Upper Eleven Point River Association, veterans of eight public hearings in fighting the dam proposed for the Eleven Point River, appeared on behalf of the Buffalo. Members of the Ozark Wilderness Waterways Club, also seasoned veterans, made statements for preservation, as did the Missouri Conservation Federation. Without listing all of those speaking for preservation, there should be mention of the Arkansas Federation of Garden Clubs, the League of Women Voters of Arkansas, and the State Audubon Society. The national conservation organizations, numerous local groups, and hundreds of individuals filed statements with the Corps.

When speakers opposing Plan D were on the podium, there was noted an annoying breakdown of decorum in the mostly local audience. People talked, laughed, got up and walked about, and scooted and scraped their chairs on the concrete floor. The very proper colonel in charge of proceedings failed to note, and made no effort, to halt that disruption. My turn came just before noon, and my address was interrupted by a lunch break. Finishing after that I felt very much like a nonperson talking about nothing at all. Afterward I learned that the PA system was off the entire time—intentionally or accidentally, we never knew.

H. Charles Johnston, Jr., finally came in and was given his turn. He stepped out in front of the microphone and talked loud enough for almost all to hear. He gave an ad-lib, emotional, but reasonable, account of the Corps' evasiveness in furnishing basic facts concerning Plan D. He pointed out that Searcy County citizens, the expectant beneficiaries of the big reservoir, didn't know just what they would be getting in the end. By then much of it would be unacceptable and disappointing to them, but then it would be too late—a great performance by Charley, but it brought not a single convert down the aisles.

But there was one there that day although we did not know it. Dick Murray was a career man on the colonel's team, a tall bald-headed fellow who occasionally had for us a friendly smile, a puzzling thing from that department. At a later date we would come to know him very well.

Whether or not there was a prior arrangement, or whether it was a spontaneous development, the president of the Buffalo River Improvement Association was permitted to function as a sort of special prosecutor of any anti-dam spokesman on the stand. He was especially good at confusing and flustering middle-aged ladies on our side, such as Dorothy Whitcomb, wife of a professor at the University. Colonel Maynard was not beyond aiding and abetting in that and did little or nothing to soften Tudor's assault on our spokesmen. There was one exception, however.

One of the last groups to be heard from was the Buffalo River Landowners Association, represented by Charles McRaven of St. Joe. McRaven stated with telling effect that his group represented the real "insiders"—those whose homes and farms would be flooded—and that all others, including Marshall residents, are the "outsiders." He said that the landowners want neither the dam nor the park, but if they had to make a choice they were definitely in favor of the National Park Service proposal.

The Buffalo River Landowners Association had sent Charley McRaven to represent them. Significantly, he was placed last on the agenda, being called at 6:30 P.M. after almost all had left the armory (most important, the press) with the exception of Ray Heady of the *Kansas City Star,* who stayed on to write up and publish what was probably the best anti-dam report of the affair.

Tudor's interrogation of Charles McRaven was especially harsh and hostile. Here was the head man of the authentic property owners along the Buffalo sitting in opposition to the "Improvement" association. He was to be tried, convicted, and sentenced on the spot.

But Charley proved to be a far better man at repartee than his antagonist, and he ran him through in that duel of words. Colonel Maynard, taking note of that, requested that the big man of the BRIA cool it. It sort of made our day.

One favorable development of the day, although not a part of the hearing in progress, was the announcement of a resolution of the Arkansas Publicity and Parks Commission supporting the free-flowing Buffalo. That agency should have been with us all along but lacked the nerve to confront many of our state legislators and the Corps. Now they had gotten up the gumption to do it, and their resolution would be made part of the hearing record later on.

Immediately following the hearing the Ozark Society sent letters to various VIPs deploring the biased treatment that we had received: to Stephen Ailes,

Secretary of the Army; to Senators Fulbright and McClellan; and to Senator William Proxmire, hoping that he would present the Corps with a golden fleece reprimand.

Whether our complaint rang any bells in Washington we never knew, but Proxmire's investigation did provoke a lengthy report from General Jackson Graham, the Corps' director of civil works, which pointed out the tortuous route that we would have to follow if we were to have any further input.

Media coverage of the Marshall event was intensive and widespread. All of the newspapers in Arkansas and most in Missouri and Oklahoma reported it. It even gained brief recognition in the *Christian Science Monitor* and the *New York Times*. Generally, the electronic media and the press were sympathetic with the free-flowing river idea, but in writing it up most emphasized audience applause, which was naturally louder for the big dam backers, thus slanting the news in that direction. Also, the hackneyed "well-off outsiders versus deprived natives" theme was given much emphasis, again tilting the news in the latter's direction.

Much of that was because of error on our part. We came to the meeting with enough copies of our statements to satisfy the procedural requirements of the affair but with no handout copies for the press. The editor of the *Marshall Mountain Wave* understood that very well and saw to it that visiting journalists got copies of the BRIA line.

As has been mentioned, the most significant anti-dam testimony was put off until the very last when all of the reporters had gone. All but one, that is. Ray Heady, outdoor editor for the *Kansas City Star*, stayed it out to cover Charley McRaven's most effective testimony. Heady's write-up of the hearing occupied a full page in the *Kansas City Star* for December 13, 1964. His introduction merits our most sentimental consideration:

> Before 1964 slips away and becomes 1965—the year that probably will decide the fate of several more beauty spots in America—we want to take one more look at the Buffalo river of Northern Arkansas. In fact, we think we will pack the canoe and make a solo trip on the river of about three days and two nights.
>
> The first night, we will build a small campfire, sit down beside it and tell the river that it probably has only a few years of life left. We will break the news as gently as we can because the Buffalo river is a stream of spirit and beauty, and it's hard to tell a friend of spirit and beauty that it is going to die.
>
> And the second night, we will build another campfire and sit beside it, and try to tell the Buffalo river why it is marked for impoundment. We are not sure we can do this in a few words because it is a complex story—involving all sorts of economic, sociological, political and regional arguments.
>
> But the two main problems are just plain economic on the part of a few persons and an over-riding don't-care attitude on the part of the majority of persons.
>
> This combination lets a small, intense, vociferous group push through a lot of projects that are called "progress."

These are difficult things to explain to a semi-wilderness river which has been flowing through the Ozark highlands for centuries, minding its own business, not familiar with the ways of modern man. The Buffalo drops from an altitude of nearly 2,400 feet down to 400 feet where it enters the White river. It is so remote and tucked away in the hills and valleys that a lot of people don't even know it is there. Only about four bridges cross its main stem which is around 125 river miles long. It's not a big river. A barefoot boy can leap across it in some places. But it is a secluded, beautiful river.

Our performance up to this time, while most annoying to the engineering clique, had not been sufficient to slow their momentum, let alone bring it to a halt. They now had up steam, and those who opposed them had better clear the track. Now it seemed that, in spite of all the effort, we were going to witness the end of the free-flowing Buffalo River.

But there was no let-up in our efforts to inform the public and to generate a grassroots wave of resentment against it all.

A communication on November 20, 1964, to Charley Johnston immediately after the hearing testifies to that and relates a bit of whimsy in regard to the Marshall affair:

> I finally have calmed down after the great fracas at Marshall. I had to go over to Yellville last night for a program with the Lions Club. My wife on discovering that this was on the schedule instituted proceedings in regards to domestic relationships which were even more ominous than those that I had been through in the National Guard Armory along with you. I made the trip to Yellville under all kinds of clouds real and otherwise, but it all turns out very well after all. There were twenty-one interested and polite citizens of the town there to hear the program and see the movie. They were most interested in what I had to say, and when the thing was over I feel sure that we had twenty-one friends. They know the Marshall folks first hand and appreciated some of the things that were said. I made no attempt to berate the Marshallites; however, one little story was worth repeating. You noticed the big photographs that we had on standards along the wall at the hearing. Craig Rosborough was assigned to keep an eye on them to prevent vandalism, and of all people to attempt some sort of thing like this was Mrs. Tudor. She came up to the photographs and started turning them around face to the wall. Craig said to her, "Lady, you shouldn't do that, we have as much right to show these as the Engineers had to show their map, and if that is the case we could go up and turn their maps around." She looked at Craig and said, "Well, they are all fakes, they are not real." I believe that this typifies the feelings and the reasoning and the situation of the Marshall folks better than any other one thing that happened. It was worth repeating and I took these photographs to the Yellville program and turned one of them just to demonstrate. It really made the point and got a good laugh. ...
>
> Once again in regard to newspapers, I received the *Marshall Mountain Wave* today. It made no mention whatever of any of the people who opposed the dam except for

Charley McRaven, and he got two sentences. This, in general, was what occurred in papers around the state. I believe that none of us got more than one or two sentences. At the end of the *Marshall Mountain Wave* article was an item stating that they had had a call from Jim Trimble late in the afternoon, and they state that Jim wished them well and that he would do what he could for them. I have been afraid of this, but it is a little strange that he wouldn't want his name announced at the meeting. We must be thinking about what possibly could be done to influence this thing from a political standpoint.

Neil

NINE

THE BOARD OF ENGINEERS FOR RIVERS AND HARBORS

Early in January 1965 there came into our hands from the halls of Congress a real hot potato: H.R. 2245.

Close on the heels of that came the announcement that the division engineer in Dallas had approved "Plan D" as presented at the November hearing and had forwarded his decision to the Board of Engineers for Rivers and Harbors in Washington.

People in the Ozark Society were not surprised at these developments. We knew that such action was inevitable and awaited with real anxiety further steps in the implementation of Plan D. But it did serve to stimulate a flurry of activity on all fronts, first an appeal to our members to write to the members of the House Committee on Public Works opposing H.R. 2245.

To set the pace I led off by sending a letter to President Lyndon Johnson, commending him for a speech he had made on the need for conserving some of our natural rivers.

One of the letters to Congressman Trimble was from A. T. Shuller, superintendent of schools in Berryville and one of Trimble's strongest supporters. In closing Shuller said:

> Let's save something of this for future generations and for those of us now living, who enjoy the rugged experience of floating and camping on a beautiful Ozark Mountain stream that remains just as God made it. The creation of the proposed national river would serve this purpose.

> A. T. Shuller

Although we never had positive proof of it, such heartfelt pleas from those so close to him as Albert T. Shuller may have dampened our congressman's big dam ardor. His behavior in that regard was sometimes puzzling. Perhaps that was due to the fact that his only son was then a colonel in the U.S. Army Corps of Engineers.

Some of our letter writers turned to with enthusiasm. There came to us in the mail an editorial from the Fort Smith *Southwest American,* which stated "Douglas Urges: Let's Keep the Buffalo!" On it was a marginal note stating that: "We are now starting on L.B.J.," signed by Robert Vater. He happened to be the president of the Landsmen's Association and a veteran and victim of the recent Maumee Massacre. The editorial in the *Southwest American* had come about from a letter written by Vater to Supreme Court Justice William O. Douglas. With his reply to Vater, Justice Douglas included a personal statement on the Buffalo River, of which a part:

> Swimming, boat fishing, canoeing, hunting, and fishing—these are the great national values in the Buffalo. Though it is in Arkansas, it belongs to all the people. All of us who have been on it love it. It would be sheer desecration to destroy it by a dam or otherwise. It should be kept in perpetuity as a remnant of the ancient Ozarks unspoiled by man. Its fast waters and its idyllic pools make it a bit of heaven on earth.
>
> William O. Douglas

Senator Fulbright, in reply to a telegram sent by me, stated that he and Senator McClellan were arranging a meeting with representatives of the Engineers and the director of the National Park Service.

His reference to a forthcoming high-level meeting aroused a nearly unendurable curiosity in our ranks. That was relieved somewhat by the senator's reply to Evangeline Archer's inquiry:

February 15, 1965

> Thank you for your letter of February 3, 1965.
>
> Senator McClellan, Congressman Trimble and I met with representatives of the Park Service and the Corps of Engineers last week to discuss the Buffalo. Secretary Udall attended the meeting and was emphatic in his support of the park plan. Colonel Maynard presented the Gilbert Reservoir proposal. The most significant development at the meeting was a statement by Secretary Udall to the effect that the Executive Branch would take a position favoring one plan or the other rather than submitting conflicting proposals to the Congress. As this came from him and followed by two days the President's proposal for a wild river system, I think this can be construed as good news for those of you who are interested in the creation of the Buffalo National River.
>
> Sincerely yours,
J. W. Fulbright

Ray Heady of the *Kansas City Star* had much better connections in Washington than the rest of us, and in the February 21 issue of the *Star* he was able to give us something of an inside view of what was going on:

Buffalo River Decision Moves to White House

> WASHINGTON—There is every indication that the fate of the Buffalo River—whether it becomes an impoundment or a national rivers park—will be decided within the next 60 to 90 days, and the decision will be made by the White House.

Probably sitting in on that decision will be President Johnson, Secretary Udall of the Interior department, and Secretary Ailes of the Army. ...

An "exploratory" meeting was held recently between Udall and the National Park service, the Engineers and Col. Charles Maynard of the Little Rock district, McClellan, Fulbright and Trimble.

While it was a closed meeting, it is understood that Udall "hit the ceiling" when the idea was advanced that the proposed dam on the Buffalo would not destroy the river's value as a national park. No decision was reached at this meeting, which was held in McClellan's office.

Udall said later that the decision rests with the administration and that a recommendation should be forthcoming.

Fulbright's strategy seems to be to wait and see how the administration is going to act, and then introduce his bill accordingly.

Conservationists and naturalists, who have been waging a strong campaign to save the Buffalo from a dam, gained hope when they learned the decision would be made at the White House level. They re-read with interest President Johnson's message of January 4, in which he said:

"We hope to make the Potomac (river) a model of beauty and recreation for the entire country—and preserve unspoiled stretches of some of our waterways with a wild rivers bill. More ideas for a beautiful America will emerge from a White House conference on natural beauty which I will call soon."

Developments outside of D.C. were not neglected. On the local scene there was help to be had, good advice from a few experienced people. That was brought out in a letter from me to Charley Johnston dated February 8, 1965:

I had an afternoon session yesterday with Bill Allen of Hot Springs, of whom you may have heard. He is a professional lobbyist for conservation interests and especially for the sporting goods interests in the United States. ...

He states that one word from Governor Faubus could stop this dam project if we can persuade him to speak it. This would not have to be a public statement on page one but could be made to any person or organization and could be announced on the q.t. if necessary. Our unfortunate position is that we don't know anybody that knows anybody, but I am suggesting that you see who you might locate down your way that could talk to Governor Faubus and get him to say just one sentence against the Gilbert dam and see that this statement is recorded in such a way that it would be official. We are working on it up here for all we are worth.

As a result of Bill Allen's suggestion Clayton Little filed a protest to Col. Edmund Lang of the Board of Engineers for Rivers and Harbors on February 22, 1965:

The purpose of our protest is to cause reasonable men to pause before accepting a decision of a single agency of government concerning a subject of much interest to

Clayton Little checking trash bags at Maumee, Ozark Society cleanup, August 1967. Clayton Little of Bentonville was the legal council for the Ozark Society for many years. Note the remains of an old ore bucket in the lower center of the photo.

the nation, not one of only "local concern." Even though the old fashioned railroad grade crossings have disappeared from our national highways, still their old familiar admonition of "STOP—LOOK—LISTEN" should apply to those considering putting a dam on the Buffalo. We want a pause, a thoughtful and thorough look, and an attentive listener.

The Corps of Engineers has, in our estimation, selected a plan of development of a river without recourse to or recognition of the laws of Nature or Congress.

They have devised the devastation of the Buffalo River. It is a virgin stream unmatched in natural beauty. It is an untamed river, unencumbered and unharnessed, enjoyed by those who understand and appreciate natural values. It is not "improved" or "developed" nor cluttered with garbage and parking lots; swimming pools or commercialized recreation; the sounds and smells of engines and pollution. In short, it is not yet an outdoor slum. It is not yet a place of mass mediocrity, regimentation, regulation, and despoliation.

The Buffalo River issue is not a question of an *unsound dam,* but the preservation of a wild river. There are subsidiary points of contention—the legal propriety of the proposal, the terms and methods of justification, and public need, but still reservoirs and public use of reservoirs per se are not at issue.

At issue is the temerity of the Corps of Engineers to ascribe its recommendation as the only solution versus *established national policy.* Reservoirs can be built; wild rivers cannot be created. The Buffalo River conflict is more than just a dam vs. a national park area. At issue are natural values associated with a clear-water stream, unique geological formations, rare and endangered plants and animals. The proposed national park will best preserve these values and at the same time fulfill a public demand for wilderness-type recreation compatible with the primary objective. It should be understood that the concept of wild river management proposed by the National Park Service is of itself unique in terms of recreation area management and is possible only with a free-flowing stream.

In the midst of these developments, Charley Johnston staged a mini-hearing of his own, more or less unbeknownst to us uplanders in Fayetteville (*Arkansas Gazette,* February 3, 1965):

Pros Clash with Cons at Little Rock Session on Buffalo Dam Issue

By Robert Shaw
Of the *Gazette* Staff

Proponents and opponents of damming the Buffalo River came together Tuesday night at Hotel Marion and tempers flared.

The occasion was a meeting called by the anti-dam group, the Ozark Society, in a blanket invitation advertised in newspapers. About 40 members of the pro-dam group, the Buffalo River Improvement Association, came down from Marshall.

The pro-dam people were allowed only to ask a few questions and wondered angrily whether the meeting was actually open.

H. Charles Johnston of Little Rock, president of the Central Arkansas chapter of the Ozark Society, told them that it was but that it was called to let "those other people" have their say. The meeting was advertised as one to get "the truth about the Buffalo from the 'Real Insiders'—those who live on the banks." ...

"We came down here expecting to say a few words," said James Tudor of Marshall, editor of the *Marshall Mountain Wave* and president of the 1,000 member Improvement Association. ...

At the conclusion of the meeting, Tudor said the Improvement Association would meet February 12 at the Searcy County Courthouse and invited Tuesday night's Ozark Society speakers to appear.

"I believe you all are good enough to prove to those people up there that they haven't got a lick of sense," Tudor said.

That meeting at the Searcy County courthouse none of us attended, and both events wound up as a lot of sound and fury meaning nothing at all.

A False Start in Harrison

At this time there transpired in Harrison an event that we hoped would be of great importance. Harrison was the business and social center for all of north central Arkansas, a place where we needed as many good supporters as we could enlist. We knew they were there, and we had tried long and hard to get them together.

It was with pleasure that I announced the event to our people in Little Rock on February 22, 1965:

As the attendant on the case I am happy to announce that the Ozark Society has gone into labor. The expected date for this blessed event is Thursday, February the 25th, and will be at the Harrison Holiday Inn at 6:00 P.M.

The new papa to be, Charley McRaven, has already picked out a name for the new arrival, The Buffalo River Chapter of the Ozark Society. If you would like to be there as a sort of uncle Charley we will be delighted.

But our hopes for the newborn chapter would soon be dashed, for it would die in its infancy of a disease endemic in that area—big damitis.

The Interim Report

As soon as the district engineer in Dallas had sent in his decision to Washington, we were generously permitted to obtain copies of that previously denied report which spelled out Plan D in detail and upon which a small fortune of public money had been expended. Its title was in keeping with the enormity of the proposal:

This *Interim Review* consisted of three volumes, each the size of a Sears Roebuck catalog, an overwhelming mass of verbiage, charts, diagrams, maps, and figures. It was biblical in scope and intent, bringing to its true believers the promise of glory and salvation. In spite of other ongoing distractions, we waded into that ponderous catechism to see if there were new revelations for those who had not heard the word. In the end, there were none that had not been delivered many times in previous releases from the offices of the Corps and thence through the various avenues of the media. The *Interim Review* was the rock upon which this edifice was to be built, spelling out everything in the greatest detail. It was a gospel not now to be questioned by the likes of us.

There was in it one stipulation, a magnanimous gesture by this great agency to those agonizing over the impending loss of the beautiful Buffalo that stands out in memory and characterizes, above all else, our differences. In discussing our oft-repeated criticism of the cold-water release from the Gilbert reservoir which would terminate body contact recreation in the river below the dam, the *Interim Review* had this to say: "As a mitigation measure we will build a $100,000.00 swimming pool at the Buffalo River State Park."

The size of the Gilbert dam reservoir was specified in it and is listed for comparisons with Bull Shoals, Beaver, and other area reservoirs.

Feature	Main Dam and Power Plant	Reregulation Dam
Location River Mile	59.5	54.8
Height above Streambed	229 feet	50 feet
Elevations		
(Feet above Mean Sea Level)		
Top of Dam	798	382
Top of Flood Control Pool	785	
Top of Power Pool	757	
Bottom of Power Drawdown	732	
Storage Acre Feet		
Flood Control	443,000	
Power Drawdown	297,000	
Area Acres		
Top of Flood Control Pool	18,200	

It is to be noted that Plan D included two dams, not one. As indicated, the pumped-storage reregulation (water goes round and round) dam would have been five miles below the main dam and would have been one thousand feet long and fifty feet high. It would have wiped out a significant section of the middle Buffalo, a fact overlooked by us and downplayed by them at the time of the hearing.

Punch and Counterpunch

During this hubbub over H.R. 2245 and the *Interim Review,* public interest concerning the Buffalo River was on the rise, and we received many requests for programs from all over Arkansas and Missouri. Difficult though it was, we managed to honor those requests, preferably in person, but if not, through the medium of our portable movie set-up operated by local program chairmen. At one such showing before the Pine Bluff Rotary Club, Ed Freeman III was in attendance. He was the editor of the *Pine Bluff Commercial* and a devotee of outdoor recreation along Arkansas's remaining free-flowing streams. As the result of the message contained in that program, Ed Freeman placed the *Pine Bluff Commercial* in the forefront of all other newspapers generating favorable publicity for the Buffalo National River. For us up in northern Arkansas, that was help from an unexpected quarter down in the delta.

The situation in the late winter of 1965 was summarized in a report to the Ozark Society membership.

> ... We have been informed that there is a further extension of time to May 12 for filing protests. So time is still available for you as an individual, or for a group to submit your views to the Board of Engineers for Rivers and Harbors, 2nd and Q Streets, S.W., Washington, D.C. 20315. They ask for "information and data not previously submitted." We have no explanation for this sudden slowing of the grinding gears of the Corps—simply give you the fact.
>
> A personal *hearing* before the Board has also been asked for. Arrangements for this are pending. The next step in Corps procedures is the preparation of a "report by the Chief of Engineers and review thereof by the affected State and Federal agencies." This process can take as long as three months. ...
>
> About twenty persons representing the Upper Eleven Point River Association, opponents of the proposed dam (Water Valley) on that spring-fed stream, appeared in an all day hearing before the Board of Engineers for Rivers and Harbors in February. Ozark Society filed a protest with the Board against Water Valley Dam and a member-representative attended the hearing.
>
> The Board is reported to have disapproved of the Corps' proposed dam, recommending instead a previously authorized higher dam which would back water into Missouri to Riverton on the Eleven Point. At the same time the Board said, "The

Ed Freeman III and his son, Andy, on the Eleven Point, August 1966. Ed was the editor and owner of the *Pine Bluff Commerical*.

power available could be generated for less cost by means of a steam generating plant." The president of the Upper Eleven Point Association is quoted as saying: "We are confused … What more is to be said if neither plan is justified by their own admission." We are confused, too.

The Ozarks Playgrounds Association, on March 20, adopted a resolution opposing a dam on the Buffalo and endorsing Buffalo National River. The resolution is reported to have passed without a dissenting vote.

The old Ozark Playgrounds Association was a regional booster group organized in the 1920s, whose policy in this case was inspired by L. B. Cook, Jr., the outgoing, unabashed owner of Theodosia Resort on Bull Shoals Lake in Missouri. The reason for L. B. Cook's attitude is best stated in his own words in a letter to the *Ozarks Mountaineer*, May 11, 1965:

> Your current issue of the MOUNTAINEER reached us yesterday and I was undecided on what to do first—mail you a check for the THEODOSIA RESORT ad that it carried, or call your attention to a few items with which I thoroughly disagree. The check will come later.
>
> In your editorial, OZARK STREAMS AND MAXIMUM DEVELOPMENT, you strongly favor the construction of a dam on the Buffalo River, in Arkansas. This, of course, is

your right. But, you are wrong. To begin with, there are many things to be considered in this proposal. We have fast become a well-dammed area. Lakes are all over the entire region, as you know. And, since lakes are not now the effective attraction to tourists that they were when first built in this region, each new one simply "thins" out the business going to existing recreational areas. Some day we may see enough business to go around, but that is in the future. Lake visitation records of the Corps of Engineers will bear this out. Existing recreational areas—and resorts—are able to retain their past business only by increasing their advertising, upgrading their facilities, or adding new facilities. Any number of smaller resorts, financially unable to do these things, are in a bad financial situation ... on all lakes throughout the region.

Our Bull Shoals Lake and White River Association, the Lake Norfork Association, the Ozark Playgrounds Association, any number of area Chambers of Commerce and Civic Clubs, and also a large number of other organizations—located in this recreational region—have voted to oppose the construction of any dam on the Buffalo River. These organizations have voted to approve the installation of a National Park on the Buffalo and surrounding area. All of us feel that a diversification of recreational interests is greatly needed for this Ozarks region to truly take its place in the national tourist, travel, and recreational picture. Another lake would not accomplish this—but rather it would detract from the existing lake facilities. A large National Park on the Buffalo, the Current River program, the lakes and scenic areas throughout—would all provide one of the most unique and interesting regions in the entire United States. This is what our thousands of tourists want—and I speak, personally, from experience in this field since 1952.

In early March the boys in the BRIA were able to deal us another jolt in our game of tit for tat (*Arkansas Gazette*, March 5, 1965):

Budget Used as Lever,
Parks Agency Shuts up in Buffalo Dam Dispute

The state Publicity and Parks Commission, already on record as opposed to the damming of the Buffalo River, announced Wednesday that it would take no further stand in the controversy.

The new stand was taken after a pro-dam legislator blocked the Commission's appropriation.

Bob Evans, director of the Commission, said Thursday that the announcement was made at the request of state Representative Jimmie Dale Myatt of Searcy County.

Myatt said he got the appropriation for the Parks Department blocked because, he said, 95 percent of the people in Searcy County favored damming the river, and he didn't think a state agency should take a stand against something in so much local favor. ...

Evans said that the latest statement did not have the effect of rescinding the earlier resolution. "It just means that the Commission won't make any more pro or con statements on the matter," he said.

He added that if more pressure were brought to bear, however, the Commission might rescind that first resolution.

How we felt about that is reflected in a letter to John Heuston on March 5, 1965, then employed by the Publicity and Parks Commission:

> I am writing you just in case I don't get to see you Sunday at Harrison. I need some advice before writing to Bob Evans.
>
> His statement in this morning's *Gazette* was about the sorriest thing that has happened since the beginning of the war on the Buffalo. Of course, the *Gazette* may have misquoted him, but he was reported as saying that if enough pressure was put on him he would reverse the stand of the Publicity and Parks Commission.
>
> If it's pressure he wants, we are getting in a position where we can put it on somebody and from all over this state too, not just Searcy County.
>
> If he is the kind of person he sounds like, he hasn't any business being where he is. If he is going to let Jim Tudor shove him around, he will find some shoving from the other direction.
>
> I don't want to write him in the frame of mind that I am now in but am asking you for advice. I will write him in the next few days and am in the mood to say a lot more than the above remarks. Give me the word on this right away.

This was especially aggravating to us because in nearby Missouri in an ongoing similar contest over the Current and Jacks Fork rivers the state conservation agencies carried the load in opposition to the dams and in favor of the scenic rivers program for those streams. In Arkansas there was to be no real help from that quarter. We would have to make the grade without them.

The Mission to D.C.

Now there was constantly before us the question of whether or not to make an appearance before the Board of Engineers for Rivers and Harbors in Washington.

Officers of the Ozark Society had been urged by various conservation groups to appear before a congressional committee, which was meeting on April 22, 1965, to present a new wild rivers bill and to plead for the inclusion of the Buffalo River in this bill. A number of high-ranking officials in the National Park Service were contacted in regard to this proposed appearance. Officials of the Bureau of Outdoor Recreation were also interviewed, along with some of the Arkansas delegation in the Senate. Following these discussions it was decided that the best policy was not to attempt to have the Buffalo River included in the Wild Rivers Bill. The reason for this is that special legislation was being prepared in the case of the Buffalo River. Those involved in its preparation urged that such legislation be permitted to continue and that its development be enhanced as much as possible by the Ozark Society and by the people of Arkansas.

After this decision, considerable argument between members of the Ozark Society ensued as to whether or not an appearance should then be made before the Board of Engineers for Rivers and Harbors. It was stated by some that this board does not constitute a legal body, that its decisions are not final, and that they are relatively unimportant. Once again, after discussing this matter with authorities in the conservation field and in the National Park Service, it was decided to request an appearance before the Board of Engineers for Rivers and Harbors.

It was also determined to work out a schedule whereby an appearance could be made before Senator McClellan of Arkansas, Congressman Trimble, and others on the same day if possible. Arrangements were made for such meetings, and efforts were then put forth to assemble a creditable delegation. The appointment before the Board of Rivers and Harbors was granted by Edmund H. Lang, colonel, resident member of the board, for May 5, 1965, in Washington, D.C.

Following the selection of a firm date for an appearance before the Board for Rivers and Harbors, efforts were made to obtain a good selection of delegates from Arkansas. It soon became apparent, however, that few, if any, of those who had been working on the Buffalo River project would be able to go. For a time it seemed that only the president of the Ozark Society would be able to attend from Arkansas. Upon contacting conservation organizations in Missouri, however, immediate response was obtained, and three delegates from Missouri agreed to appear.

Late in the effort telegrams were sent to the various conservation organizations in Washington, D.C., and representatives from some of these were also obtained.

The final delegation was as follows: Neil Compton, M.D., Bentonville, Arkansas, president of the Ozark Society; Charles McRaven, St. Joe, Arkansas, spokesman for the Buffalo River Landowners Association; Larry Burns, Bentonville, Arkansas, electrical engineer representing no organization; Ray Lockard, representing the Ozark Wilderness Waterways Club, Kansas City, Missouri; Ed Stegner, executive secretary of the Conservation Federation of Missouri, Columbia, Missouri; L. B. Cook, president of the Bull Shoals and White River Association, Theodosia, Missouri; Ben Ferrier, outdoor writer, home state Wisconsin; Rupert Cutler, representing the Wilderness Society, Washington, D.C.; Robert T. Dennis, representing the Izaak Walton League,

Washington, D.C.; and Harry Graham, representing the National Grange, Washington, D.C.

Compton, McRaven, Cook, and Burns traveled by automobile taking along maps, printed material, display boards, projectors, and film. For these delegates the Ozark Society provided gasoline, oil, and two nights' lodging. Lockard, Stegner, and Ferrier traveled by air. The Ozark Society provided expenses in the case of Ferrier. The other delegates were present already in Washington and required no transportation.

Testimonials

The meeting was convened by Col. Edmund H. Lang. No other military personnel were present. Three civilians, members of Lang's staff, were in attendance. These civilians were senior engineers who draw plans for dams such as those now proposed for the Buffalo. The entire proceedings were taped by the Corps of Engineers.

Introductory remarks were made by Dr. Compton of Bentonville, who then proceeded to introduce the speakers in turn. The first speaker, Charles McRaven, gave an eloquent statement similar to his previous statement at the hearing in Marshall on November 18 on behalf of the actual landowners involved in the proposed Gilbert reservoir.

Ben Ferrier, next to speak, first described his experience as an outdoorsman. Mr. Ferrier was a trained biologist and had taught on the college level. He had also been an athletic coach and was also interested in outdoor activities generally, including canoeing. His plea for the Buffalo is worth recording:

> The Buffalo River is tops in the United States as a canoeing, floating, kayaking and jon-boating stream. I make this statement because I have been in every state in the United States and nearly every county during the last twenty-six years. I have seen nearly every stream of importance and have placed a canoe in the waters of many in every section of the country. I am a professional film narrator and my subject is canoeing in the wilderness. I have given over four thousand programs. I am a wilderness biologist with an MS degree. I have been engaged in professional exploration by paddle and portage in Canada and Alaska since 1933. Since 1920 I have traveled by canoe more than 40,000 miles in Canada, Alaska and the United States, which is considerably more mileage than any other known person. For the last five years I have averaged a month of canoeing the Ozark rivers of Arkansas and Missouri.
>
> The Buffalo has qualities which make it ideal for classification and preservation as one of our National Wild Rivers. In 130 miles of its length from Ponca to the White River there are no towns on its banks. In this whole distance there are very few cabins. If compared with Canadian and Alaskan Rivers such as the Peace, Mackenzie,

Ben Ferrier, *right*, tells Sierra Club floaters of his past canoeing adventures.
(*Courtesy of Kenneth L. Smith*)

Churchill and even the Yukon there are fewer cabins and less people. On its entire length it is a scenic gem. The Buffalo has cliffs rising to the height of seven hundred feet. It is heavily wooded. There are large clean gravel bars which are excellent for camping. The water is very clear and some people still drink the water from the river. …

Wilderness recreation is far more valuable to the future of the United States than artificial lakes. Let's get on the ball and see to it that this generation and the next and the next will have an opportunity for recreational experiences that develop resourcefulness, stamina and muscular physique. Outboard motors, cabin cruisers and soft living won't do it.

Ed Stegner, of the Missouri Conservation Federation, gave a brief and convincing statement and emphasized that the Buffalo River is the most beautiful river in the Ozark area and that it transcends the Current River in his own state of Missouri in this respect. He emphasized the need for the preservation of more than one stream in the Ozarks and made a special plea that the Buffalo be this other one.

Ray Lockard made a moving plea for the salvation of the Buffalo; he, too, emphasized that it is a much more beautiful stream than the Current River. Mr. Lockard finished his statement with a short poem about the river to add an appropriate bit of sentiment.

L. B. Cook, president of the Bull Shoals and White River Association, made a strong argument for the preservation of the Buffalo in its natural condition, stating that the area was already over-supplied with large reservoirs and that the tourist industry based on these reservoirs is in considerable economic distress. His statements brought forth positive interest from Colonel Lang, who interrupted Mr. Cook to request further information on this subject.

Mr. Rupert Cutler of the Wilderness Society gave a short statement extolling the Buffalo River for its wilderness qualities.

Robert Dennis spoke briefly for the Izaak Walton League and pointed out the desirability of retaining the Buffalo River in its natural state because of the need to preserve some streams as a natural habitat for native fish.

Finally, Larry Burns spoke on his own behalf, stating that he had visited many of the national parks in the United States and that the Buffalo River compares very favorably with any of these from the standpoint of scenic beauty and attractiveness.

Following the statements by the speakers, the twelve-minute movie was shown to the board demonstrating the scenic beauty of the Buffalo and showing action on the whitewater sections of the stream.

Following the presentation, Colonel Lang made a short address. He stated that the reason the Water Valley dam proposed for the Eleven Point River was de-authorized was that it was noncompetitive from the standpoint of hydroelectric power generation versus the generation of power by steam plants. He pointedly added, though, that this would not be true in the case of the Buffalo and that the generation of power on that stream would be economically feasible according to the Corps of Engineers. He did admit, however, that there were many intangible factors involved in this case which might have a bearing on the ultimate decision.

After adjournment of the meeting of the board, much interest was shown by its members in the statements that were made, and a question and answer session on an informal basis ensued for more than an hour. The relationship between both groups was cordial and friendly. No effort whatever to embarrass or discredit those offering testimony was made by the board, and the chief regret by those presenting the case for the Buffalo was that more could not have been present on both sides to witness the presentation.

As we prepared to leave, one of the Corps' civilian staff members approached me with some most gratifying information. This bald-headed gentleman we had seen at the Marshall hearing. He said that he was soon to retire from the Corps and that he had selected Fayetteville, Arkansas, as his new home. Dick Murray, an engineer from Georgia Tech, was now completing his career as a civilian employee of the Corps. He wanted me to know that as soon as he was settled in Fayetteville he would like to become a member of the Ozark Society and participate in all of our activity, especially hiking. That was as pleasant a surprise as we had had in this lengthy controversy, a convert from the Corps' own headquarters. During the next twenty years of his retirement, Dick Murray was to fulfill his desire and become one of the principal cogs in our machinery. It was he who initiated the hiking trail program in Arkansas, now undergoing significant development, an endeavor for which we should always remember Richard Murray. During those years with us, Dick did not demean the Corps of Engineers, respecting them for the efficient organization that they were; but if the Corps was in error on a project, he was not afraid to acknowledge it.

Other Meetings and Interviews

Following this meeting a few minutes of spare time were available, and Mr. L. B. Cook, Jr., made an effort to see Senator Symington of Missouri. Senator Symington was not in; but his chief aide, Stanley Fike, was, and we were granted a short but cordial interview by him. Later acknowledgment of this meeting was received from Senator Symington, and he stated that he would give the case of the Buffalo consideration if and when the time came to do so.

Senator McClellan of Arkansas had granted an audience to the delegation but was short of time. Attempts to make oral statements were not regarded as satisfactory, and it was elected to show the senator forty colored slides of the Buffalo River. The senator displayed considerable interest in these and stated at the end of the series that he had never had an opportunity to see the Buffalo River.

As we were preparing to leave, Senator McClellan reminded us that the proposed dam on the Buffalo was Congressman Trimble's pet project and that he (McClellan) would find it hard to oppose him. Of Jim Trimble he had this to say: "He keeps getting elected, doesn't he?" With those words in our ears, we looked at each other, swallowed hard, and took leave of Senator McClellan.

Following this discussion, arrangements were made to meet Congressman Trimble in his office. Mr. Trimble appeared to be concerned and almost anxious about the controversy over the river. He reiterated his previously stated stand for the building of dams on the river but at the same time seemed to be interested in the proposed national river projects as well. On that subject he repeated something that he had said to me a year before, which got the attention of all: "I don't understand what you want with that narrow rope of land along the river. It will be hard to administer in that shape, and a lot of interesting things have been left out, like Diamond Cave and Marble Falls."

He should have seen our original proposal, which included those places and more. But the National Park Service set the boundaries. It was all their financial resources would permit, and we knew that we were lucky to have that much under consideration. But it was tantalizing to think how easy it would have been, and how large the area, if Jim Trimble had been with us from the start.

We then proceeded to the National Park Service headquarters in the Interior Department building where various members of the service were interviewed in informal discussions. At the same time, members of the press were contacted on the telephone and activities carried out up to that time were described. These reporters were identified with the *Southwest American* of Fort Smith, Arkansas, the *Post Dispatch* of St. Louis, Missouri, and the *Kansas City Star* of Kansas City, Missouri.

At about 4:30 P.M. an interview was granted by Stewart Udall, Secretary of the Interior. Mr. Udall's first emphasis was upon the need to convince the Arkansas delegation of the desirability of the Buffalo National River. He made it clear that no department of the government wished to override Mr. Trimble in his position for dams on the Buffalo. It was obvious that this was because of Mr. Trimble's seniority in Congress and that he (Trimble) exerted considerable power on certain committees in Congress. Mr. Udall did make it clear, however, that if positive steps were taken to implement the program for damming the Buffalo, his department of government would then make efforts to stop such a program.

The idea of an expanded area for the national riverway which had been brought up by Mr. Trimble was mentioned to the secretary. He displayed immediate interest, saying: "Oh! in that case we could have a real national park, couldn't we?" But we all knew it was best to stay with the present game plan. If Mr. Trimble had showed a definite interest in the expanded park area, this could then be developed later. Secretary Udall made it clear that more interest by the people of Arkansas in the proposed Buffalo National River was going to be necessary before this proposal could become an accomplished fact.

After the interview with Secretary Udall, a long telephone conversation was had with Spencer Smith, director of the Citizens Committee on Natural Resources. This organization proved to be a most knowledgeable group concerning conservation legislation. Most of what was stated was of a confidential nature, and it was recommended that closer cooperation between the Ozark Society and the Citizens Committee for Natural Resources be practiced in the future.

Following this series of appearances, all of those involved felt certain that a measurable good had been accomplished. We felt reassured in the words of Spencer Smith of the Citizens Committee for Natural Resources, who stated that had we not made the appearance, it would have been fatal.

TEN

A RUMBLE
ON THE RIVER

In the spring of 1965 our principal activity was not in the nation's capital but in Newton and Searcy and sometimes Marion counties through which the Buffalo flowed. The Ozark Society, the Ozark Wilderness Waterways Club, and occasionally Sierra Club visitors were out every weekend cruising its waters, photographing its scenery, and taking note of its geology, topography, wildlife, and botany. Adventures and misadventures were inevitable as our efforts to better know the Buffalo became more intense.

That activity did not go unnoticed in Marshall where shenanigans of dire import for us "outsiders" were in the process of incubation.

But to begin with, many of our difficulties on the river were the result of natural conditions and the lack of experience and judgment on the part of us would-be voyageurs.

A Ride down Buckin' Shoals

As a starter, on the Mt. Hersey to Gilbert run, I very nearly lost Laurene, that pretty girl who had watched the eclipse of the moon with me from White River's panorama in the summer of '35.

We had scheduled that long float for two days, April 10 and 11, and in order to get to Mt. Hersey for an early put-in on Saturday had checked in at the Crest Haven Motel in Harrison Friday night. There we were joined by Ken Smith, who hadn't yet had his supper. He produced it from a small packing box—canned peaches and Post Toasties—a lesson in frugality to the rest of us as to how to make out on a modest park service salary and how to research, write, and produce a book at the same time.

While we were contemplating that, a Volkswagen beetle drove up and parked

Harry Pearson, investigative reporter for the *Pine Bluff Commercial*, roping in to Indian Creek Canyon.

alongside my canoe-topped International Scout. I went out to investigate the newcomer and there beheld in the window on the driver's side the anxious countenance of a young man. He looked, and what's more acted, like a junior-grade edition of Charles Laughton, prominent British movie star of the era (Captain Bligh and all that).

"I have just been through the most traumatic experience of my life," he had me know in dramatic tones.

He was pleased to find that I was the person he had been looking for, Dr. Compton of the Ozark Society. His name, he said, was Harry Pearson, an investigative reporter sent out by Ed Freeman III of the *Pine Bluff Commercial,* to do an in-depth story on the intensifying struggle for the Buffalo River. I invited him inside where we all sat listening in amazement to his theatrical recitation of his "most traumatic experience." He had just come from the town of Marshall where he had attended a meeting of the BRIA and where he had heard lurid threats of violence, couched in the most revolting Arkansas-style vituperations that the human ear could tolerate, all directed toward us peace-loving, save-the-Buffalo types. During his delivery of that chilling news, he paced back and forth, every word a revelation of what evil might be in store.

Laurene, my wife, whispered in my ear: "Get that guy out of here! I can't stand him another minute!" But what he had learned in Marshall I wanted to know, and we let Harry carry on.

After several months, when we all knew Harry better, Laurene, who always appreciated art and drama, decided that Harry was the smartest thing on two legs.

After hearing what he had in Marshall, Harry was already on our side, but the BRIA people, not knowing, expectantly delivered to him various of their papers, material for a favorable write-up in the *Pine Bluff Commercial,* they hoped. But instead, he turned over to us that journalistic gem, just hot off the press, Tudor's broadside: OH MERCY, WE'VE BEEN GORED!, some DAM THE BUFFALO bumper stickers, and other trivia, which we have saved for posterity.

Harry wanted to go with us that weekend so that he could write a firsthand report on our unspoiled river. We were glad to accommodate, and he was assigned as bow paddler in my canoe, a position originally intended for Laurene. She would serve as bow paddler for an expert canoeist, Joe Mills, a Sierra Club member from Minnesota. Thus I would be able to deliver a running commentary concerning the Buffalo to our newfound reporter as we went along.

With us Saturday morning was Ben Ferrier and his wife; Boyd Evison, a park naturalist at the Hot Springs National Park, and his wife; Larry Burns, who would travel with Ken Smith; and Joe and Maxine Clark of Fayetteville. We would meet H. Charles Johnston and a few others at Mt. Hersey.

Joe Clark was a large, friendly fellow, full of fun and always ready to help when there was need for it. He always had that tool or that item, to take care of almost any emergency. He had recently moved from Tulsa to Fayetteville where he was employed by the Arkansas Western Gas Company as a field geologist.

Harry was no experienced river runner and came without protective containers or covers for his gear, which consisted mainly of a Leica camera and an old sleeping bag. Those he deposited in the bottom of the canoe where they would surely get a wetting. Somehow he kept the camera dry but not the sleeping bag. The river was up, not in flood but rolling and heaving and out in the willow thickets that grew down to the water on the banks and gravel bars.

Upon launching, we immediately hit the long rapids below Mt. Hersey; and when Harry felt the surges of that watery roller coaster, he let forth with whoops and yells that could be heard all the way to Woolum.

On account of the late start, I decided to make camp at the Narrows, only about five or six miles below Mt. Hersey, in order to get settled down before dark. The Narrows bluff is a limestone spine standing up above the river on the west and Richland valley on the east. It rises to seventy-five feet straight above the water in the middle where it is less than two feet wide in one place, then widening and sweeping upwards for another seventy-five or one hundred feet on either end. By being careful it is possible to climb the Narrows bluff on the river side, and that we did for the delightful view of the two valleys from the narrow saddle back. It was April, the month of storms, and we spent a rather fitful night after learning on Joe Clark's radio that there had been a tornado in Conway, but all we got was a splattering with a few king-sized raindrops. Sunday dawned clear and warm.

While busy preparing breakfast, I noticed Larry Burns chuckling to himself. Upon asking what was so funny, he replied: "Well, it was Harry. I couldn't tell whether he was praying or cursing. You know when we started to turn in last night, the lower half of his sleeping bag turned out to be wet. When he put his feet down into it he gave a great groan and said, 'Oh! Jesus God!'"

We got off that morning in good order, passing by the spooky eye sockets of Skull Bluff and on past Woolum Ford, past Margaret White Bluff and Lookoff Bluff. Then came early afternoon and time for a rest stop. With an eye out for an easy place to beach canoes, I spotted on the right an embayment off of the main channel. A low bluff stood on its upper end and a long gravel bar at the lower, covered with a willow jungle through which the rushing waters of the swollen river roared as if over a waterfall. We all pulled in past the upper ledge, got out, stretched legs, partook of cokes and sandwiches, admired the lively waters, and savored the zestful air. Harry and I were the first to depart, hoping to get some good action shots with the movie camera as the rest of the flotilla came down the river. We dawdled along but saw nothing of them, and finally I decided to put in to a low ledge out of the fast water where we tied up to wait.

Soon Boyd Evison and his wife came flying by, in a hurry to get to Gilbert so they could get home to relieve a babysitter. Presently, Charley Johnston came sailing along and turned in next to us. We wondered where everybody was. Then down the swift current came Ben Ferrier and his wife. They hauled up for a moment to tell us something. There had been a canoe wreck up river. Someone was jumping up and down in a canoe to straighten out the dents. It was assumed that the crew had been saved. With that, the Ferriers were on their way; they had apparently seen that activity while passing by.

Finally down the river came my second canoe, moving fast but low in the water. Sitting low in front of the middle thwart was Laurene, the ribbon on her hat streaming out behind, with Larry and Ken plying bow and stern paddles to keep the craft straight in the current. There was extra gear stacked amidships, obviously salvaged from the reported wreck. They called to us to keep them in sight since they were so heavily loaded and their steerage poor. Before we could get under way, around the bend came Joe Mills, alone in his new lightweight Grumman, the bow riding high a few inches out of the water. The reason was to be seen. There was a six-inch gash in the hull a few inches above the water. Joe's weight in the stern kept the rent up so it couldn't take in water. Then bringing up the rear, with his canoe also full of extra gear, came Joe Clark and Maxine. It was all too obvious that it was Joe Mills' canoe that had been wrecked, and I was anxious to get to our destination to learn the details.

As for Laurene, there was no opportunity for me to fear for her. She had been rescued by others before I even knew that she had been in danger. It was all so unlikely. Joe Mills was an expert canoe operator. But there they were, safe but not exactly sound. I was thankful for that. It was upsetting to imagine what her reaction might be to this experience. She had come to enjoy some of our outings on the river, but her real preference was Oaklawn and the races for rest and recreation, something that we weren't having on the Buffalo that day. However, she was not turned off by it all and later didn't mind telling about her canoe wreck on the Buffalo.

Still we were not home free. It was ten miles yet to Gilbert when we rounded the bend at Peter Cave Bluff. There we entered the Buckin' Shoals, a long straight rapids racing for more than a quarter of a mile over a hard bedrock surface. There were enough loose, melon-sized boulders scattered on the rock bottom to produce an erratic turbulence with waves up to two feet high. Whether Ken, Larry, and Laurene could run that gauntlet in their laden craft without swamping was doubtful. We watched helplessly as bucketfuls of water came over their bow, their canoe riding ever lower in the water.

I tried to prepare my spirits for the sight of Laurene being gobbled up by that wild and scenic river, calculating how best to effect a rescue. But Larry and Ken amidst much yelling and mighty digging in with their paddles managed to get the foundering craft into the right-hand bank at the end of Buckin' Shoals where Harry and I, who were a few moments ahead, waited to give a hand. The canoe was full of water with Laurene sitting in the middle as if in a bathtub, but at least she was not floating off down the river alone. Very quickly everything was out on the bank, and with all of us straining mightily we lifted the canoe up and turned it upside down, dumping out the water, then putting it back down right side up and ready to go.

With the weight distributed better, some of the gear in my canoe, we were off in better trim for Gilbert. On past Blue Bluff, Red Bluff, Tyler Bend, the 65 bridge, and the exciting curving rapids below it we went, tired now and hungry but arriving at Gilbert with still enough light to get our vehicles loaded for the long drive home.

Then there was opportunity to hear about what really happened at the rest stop.

Joe Mills and Laurene started to leave the place ahead of Larry and Ken and Joe Clark and Maxine, luckily. If they had been last, none could have gone back to help. Instead of entering the main current upstream from the willow break, Joe Mills came straight out at right angles and not far enough above the willows. The current caught him and slammed his canoe against the first tree. Laurene instinctively leaped out of the bow, clutching for the nearest tree, which she managed to hang onto. She said that the scariest part of it all was the awful force of the water on her lower torso pulling her down from the tree. But she managed to climb up far enough where she found a perch above the water, a place where she could wait safely for that old river man she was married to or whoever else might undertake it to come and rescue her. Joe Mills, being taller and stronger, was able to stand against the current while hanging onto a tree with one hand. He managed to reach the tree where his canoe was impaled with its back broken, the bow and stern being folded backwards against each other. But it was a nearly impossible task.

Joe Clark and Larry Burns, who were following Mills, quickly turned back before making the same mistake, beached their canoes, and initiated rescue operations. After determining that Mills and Laurene were safe for the moment in their trees, the rescuers took off down the gravel bar after the floating gear and somehow rounded up all of it. Then Joe Clark produced an essential item, a long piece of nylon line. After unloading his canoe, Joe tied the nylon rope to his canoe, and he, Ken, and Larry lined it out into the willow thicket, positioning it under Laurene's tree. She dropped down into the canoe and was pulled to the shore by those instant coast guardsmen. Joe Mills had meanwhile been able to free his bent-double canoe and with Joe Clark's line attached was likewise hauled to safety. There it was stomped back into a semblance of itself so as to at least carry its owner on down river.

Now it was over, and we were all alive and on our way home across the familiar hills, thankful that somehow it had been no worse. It had been a strenuous day mentally, physically, and emotionally, more so than almost any in our lives before the struggle to save that river began. It was enough to make a tired soul wonder if it was worth it, if it would ever be worth it—even if one day we really did save the Buffalo. With that question in mind, and Larry spelling me at the wheel, I leaned back against the hard seat in that hard-riding International Scout and went sound asleep.

As for whether the river saved would ever be worth it, there would come a day—and it was.

The Pearson Series

Immediately, dividends from the Mt. Hersey/Gilbert trip poured forth from Harry Pearson's typewriter, with more to come. One of our greatest needs was now in the process of fulfillment—comprehensive and sustained newspaper coverage of the battle of the Buffalo.

In a period of a few weeks that spring, the Harry Pearson series appeared in six installments, in the *Pine Bluff Commercial*. They covered the contest from all angles: the river in being, the local big dam promoters, the Corps of Engineers, the National Park Service, the regional river savers, and the politics of the affair.

When the series was completed, the Ozark Society had fifteen thousand copies reprinted, along with related items from other newspapers. Over the next seven years they were distributed far and wide, a piece of handout information of incalculable value.

Harry came to be a party to all of our activities from then on, chapter meetings, public programs, and various outdoor activities. In addition to his series, he reported all ongoing news about the Buffalo controversy from day to day. The *Pine Bluff Commercial* was the champion speaker for the salvation of the Buffalo in all of the news media from then on.

The Memorial Day Blockade, May 24, 1965

But as chance would have it, neither Harry, the indispensable reporter, nor I, the apostle of the park service, would be present to witness firsthand the most distressing set-to between the river dammers and the river savers during the entire conflict. Our comment about it would have to be after the fact and secondhand.

The Ozark Wilderness Waterways Club had always taken advantage of the long Memorial Day weekend to get out of Kansas City, usually coming to the Buffalo. The event was of special importance in 1965, a crisis year for the Buffalo, and the O.W.W.C. turned out in strength. They would be joined by a few from Arkansas, notably attorney and later Circuit Judge William Enfield and his wife, Miriam, from Bentonville. They were all going to float from Ponca to Pruitt, 130 or 140 men, women, and children. But that was not all.

Charley McRaven had thought up the canoe race as an attention getter for the same weekend. Eighteen canoes were registered for the race. Those of us in the Ozark Society headquarters were not exactly sure that it was the sort of thing that we should be engaged in, but we gave in to Charley's enthusiasm. But even as we did so, things were happening that would scare off most of the entrants.

The BRIA had learned through various channels of our Memorial Day schedule and were determined to bring it all to a halt.

They had already started a campaign of telephone harassment against the Buffalo River Chapter of the Ozark Society in Harrison. Members of the chapter were business and professional people, mostly in Harrison, but whose customers lived all over Boone and Searcy counties. Their delivery trucks and service people made calls throughout the area, and some maintained small business outlets in places like St. Joe, Everton, and Pindall.

Those members were subjected to abusive phone calls night and day, being threatened with all sorts of violence if they did not cease and desist in their opposition to the dams on the Buffalo. They were told that their trucks would not return from service trips out into the country, their in-town businesses would suffer damage, and they themselves might not be spared.

The big dam faction had influence with many officials in the counties and towns in the area. Most of those in elective office respected and catered to Congressman Trimble. They would go along with him, and that attitude percolated on down to justices of the peace and local school boards. Most of them didn't engage in personal threats by phone, or otherwise, but they could be depended upon to embarrass anti-dam citizens if called upon.

An example was Charley McRaven. Originally he had been an employee of the Arkansas Power and Light Company, but as the Buffalo River controversy heated up it became necessary for him to terminate with AP&L due to hostility of the company's customers in Searcy County. Charley, who was no real farmer, then got on as a schoolteacher at Valley Spring but soon gave up that job for that of announcer for radio station KHOZ in Harrison. He left that position rather suddenly because he was also chairman of the Buffalo River Chapter of the Ozark Society and a prime target for the people in the BRIA (or so we all believed).

Trees felled across the Buffalo River by anti-park, pro-dam activists, Memorial Day, 1965. *(Courtesy of Nancy Jack)*

The other members of our Harrison chapter soon became completely unnerved by the viciousness of the assault. If they were to stay in business in their hometown, they had best no longer affiliate with an anti-dam organization, and the Buffalo River Chapter discreetly disappeared from the scene.

Fortified by what seemed a significant victory, the pro-dam faction prepared for a hands-on confrontation on the river itself.

On Monday morning (Memorial Day) 1965, I was busy in my office when word came that the O.W.W.C. expedition on the Buffalo was in serious trouble, the river blocked by downed trees and by fences, and the canoe travelers under rifle fire from the river bank. Knowing that the local constabulary could not be counted upon to intervene, my next thought was to see if we could get some good reporters up there to tell it like it was. Harry Pearson was out of pocket at the moment, so I called the editor of the *Southwest American* in Fort Smith, Jim Crow, who had succeeded Bill Barksdale. He declined to send anyone. There would be something in the papers, without doubt, and I would simply have to wait to see how the AP would tell the story.

But there was a crusty news-hen on the river that day, a long-time member of the O.W.W.C. She was Nancy Jack, whose column appeared regularly in the *Kansas City Kansan.* Her account of what happened to the O.W.W.C. floaters was reprinted in the June 12 issue of the *Pine Bluff Commercial*:

They Say … "Did You Get Shot At?"
Kansas City Wants to Know

"Well, did you get shot at?" I was asked often after returning from nine days in the wilds, the first three of them spent canoeing the Buffalo river in north central Arkansas.

For the record, no firearms were seen or heard by our party of 121 people traveling in 55 canoes on the Ozark Wilderness Waterways Club's annual 3-day Memorial day weekend trip on the Buffalo. There were two reports of firings and one of rock-throwing and a barbed wire injury involving other people on other parts of the river.

What the friendly, tourist-dollar-hungry Arkansas natives did for us was to block the channel by felling 18 magnificent trees across the river.

… but some of us could not resist the challenge. We took empty canoes and went straight through the jams. …

Trees that were cut down in our path were the largest, most majestic sycamores, sweet gums, maples and elms nearest the river. It took at least 100 years for them to grow, and they won't be replaced tomorrow. …

The *Arkansas Gazette* significantly reported the events from Marshall the next day (June 1) in a manner that the BRIA people would approve:

Barbed Wire Strands, Felled Trees Mar
First Canoe Race on Buffalo

A series of incidents Sunday marred the first canoe race, narrowed to six teams because of reported threats, on the Buffalo River.

The canoers found three strands of barbed wire stretched across the River at one point along the 11-mile course south of St. Joe, a stretch of the River that would be inundated if a proposed multi-purpose, and controversial, dam is constructed at Gilbert.

Jack Riggs of Little Rock was scratched on his upper arm by the wire as he and his partner attempted to go under it. He said two men were standing on the bank whipping the wires up and down. …

Four men in a canoe were spilled when their boat hit two freshly felled trees across a rapids, and one said gun shots peppered the water around them. …

James Tudor, a Marshall newspaper man, said that Homer Blythe, who owns a farm in the Woolum community and whose property is on both sides of the River, had fenced his land and told some canoers they were trespassing.

Tudor said there had been quite a bit of destruction in the area and that Blythe had had farm machinery chopped up for firewood.

"There are a lot of these canoe riders floating down the River and usually one of them sits in the front with a high-powered rifle and one sits in the back with a high-powered rifle," Tudor said. He said some of the residents were getting "tired of them running up and down shooting."

Tudor said several persons had had cows killed. One farmer, he said, found his dead cow by tracking the blood from the river bank to a thicket. Her calf was almost starved, Tudor said.

Sheriff Beal Sutterfield of Searcy County said one of the landowners had telephoned him Saturday saying he had fenced part of the River and didn't "aim for canoeists to go across it."

The sheriff said he deputized two men to go to the farm and serve warrants on any trespassers but that he hadn't heard from the deputies.

Sheriff Loy Waggoner of Newton County said he knew about the felled trees but that the state Game and Fish Commission would have to investigate since obstruction of a public waterway was involved. The Commission said it had no jurisdiction.

Upon becoming aware of the outbreak of hostilities on the Buffalo, Harry Pearson got hot on the phone and made in-person contacts with as many as he could, his account appearing in the *Pine Bluff Commercial,* June 1, 1965:

On the Buffalo, They Worry: A New Vietnam?

"If things keep going on like this, we're gonna have another Vietnam up here," Don Gronwaldt said last night.

What Gronwaldt, a Newton County man, was talking about were the shootings, tree fellings, and other incidents which occurred on the Buffalo River Memorial Day weekend. …

Gronwaldt, the owner of the now-defunct Buffalo Basin Guest Ranch, was one of the many people involved. He was acting as a guide for an elderly Kansas City couple who were float fishing Sunday afternoon on the river about two miles above Pruitt in Newton County.

As Gronwaldt's john-boat rounded a blind curve (called Crisis Curve by the canoeists), Gronwaldt saw felled trees across the stream.

"There were two large sycamores right across the river," Gronwaldt said. "They were dropped in very strategic places so they were impossible to miss. The couple were up in years—they were in their sixties and neither one of them could swim. The water was moving about 20 miles per hour and was about chest deep."

The boat hit the trees and swamped.

"It could have been disastrous if they hadn't been able to get up on the logs," Gronwaldt said. "They could have been swept under."

Gronwaldt, who along with his customers lost $250 worth of equipment, was enraged. He made his way upstream to warn the Ozark Wilderness Waterways Club about the logs, since the club was floating down river from Ponca to Pruitt. …

The shooting incident also took place Sunday at the spot where the trees were felled.

John Saxon, vice president of the Buffalo River Chapter of the Ozark Society … , said he had decided not to enter a canoe race downstream in Searcy County after receiving threatening telephone calls.

Saxon, a Harrison businessman, entered the river at Erbie Ford upstream from the log jam. When he got to the trees and was trying to get the boat under the obstacles, he said someone began firing at his party with a gun.

The River War
Pine Bluff Commercial
June 13, 1965

The recent outbreak of gunfire, felled trees and barbed wire on the Buffalo River—treated with levity in some quarters—has more serious implications for thoughtful Arkansans.

Most of these implications are reflected through the story of Bob Crow, the owner of the Lost Valley Lodge in the hamlet of Ponca. ...

Up until the gunfire, he was doing a fair business that promised to become better as the fame of the Buffalo spread far and wide. Mr. Crow, in keeping with the feelings of his clientele, is opposed to damming the river.

The gunfire and other harassment of tourists that occurred in Newton County on Memorial Day weekend sent Mr. Crow to the local law enforcement officials seeking some satisfaction.

He tells it this way:

"I went to the State Police. They told me it was the sheriff's responsibility.

"I went to the sheriff. He told me to see the game warden.

"I went to the game warden. He told me it was the sheriff's problem."

Any minister in Churches United Against Gambling knows about how Mr. Crow felt. ...

And what this attitude does in turn is cripple an industry Arkansas wants and needs: Tourism. Meeting visitors with shotgun blasts is not, we think, one of the better ways to attract tourists to Arkansas.

In summary, to clarify the sequence of events, we should relate that the large O.W.W.C. flotilla had encountered only minimal difficulties until they were about halfway between Erbie and Pruitt. There they found newly felled trees partially blocking the river and pulled up on a small gravel bar (not where they had intended to spend the night) to investigate. There they soon learned from Don Gronwaldt, who had been wrecked in a big log jam further down river, that the river was not passable and that they had better take out where they were. They were on Cooper's gravel bar, as was reported, from which point there was a rough access road out and which they took advantage of the next day, except for rough-and-ready Nancy Jack and a few others.

In addition to all those children, there was one family of four, dad and mom in two canoes, each with one of the kids in the bow. Mom, in this case, was expecting and was well along, much to the anxiety of all present, especially since their only obstetrician (this writer) was home in his office that day. She had already had to navigate her craft through a downed treetop by herself, but that she did without having a single sign of labor, something that could not be said of the worried trip leader.

After the unpleasant argument with Mrs. Cooper, a general state of anxiety pervaded the camp. Everyone was quite concerned about what might happen next, especially after hearing that floaters had been fired upon down river. They felt that they had been trapped without an avenue of escape and that worse things were in store. They were glad to find that rough route out the next day, leaving the big dam skirmishers in possession of the field. The BRIA and their cohorts had been successful. They had stopped the big O.W.W.C. Memorial Day float, but it was a sorry way to sell Plan D to the general public.

Down river the situation was, if anything, more chaotic. There were no organized floaters there, but a number of individual fishermen and boat paddlers were out. They were the ones who encountered the worst of the felled tree blockade. In two or three cases near drownings occurred when people became unexpectedly trapped in swift water rushing through the tree tops. In that area most of them received rifle fire from the riverbank, one of whom was John Saxon, vice president of the by-now-moribund Buffalo River Chapter of the Ozark Society. He was not hit, but a fisherman nearby had a hole shot in his boat.

The canoe race farther down below Woolum had received regional publicity, and the BRIA was ready with retaliative measures. Barbed-wire fences had been strung across the river as a mark of ownership, and anyone crossing them would be "trespassing." That would require sheriff's deputies to make arrests, which personnel the sheriff of Searcy County was pleased to provide. One of those deputies turned out to be a son of the president of the BRIA. He and another instant lawman were stationed at the newly built fence, like spiders on a web, to catch the canoe racers as they came through. As it turned out, they would catch more than they bargained for.

After the new deputies had harassed several of the frontrunners in the race, along came a pair who had taken a wrong turn up river and were trying hard to catch up. One was Don Stanley, a local boy, raised in the Pruitt area, whom we would come to know well several years later. His partner, Jerry Heard, was from St. Joe (no "outsider" he), a six-foot-three, ham-handed fellow with the strength of a full-grown bear.

They arrived at the barbed-wire barricade alone, and as they approached one of the deputies waved a .38 calibre pistol and yelled: "If you guys try to go through the fence, I'll shoot!"

Jerry Heard shouted back: "The hell you will!" And with that turned the canoe hard for the bank uttering some of the best country boy cuss words that had ever been heard along the Buffalo. Fortunately, the "deputies" chose not to use their artillery. Don Stanley said, "They turned tail and run like turkeys."

Kerm Powers, editor of the Yellville *Mountain Echo,* had been favorable to the national park proposal from the time he first heard of it. He covered the canoe race in person, reporting it in the *Mountain Echo* for June 3:

Barb Wire and Threats Fail to Halt
Buffalo River Canoe Race

James Taylor and Jack Hensley, St. Joe guides, won the first Buffalo River canoe race sponsored Sunday by the Ozarks Society.

Taylor and Hensley got an assist from barb wire strung across the river about halfway through the course.

Bruce Hammock of Little Rock and Wayne Johnston of Stuttgart were leading the race when they were warned by fishermen in boats tied up along the bank that there was barb wire ahead and they could not continue. That was why the fishing boats had stopped.

Uncertain what to do, the two pulled over to the edge of the river but the two St. Joe guides, who were in second place at the time, continued on course and forced their canoe through the strands of barb wire strung across the river.

When they did, two men, standing at the wire on the river bank, began to curse and threaten them, they reported. They said they recognized the two men on the bank, one of whom was a Marshall resident.

Vernon Ginn of Little Rock and Ted Reeves of St. Joe were right behind Taylor and Hensley and also forced their canoe through the wire and Hammock and Johnston followed. The boats finished the race in that order.

When the two on the bank saw the barb wire was not going to stop the canoes from continuing down the river, they became wildly abusive, the canoeists said, and began to shake the barb wire every time a canoe went through it. Jack Riggs of Little Rock, an employee in the Secretary of State's office, received two bad cuts on his arm as a result of the actions of the two on the bank, and another canoeist was slightly cut.

One canoe turned over in getting through the barb wire, but its occupants righted the craft and continued the race.

Some of the canoeists had traveled the course Saturday afternoon, and the new wire was not across the river at that time.

The Ozark Society, as a better part of judgment, chose to avoid the mid-section of the Buffalo for a while after the Memorial Day incident. Our basic position in the conflict was not bad, and it would have been foolhardy to further aggravate the hotheads around Marshall and Pruitt. Some of them were not beyond perpetrating physical injury upon anyone opposing the Buffalo dams who might wander into their territory. That was one reason why we did not schedule any more canoe races. But, curiously, the Marshall Jaycees took up the sport and held races on the river for a few years thereafter.

The Memorial Day conflict naturally shook us up, posing some serious questions. How best to continue our program without more such confrontation and without people getting hurt or even killed?

There was extensive exchange of information about just what did really happen. In some instances the aftermath was better than the main event as was explained in a memo to Bernie Campbell on June 11, 1965:

> ... We have all been considerably distressed by the tendency of the Associated Press to play up the sensational side of this affair. They have treated it as if it were a cowboy and Indian episode in the movies. ...
>
> None of these accounts have been accurate and have all given the wrong impression that perhaps members of the Ozark Wilderness Waterways Club or that the Ozark Society members fired back at farmers or whoever it was that shot at some of our people that Sunday.
>
> Fortunately, Mr. Kerm Powers of the Yellville *Mountain Echo* did appear on the scene and did write a firsthand account which he published in the *Mountain Echo*. ...
>
> In spite of how all of this may sound in the papers, I can assure you that the large majority of the people in Newton and Searcy counties resented the building of a fence across the Buffalo River and the blockading of it by the felling of large trees. I was over in the area last week and discovered that feelings against the Tudors are bitter. You will note that the two winners of the canoe race were local boys from St. Joe. You will be interested to know that they returned to the place on the river where the fence had been built and engaged the two deputies of the BRIA in an argument after the race was over. One of these deputies ... became belligerent, and Mr. Taylor, one of the canoe race winners, slapped him down and tore off his shirt. After this hostilities subsided rapidly, and a lot of people in Searcy County are laughing about how Donnie Tudor got beat up after the canoe race.
>
> Another bit of interesting news is that Boy Scouts from Bartlesville, Oklahoma appeared on the river on Friday after the above incident with chain saws and sawed their way through the log jam. The river is being used again by the commercial boatmen and others without further disturbance.

Especially galling was the tendency of such authoritative sources as the Associated Press to sensationalize the incident. Much of their reporting had come from people in the BRIA, in which case the O.W.W.C. came out looking like the Kansas Jayhawkers and the Ozark Society like Quantrill's Raiders.

In an effort to counter that we tried to get a position paper into the news channels, but eventually we just had to live it down. We appealed to our number-one newsman in Pine Bluff for advice and assistance.

Joe Acuff, from Kansas City, a big, easygoing outdoorsman, was trip leader for the O.W.W.C. on that Memorial Day float. Puck, his wife, often brought along a pinwheel to twirl in the breeze on the bow of their canoe and a harmonium for music and song by the evening campfire. The Acuffs were fun people on any trip. Three weeks after the Memorial Day business I got up enough gumption to go over into Newton County to find out how the local folks felt about what had happened. There was information that Joe Acuff needed to know about, and I forwarded that to him on June 21:

I had an experience yesterday that you need to know about. It is also important that Harold and Margaret know, and I am also sending them a copy of this letter.

Dave McDonald and I did something which we certainly should do more of, and will do if we can find the time, and that is we conducted a public opinion survey in and around Jasper and Pruitt. We spent all afternoon in talking to people such as Pearl Holland, R. E. Bickerton, Don Gronwaldt, Ross Atkinson and the banker Nance in Jasper. On the way down to Gronwaldts we encountered a log jam. It seemed that someone was cutting a lot of big cedar trees and other trees to widen the road, and we had to wait several minutes to get by this obstacle. While we were parked there a large man with a crew cut came up, and we introduced ourselves. He was most delighted to meet us and informed me that he had been intending to call me on the telephone to try to straighten out certain bad impressions that had been published about him in the paper. He informed me that he was Major Bob Cooper, a retired army officer, and that he owned the land on which the Ozark Wilderness Waterways Club was forcibly compelled to camp on Memorial Day weekend. He stated that he had been in California during the time of this incident and that had he been at home it never would have happened. He said that his wife had guests to arrive on that afternoon and that she had decided to take them swimming in the Buffalo River. When they walked down to their gravel bar to swim, they found it inhabited by the O.W.W.C. plus a few campfiring Ozark Society folks. ... She never did understand the sort of distress that you folks were in and at least for some time was under the impression that the camp had been there intentionally and without asking permission. Mr. Cooper said that they had been having a lot of difficulty lately with campers and fishermen who were constantly littering the gravel bar and that he had to take his pick-up truck down about once every week or two to clean up the mess. ... Mr. Cooper actually favors the National Riverway proposal, but he is not trying to do anything to bring about its accomplishment. He did seem to be interested in selling his land to the government and expressed the idea that the Park Service might want his big house for some sort of headquarters. I did notice that he has a for sale sign on his gate. He is in no way allied with the Shaddoxes who did the shooting and the tree cutting and has about the same opinion of Jim Tudor ... that we do.

On this trip we found that there is all sorts of misunderstanding as to what happened among the local people. Some of them said Cooper had done the tree cutting and the shooting, and we are still hearing all sorts of wild statements about what the National River program would really do to the area.

Joe and Puck Acuff singing with harmonium after a hike into Indian Creek; Bernie Campbell *(sitting on tailgate)* and Ken Smith with the camera.

... Everyone that we talked to was down in the mouth about the incidents of Memorial Day. It has hurt their business in that area, and they wish that it had never happened. Everyone's main interest seemed to be in just where the park boundaries are going to be and just how this project will affect them. It is going to be up to us to get some specific information into their hands, and we are making plans on how to do this now.

You are probably wondering about what is going on on the river right now. ... I do think that it would still be dangerous to antagonize the Shaddoxes in any way. Don Gronwaldt, who lost his gear and equipment, knows where it is in the river and is afraid to go and get it for fear that they might attack him. ...

THE EVIL AFTERMATH

Although I was not present on the Buffalo River shoot-out and chop-down on the Memorial Day weekend of 1965, I was to have a taste of it in June 1985. We were on a family float from Erbie to Pruitt and were a mile or so above Pruitt when I heard a loud "splat" in the water. Looking astern I saw a geyser of water about two feet high and four feet away and then there was another and another and another bracketing my canoe and those ahead. It was gunfire from a weapon with a silencer on it. My daughter Ellen, and a lady from St. Louis who was in another party, got out on a nearby gravel bar, faced about, and shouted at the gunmen who then went crashing off through the underbrush without us seeing them clearly. We reported the incident to the park service personnel, but they were occupied with more serious things. A diver up river had hit a rock and suffered a broken neck. If one of us had been shot, we might have gotten more attention. I went home that day with no fear whatever of floating the river again but beset with curiosity as to who the shooters were. How did they learn that an old national park guru was on the water that day?

But that was a triviality compared to what was in store for Harold and Margaret Hedges.

In the rugged valley of the Buffalo three miles above Boxley, the Hedges had purchased seven hundred acres of land, there to build their dream home in 1968. The house was as charming a habitation as one could imagine. It nestled on an eminence above the river and presented on the exterior unpainted clapboard siding. The interior was finished with weathered gray oak barn siding that Harold had collected locally. It was two miles from Highway 21 on an old neglected wagon road, which wound through beech woods past stands of native umbrella magnolia.

The Hedges were devout members of the Boxley Baptist Church and had good friends in the valley—but there was a discordant note. There were some who resented pro-park people in their midst. Because of that, after a few years, Harold and Margaret were forced to leave the Boxley church to drive into Harrison for services.

With the creation of the Buffalo National River in 1972 the Hedges' land was encompassed by it, and they sold to the government. They then rented back the house and its grounds, paying in advance for twenty-five years. Then with the extension of the Wilderness Act in 1974 they found themselves living in the Upper Buffalo Wilderness Area of the National River. There no roads, vehicles, or human habitations were allowed, but they were granted the right to remain until their rent expired. A No-Admittance cable was placed across the old road for which they were given a key to the lock.

All of this infuriated some of the locals who felt they had the right to drive any road any time for any purpose. Someone shot up Harold's mailbox as a starter and then knocked down the posts that held up the cable, after which the park service put up a heavy steel gate with immovable supports.

Harold and Margaret had always liked to travel, going to the Grand Canyon, Canada, and Old Mexico for lengthy stays. Once in their absence someone tried to burn an old Jeep and shed on their place but failed.

Then in the spring of 1990 the vacant early-day Lewallen house nearby on park service land, which house was being saved for its architectural value, was mysteriously burned.

For Christmas 1990 the Hedges went to Old Mexico, and on New Year's Eve their lovely home burned to the ground while they were gone. The next day friends of theirs, national park officials, and a local fire marshall inspected the site. The latter officially reported that it was caused by faulty wiring. The superintendent of the Buffalo National River said that an "accelerator" had been used.

There were fresh tire tracks in the snow as far in as the gate and from there on the tracks of one man. On February 5 the acting superintendent of the Buffalo National River offered a five-thousand-dollar reward for the apprehension of the arsonist. Even if that should happen nothing can restore the loss, the charming house, the personal belongings, the records that Harold had kept, and the natural science research headquarters that the house might have become one day.

Ben Ferrier—RIP

The summer of '65 marked the passage of an irreplaceable member of our save-the-Buffalo team. What Ben Ferrier meant to us is revealed in a letter to Harry on June 18, 1965:

> … I am enclosing the statement made by Ben Ferrier before the Board of Engineers for Rivers and Harbors. You will recollect that Mr. Ferrier as being the man who brought us the news of my wife's canoe wreck during the Mt. Hersey to Gilbert float. He is for a fact the most experienced canoe man in the United States today, and I believe that his statement in some ways is the most important that we have had about the Buffalo River. The fact that he maintains that it even transcends the rivers of Canada in quality surprises even me.
>
> Mr. Ferrier is going to come to Arkansas and spend six months each winter during the next several years. He intends to write a story or a series of articles on canoeing southern rivers during the winter time. When he does this, you will have an opportunity to interview him and perhaps write a story about Ben Ferrier himself. I think that it might turn out to be a most interesting article. You know he claims to be the last living white man to have lived with the canoe culture Indians in Canada and who traveled with them. …

Ben Ferrier and his wife had repaired to an island in one of Minnesota's lakes that summer as part of their outdoor-living routine. There Ben became ill with acute indigestion, which turned out to be a ruptured appendix and from which he died before they could get him to a hospital. Could it have been otherwise, we would have had some outstanding reports on the wintertime recreation potential of our southern rivers, something we are lacking to this day.

Thomas Hart Benton pauses with his sketch book before embarking on a float trip down the Buffalo River. *(Photo by Harold Phelps. Courtesy of the Arkansas Publicity and Parks Commission)*

Thomas Hart Benton on the River

But we were pleased that season to have as a visitor to the Buffalo a famous personality from the world of art, Thomas Hart Benton.

From John Heuston's report in the *Arkansas Gazette,* May 30, 1965:

... "The Buffalo River is no stranger to me," Benton said last week. "I hiked along its banks from Ponca to Jasper back in 1926 when I made a series of sketches about rural American life. Some of these sketches appeared in my book, *An Artist in America.*"

Benton floated the mid-section of the Buffalo River from state Highway 123 to U.S. Highway 65 last summer and said he was so impressed with the river's pristine beauty that he decided to return. One of the paintings that resulted from his last trip sold recently for $8,000.

"I love this country," Benton said. "I traveled all through it in the 1920s with a candy salesman. While he sold his wares, I made sketches of everything that interested me. The people were very friendly. I especially remember a young farm couple near Jasper who took me in, a total stranger, and gave me a warm place to spend the night. ..."

Benton and his party arrived at Al Gaston's White River Resort at Lakeview a week ago Saturday and will float various sections of the Buffalo River for an undetermined length of time. His trip was jointly arranged by the Missouri Conservation Commission and the Arkansas Publicity and Parks Commission. ...

Benton made two sketches and several studies during the first part of his trip. He was especially interested in portraying an unusual rock formation above Woolum in Searcy County known locally as "the Narrows."

Harry Pearson continued to provide the latest news, most of it encouraging, if not sometimes overoptimistic. But some of it would turn out to be factual enough to make the big difference in the long run (*Pine Bluff Commercial,* June 13, 1965):

Washington Ready to Act on Buffalo "National River"

The Johnson Administration has decided to sponsor legislation to make the Buffalo River a national river.

This decision, along with two other new developments, has decidedly brightened the picture for those dedicated to preserving the river as a free-flowing stream.

The other new developments:

—Governor Faubus, in letters to two Pine Bluff proponents of a free-flowing river, has endorsed the national river idea.

—Congressman Wilbur Mills of Kensett, who favors the national river concept, is now directly involved in the fight as a result of congressional redistricting. ...

The entrance of Mills into the picture may give Trimble just the excuse he needs to back down somewhat on the issue. All the Washington proponents of the river want is simply a lack of active opposition from Trimble.

Another avenue of withdrawal for the congressman may come from the Engineers themselves. Historically the Engineers have never built a dam which the governor of the state involved opposes.

Here is what Governor Faubus said in a letter to Anthony T. Schimmel of 1919 West 11th Avenue in Pine Bluff:

"I agree with you that this is one of the most beautiful scenic spots in America. It is also my favorite vacation area, and I get there for fishing trips whenever I have the opportunity. I would regret to see anything done which would destroy the natural beauty of this great free-flowing river. ..."

In the second letter, the Governor told Mrs. C. W. Horwedel of 2701 Howard Drive in Pine Bluff:

"I have made very few public statements on the matter.

"However, it is my viewpoint that the Buffalo should be preserved as a national recreation area. We already have so many dams and beautiful lakes in that region that I believe it will be to the advantage of Arkansas economically and tourist wise to have the national river, rather than one or two more dams and lakes."

The use of President Johnson's name in the controversy may have been presumptuous since there was no public word from that source on the matter. But we did have reason.

There had recently appeared on the scene a knowledgeable gentleman from Washington, D.C., Dr. Spencer M. Smith, Jr., secretary of the Citizens Committee on Natural Resources. He was here, he said, as an agent for President Johnson to investigate and to evaluate the Buffalo as a possible national river prior to a firm decision by the president. He was not a secret agent but did not desire publicity concerning his activities. We were glad to accommodate him in that and went all out to see that he got to float various sections of the river. The Citizens Committee on Natural Resources was a semi-private organization that kept tabs on all legislation concerning natural resources being processed in the congressional mill at any one time. That's where the bill for the construction of Gilbert dam was at the moment. We certainly needed contact with people like the Citizens Committee because the big Corps of Engineers' battlewagon was by no means dead in the water. Knowledge of its next move was of prime importance.

We didn't have to wait long for that. The Corps let go a salvo in the general direction of our friends and fellow combatants on the Eleven Point. They received the full backing of the entire Arkansas delegation in both the House and Senate, to the dismay of all of us would-be dam busters.

Almost simultaneously a similar blast was aimed at the Buffalo River defenders (*Pine Bluff Commercial,* July 18, 1965):

Move Stuns Park Hopes; Congressmen Back Down

All four of Arkansas's congressmen have signed a letter urging a dam be built on the Buffalo River.

The move has both stunned and angered those who favor preserving the river as a national park.

The letter was drafted by Representative James W. Trimble of Berryville, a long-time advocate of damming the Buffalo. Arkansas's two senators, John L. McClellan and J. William Fulbright, refused to sign Trimble's letter.

Trimble's letter is a simple one-sentence request. The text:

"We the undersigned recommend to the Board of Engineers for Rivers and Harbors that the Gilbert Dam on the Buffalo in Arkansas be reported favorable."

So far, the Army Engineer's review board has reported nothing.

It has had the issue of damming the Buffalo under advisement since early May, and the long delay caused many to wonder if the board were going to recommend a dam on the river at all.

The Northwest Arkansas congressman mailed his letter to the board June 29.

Dr. Neil Compton of Bentonville called the move by the congressmen "sneaking and conniving." Compton, president of the Ozark Society said: "We resent it because they did not inform us of this action. We feel like we have a right to know what they're up to, since we've been vitally interested in this for years."

That Arkansas's congressmen did sign such a letter first came to light this week, when McClellan wrote to a constituent interested in preserving the Buffalo. He wrote:

"As you may know, the entire Arkansas delegation in the House of Representatives signed a letter recently urging that the dam be built. Senator Fulbright and I declined to sign the letter."

One of Fulbright's aides further confirmed this: "All the (Arkansas) House members signed a letter asking that the Buffalo be dammed. The senator (Fulbright) would not."

When Trimble was asked if this were true, he said: "That's one of those things I hate to answer."

Compton said, "What I want to know is why the delegation is flying in the face of the wishes of the overwhelming majority of Arkansas's citizens. ..."

The best guess about what this means (culled from a variety of state and Washington observers) is this:

The Johnson Administration, which took the feud over the Buffalo under advisement some months ago, was reported ready to have a bill introduced which would make the river a national park—along the lines suggested by the National Park Service several years ago.

Prodded to Action

This news, the theory goes, prodded the dam proponents into action, and it was thought that about the only effective action on behalf of the dam would be a blanket endorsement from the Arkansas delegation.

John Pickett, president of the Upper Eleven Point Association, said he learned that the entire Arkansas delegation (including both senators) "signed and sent a letter to the Chief of Army Engineers requesting that Water Valley Dam be built."

Pickett, a Dalton farmer who bitterly opposes damming the Eleven Point, wrote in an "election alert" to association members:

Attempt to Persuade

"This is an attempt to persuade the chief of Army Engineers to overrule the Board of Engineers for Rivers and Harbors who recently recommended that Congress deauthorize this project. We feel that, should the Chief of Engineers overrule the Board, an immediate effort will be made to ram authorization of this project through Congress.

"It is shameful that we must fight our own Congressmen and Senators in order to preserve our homes and farms. We have had no peace.

"We get no assistance from the elected public servants who are supposed to protect and serve the people of the Eleven Point Valley, as well as a handful of land speculators and proponents from Pocahontas, Arkansas. There has been no widespread and public demand for this foolish and worthless project. ..."

To say that we were confused by the murky situation would be putting it mildly. Just what Congressman Trimble, the Board of Engineers for Rivers and Harbors, Lieutenant General William F. Cassidy, Wilbur Mills, President Lyndon Johnson, Secretary Udall, the National Park Service, and Governor Faubus were doing about the Buffalo and the Eleven Point was anybody's guess.

Bernie Campbell gave me an on-the-level report from the Park Service:

Hot Springs National Park

I have read so much lately in the Associated Press and other news reports about the Buffalo becoming a National River that I called our Washington Office this morning to clear up some of the items that have been puzzling me and a number of people who have been calling me for information.

There appears to be a misunderstanding that the Wild Rivers Bill is to include the Buffalo in Arkansas. This is not true; rather, the Bill includes the Buffalo in Tennessee. Further, the Associated Press recently carried a release pointing out that *separate* legislation to make the Buffalo a National River is pending in Congress.

Frank Harrison, who keeps abreast of legislation in our Washington Office, told me this morning that no legislation is pending in Congress to make the Buffalo a National River. He said rather that the Buffalo is still under consideration as a National River by the Department and by Senator Fulbright. Harrison further says he has no knowledge of proposed legislation by the Johnson Administration to make the River a National River.

All of which leads me to believe that legislatively we are pretty much in the same status as we have been for the past several years. The cold fact is that Trimble's Bill for a dam is still the only legislation before Congress. Further, it currently appears as if none of our local Congressmen plan to take issue with Trimble. This, of course, is what is needed or else to have Congressman Trimble back off from his present position.

The other day, when I talked with Harry Pearson, he seemed very confident that the White House would take action shortly in favor of a Buffalo National River. I hope he has the inside word. Obviously, our Bureau does not. ...

From Spencer Smith the latest word and some good advice (July 21, 1965):

Confidential from Washington

The Senate has reported out S.B. 2300—(Omnibus bill). The Eleven Point and Buffalo dams are not in it.

The House will soon also report its Omnibus bill, and if there are conflicts a conference committee will convene at which differences will be received. Results from this committee are unpredictable and beyond any control.

If the House Omnibus bill contains authorization for Water Valley and Gilbert dams, this is very bad because it will mean that the President will have given his OK for them to be included.

President Johnson made a decision against these two dams some time ago. The Corps of Engineers is fearful of him, and for this reason they recommended abruptly the de-authorization of Water Valley dam and have delayed for two months a decision on the Gilbert dam although they are eager to build both.

The recent letters concerning these dams submitted by the Arkansas delegation to the Chief of Engineers is an all-out effort by Mr. Trimble to override the President in his position.

The evasive attitude by Mr. Mills is not understood since he is usually forthright, blunt and uncompromising. He emphatically denies that he signed the letter for the Gilbert dam or states that what he did sign was not a letter for its approval. However, the Corps states that his signature is on a letter to them requesting the approval of Gilbert dam, and this signature has been witnessed by others not of the Corps of Engineers.

The maneuver by the Arkansas delegation in reference to these dams is in itself extraordinary and is not in accord with the usual procedures for the development of such projects. It represents an extreme anxiety on their part or on the part of the Corps to secure legislation for these projects.

Then, on the heels of that, one of those tangible intangibles from the high command that we had heard how many times before:

DEPARTMENT OF THE ARMY
OFFICE OF THE CHIEF OF ENGINEERS
TECHNICAL LIASON OFFICE
FOR RELEASE FRIDAY, 30 JULY 1965
ARMY ENGINEERS PROPOSE SINGLE DAM
ON BUFFALO RIVER, ARKANSAS

A solution for development of the controversial Buffalo River in Arkansas that would preserve scenic and wild-river areas and yet permit reaches of the river to be shared for flood control, hydroelectric power, water conservation, and public recreational uses essential to regional growth has been proposed by the U.S. Army Corps of Engineers.

The proposals were presented in a draft report sent to the State of Arkansas and Federal agencies concerned by Lt. General William F. Cassidy, Chief of Engineers, for comment in advance of preparing his final recommendations for transmittal to the Congress.

The plan calls for construction of a dam at the Gilbert site.

A dam already authorized for the Lone Rock site as part of the flood control plan for the White River Basin is recommended for deauthorization because it would back water up into the reach of the river considered suitable for preservation as a wild river. ...

The Buffalo River is the principal contributor of uncontrolled flood flows in the yet untamed portion of the White River Valley. Even if all other authorized flood-control projects in the basin were built, flood damage would still average about $3 million a year unless flood control were also provided on the Buffalo, according to Corps estimates. Gilbert Dam alone would eliminate about two-thirds of this damage potential.

A dam at the Gilbert site would provide lake recreation to complement the recreational and scenic values of the National River area. ...

In the face of these developments, we redoubled our efforts, as an example a letter from me to Secretary Udall, July 26, 1965:

The recent action taken by the Arkansas delegation requesting that the Bureau of Engineers for Rivers and Harbors report favorably on the Gilbert dam for the Buffalo and the Water Valley dam for the Eleven Point River in Arkansas does not reflect the desires of the people of this state. I am sure that you understand that this action is the result of commitments by one and perhaps two of our congressmen to carry on further pork barrel projects in our state. ...

We sincerely hope that no compromise will be made in regard to either of these projects since such a compromise will only result in a face saving measure for the incumbent congressmen who would then be able to claim credit for a dam and a park on the Buffalo River.

You will recall our visit with you in Washington on May the 5th of this year. At that time you stated that something could be done to thwart the development of the Corps of Engineers project and that you would take steps to prevent it if it became necessary. It is obvious that Representative Trimble is deadly serious in his intentions to secure the Gilbert Dam on the Buffalo River. It appears that such steps are now in order. If he can be stalled for the time being, we firmly believe that we will have a new and favorable representative from this district after the next election.

We sincerely appreciate your splendid performance as Secretary of the Interior and the fine accomplishments that have been made in the cause of conservation during your tenure in office.

The importance of Governor Faubus had not been overlooked. From us he received one of my communiqués on June 23, 1965:

Those of us who have been working for the preservation of the Buffalo River have known for some time of your interest in this beautiful stream. We have always hoped that you would see fit to make a public statement in its behalf and were most gratified to read in the newspapers recently that you had expressed such sentiments although not in an official capacity. ...

Lately we have heard from some of our friends in Washington that the Board [of

Engineers for Rivers and Harbors] has been having difficulty in making up its mind on what to do about the Buffalo River. We believe that submission of copies of your letters to the ladies of Pine Bluff to the Board of Engineers at this time might have a decisive effect in preserving the Buffalo River. For that reason we would like to beg your permission to submit these letters to the above mentioned board as soon as possible.

You may know that the news item in question was reprinted in several of the newspapers in the state and did appear in the Harrison paper last week.

That time to bat he walked.

July 14, 1965

Dear Dr. Compton:

This acknowledges your letter of June 23rd in reference to the Buffalo River project.

My position is known to Secretary of the Interior Stewart Udall. He told me some time ago in Washington that we would be successful in getting the Buffalo River project. It seems the only obstacle in the way at the present time is Congressman Trimble's commitment to the construction of the dam on the river.

…I think it is not wise for me to become engaged in a controversy which I cannot decide, and in which my attitude would not be decisive, and, thus, jeopardize the possibility of passage of worthwhile and progressive programs for the State of Arkansas.

I am personally of the opinion that the National River Recreational Area will be the ultimate decision of the federal government. I certainly hope so. I am strongly in favor of it, and now I must remind you that this letter to you is confidential and is not to be used publicly. I merely wanted to explain to you how I could suffer more loss than gain by taking an open position on this problem.

Evangeline Archer had made an appeal to Leonard Hall, the veteran big dam fighter in Missouri, and received this interesting reply on July 20, 1965:

I think you can blame the Buffalo situation entirely on Congressman Trimble. He is a very powerful member of the Rules Committee and can, in company with Wilbur Mills, hamstring most of the Administration's legislative program.

My understanding, from sources I am not at liberty to quote, is that Trimble says he is not particularly in favor of the Buffalo River dams. On the other hand, he says that ever since he has been in the Congress, he has been promising his people to build these dams and that this is now a matter of his word of honor which he is not going to break under any circumstances.

Again, this is more than rumor, even though I can't set down the sources in black and white.

What Leonard Hall's source of information concerning Mr. Trimble was we would never know, but it was most intriguing. We, as well as the BRIA activists, were aware of Trimble's sometimes laggard effort to move the Gilbert dam proposal along.

Was he old and tired?

Why didn't he come, full of enthusiasm, to the last hearing at Marshall?

Had the BRIA boys directed some of their harsh tactics toward him for his slowness? We had heard such rumors, that they had subjected him to critical phone calls at all hours, just as they had some of our people here at home.

Borborygmus* in Little Rock

While pondering our problem with Congressman Trimble, those of us in northwest Arkansas were astonished beyond words to read in the state papers a sure and simple solution to our problem. That news, naturally, had first appeared in the *Pine Bluff Commercial* under Harry Pearson's byline, July 31, 1965:

Foes of Dam Vow Revenge on Trimble

Forces opposing the construction of Gilbert Dam on the Buffalo River said yesterday that they would raise $50,000 to help a candidate defeat Congressman James W. Trimble of Berryville when he runs for reelection.

H. Charles Johnston, head of the Little Rock chapter of the anti-dam Ozark Society, made the statement. He did not say who the candidate would be. Johnston denounced Trimble for his efforts to get the Gilbert Dam built.

Johnston said the approval of the proposed dam at Gilbert by the Army Engineers was the result of a "power play" by Trimble.

He said the candidate would be a young office-holder in the Third Congressional District "who is a proven vote getter" and who promised to run against Trimble in 1966.

We were struck dumb by the very idea. Fifty thousand dollars we didn't have and never would have. Besides, that was not the way we played ball. Charley Johnston in his unbounded exuberance and irrepressible impulsiveness had been taken with the idea on his own and had let the cat out of the bag to Harry, and there it was on the front page of the *Commercial,* the *Gazette,* and all of our other dailies and weeklies. Being the heir apparent to the First Federal Savings and Loan Association of Little Rock, the sum may have seemed within reason to him, but not to us and our now two-dollars-per-year membership.

It had never been our intention for the Ozark Society to become involved in personal political contests as an organization. It had been my observation that one could never be absolutely sure how the dice would fall in that game. Our policy should be to try to convince the various candidates of the rightfulness of our cause or officeholders once they were elected. Our members, as individuals,

A rumbling of the guts

232　*Orval Faubus Comes to Bat*

could work for and donate to any politician they might choose. But the Ozark Society as an organization—no. If possible, the Society should maintain a proper relationship with any candidate or any public official. There would always be that possibility that an opponent would convert to our side if we remained on speaking terms. Before it was over some did—but not Jim Trimble.

We didn't berate Charley Johnston for his gaffe. There was no point in arousing an in-house ruckus over it. The BRIA would gloat over that evidence that we had millions of dollars in "vested interests" behind us and that we were going to buy a congressman. We just went ahead, kept our mouths shut, and let the old cat die.

That news report was not all fiction. There was a young officeholder coming on the field to challenge Jim Trimble the next summer. He was a veteran member of the state legislature, representing Washington County. He, along with Washington County's other member of the state legislature, Charley Stewart, were charter members of the Ozark Society. His name was David Burleson, and he was all that we could ask for in the way of an anti-dam, environmental-minded congressman. As individuals, we would help him in every way we knew in the Democrat primary that next year. But it was to be all for naught. He would be handily defeated by Congressman Trimble in that contest.

Darkness before the Dawn

As the summer wore on, the fate of the Buffalo River reached the crisis stage. From Washington came word that the Corps of Engineers had thrown the big dam switch to ON and that the machinery for the building of Gilbert dam was running according to plan.

That fact was reported in the *Arkansas Gazette*, August 2, 1965:

Approval of Dam on Buffalo River Greeted with Joy

MARSHALL—Joy reigned at Marshall Saturday because the chief of Army Engineers had given the go ahead for a dam on the Buffalo River.

Although this necessarily doesn't guarantee that the dam will be built (there still must be some congressional action and the presidential signature), it was what thousands along the controversial river had waited for.

"We're going to have a big dam rally at Marshall on Saturday, August 14," James Tudor of Marshall, the editor of the weekly *Marshall Mountain Wave* and a leader in the fight for the dam, said Saturday.

"It'll be a big get-together," he said. "We'll serve free watermelon and we're inviting Jim Trimble, the governor, Colonel [Charles D.] Maynard [District Engineer] and the senators."

"The dam'll be the salvation of the whole area," said Harold H. Reid, president of the Marshall Chamber of Commerce, which will sponsor the celebration. ...

Tudor said a "sort of celebration" had been going on ever since Friday morning when they got word that the Engineers had approved a $55.3 million dam and reservoir at Gilbert, about halfway up the river and about 15 miles north of here.

Representative James W. Trimble of Berryville, who has fought for dams in his district for some 20 years, gave the first word. He telephoned Tudor early Friday. Word of the action spread through the county as fast as if it had been broadcast, Tudor said. "I doubt if there's anybody left who doesn't know about it."

The hope now is that there's still time for Congress to put the project in the omnibus water projects bill and get an appropriation, at least for planning this year. ...

Tudor said he thought Trimble wouldn't have any trouble getting re-elected, and added, "We can't understand how those people in Little Rock and Kansas City, Missouri think they are the ones who know what's best for us. I certainly don't think Charlie Johnston is qualified to tell us what to do up here. ..."

National River Called Not "Worth a Dime"

Tudor said he felt that a National River, which has the backing of the anti-dam groups, would not be "worth a dime."

"The Buffalo goes dry every summer," he said.

At the same time Congressman Trimble personally contributed to that move with words of his own, a rare occurrence in the drawn-out controversy (*Arkansas Gazette*, August 4, 1965):

Trimble Pleased, Says He'll Push
for Recommended Dam on Buffalo

WASHINGTON—Failure to put a dam on the Buffalo River would be as senseless as completing only 90 percent of a roof on a house, Representative James W. Trimble of Berryville contended Tuesday.

Trimble said he intended to go ahead with full support on the heated issue of flood control for the free-flowing Buffalo. ...

"I was heartened by the Engineers' report because when I came to Congress 20 years ago White River flood control was just beginning," Trimble said in his first comment on the new plan.

The engineers say that the Gilbert dam is needed to approximate full flood control in the White River Valley.

"This is just like a new roof on a house; it's not wise for me to quit building it until it is finished," Trimble said. ...

In response to all that, the BRIA put on what they called: "A real, red hot, never-to-be-forgotten, V-J day rally for the dam"—in Marshall all day August 14. It was announced in the *Mountain Wave* on a full-page insert urging all and sundry to:

Come!!! Help Us Make this BLAST So Positive that our Support for the Great New "Trimble" Lake on Buffalo River Will Be Forever Beyond Question

Meet your Congressmen and Other Officials

BIG Hootenanny and Other Entertainment

Many state and national figures will be here to hear what you have to say about the Buffalo. The event will be covered by radio, TV, and newspapers throughout the midwest.

Activity Schedule

10 A.M.—Free watermelons served from 10 A.M., as long as 200 big melons last.

10:30 A.M.—Arrival of Congressman Jim Trimble and other dignitaries

10:30—Opening of Booth on Court Square for collecting letters, cards, and telegrams to Washington. Also a headquarters for Press, Radio and TV

10:30—First guided tour to Buffalo River for dignitaries, by B.R.I.A. Officials.

Noon—Luncheon for Guests, Courtesy Marshall Rotary Club

2:30 P.M.—Public Speaking

SKIT: 15 minutes of humor and satire by Marshall B. & P. W. Women

"That Was the Year That Was"

Followed by home-made music, visiting, shopping, or what have you 'till 7 P.M.

BIG Hootenanny Starts at 5 P.M.

At that affair they had a "Hot Line" booth on the square with special low rates for telegrams to the president, governor, congressmen, or senators. For that purpose typewriters, stationery, and girls to type the message were provided.

The main event of that "never-to-be-forgotten rally" was the appearance of Congressman Trimble in person as was reported by the Fayetteville newspaper, the *Northwest Arkansas Times*, August 16, 1965:

Buffalo Dispute Nears the Boiling Point

... Rep. J. W. Trimble of the Third Arkansas district visited Searcy County and had his picture taken on a stretch of the Buffalo River; and both sides in "The Battle of the Buffalo" over the weekend marshalled strength for an all or nothing fight.

Congressman Trimble flew to Arkansas from Washington to appear at a Saturday rally in Searcy County. He made it clear he intends to ask for reelection in 1966, and said that it will take construction of the Gilbert Dam on the Buffalo to complete the planned White River flood control program which he said "is 90 percent complete." He pledged all he could do to get the dam built. ...

Congressman Trimble told the crowd, "I came here to pledge to you again that everything I can possibly do to get this project done, I'll do." The dams on the White River cost the taxpayers a half billion dollars, he said. "Shall we quit now and leave the job undone after spending that much money? I think not."

That Jim Trimble intended to get that dam built should not now be doubted by anyone.

At the time of the watermelon feast, the officials of the BRIA, still whipping a dead horse, took Congressman Trimble down to the Buffalo, found a big gravel bar where a picture could be made without showing the water, and there photographed Trimble with his big dam buddies.

But the Ozark Society was not without support in the high echelons of government that summer (*Arkansas Gazette,* August 22, 1965):

Udall Scores Dam Decision

Claims Buffalo Site Would Flood Park

WASHINGTON—Interior Secretary Stewart L. Udall said Friday that he regarded a proposed dam on the Buffalo River near Gilbert "no compromise at all."

… "I see the dam as directly counter to a national park plan. If this goes ahead, the dam and the park will be on a collision course and a decision is going to have to be made.

"I was surprised," Udall noted Friday, "to see this [the Gilbert plan] described as a compromise. We don't regard it as such."

The Ozark Society's annual meeting was again held at Petit Jean State Park. Several speakers were present to bolster our opposition to the dam. We were especially happy to hear from Will Goggin, one of the real Buffalo River landowners whom we had not heard much of since the departure of Charley McRaven; Elbert Cox, southeast regional director of the National Park Service; L. B. Cook, Jr.; and Don Cullimore.

John Heuston reported the event in the *Arkansas Democrat,* August 16, 1965:

Ozark Society Reaffirms Opposition to Buffalo Dam

PETIT JEAN STATE PARK—"I've spent my life on the Buffalo River and don't intend to see it dammed without a fight. Don't let the pro-dam boys at Marshall fool you. I know of at least 17 of my neighbors along the Buffalo, who own thousands of acres where this useless lake will be, and none of them wants their homes and farms flooded. No sir. We'd rather have the national river. It would bring more money into our county than a dam, and we wouldn't have to move from our land."

So spoke Will Goggins of St. Joe, president of the Buffalo River Landowners Association, a group composed of farmers and landowners within the site of the proposed Gilbert Reservoir on the scenic Buffalo River of north central Arkansas. Goggins was one of 60 members of the Ozark Society who gathered at Mather Lodge Saturday and Sunday for the conservation organization's third annual meeting.

Ozark Society members assembled at Petit Jean heard Elbert Cox, Southeast Regional Director of the National Park Service, reaffirm an earlier stand by the Park Service and Department of Interior opposing the construction of any dams on the Buffalo River.

"The Gilbert Dam plan proposed by the Army Corps of Engineers is not compatible with the Buffalo National River plan of the National Park Service," Cox said. "A dam anywhere on the Buffalo River would destroy the river's national significance to all the people of the United States."

… Don Cullimore of Columbia, Mo., executive director of the Outdoor Writers Association of America, explained the tactics Missourians used to defeat a dam on the Current River and urged Ozark Society members not to "give up the fight."

My address to those assembled ended with a summary of the situation and a plea for everyone to do more than we had yet done to stay the big dam juggernaut.

Just before the annual meeting, we had had a letter from Senator Fulbright which was the subject of much discussion and concern. Whether or not his language might imply an upcoming change in his policy was the question. Whatever he meant by "preservation of the water in the river basin while at the same time preserving the scenic and aesthetic value" was considered with no little dismay. To cap it off was that word, "compromise":

August 11, 1965

Dear Dr. Compton:

I have delayed an answer to your telegram of July 24 because the recent action of the Corps of Engineers was imminent. I am sure you have read about the "proposed" report issued by the Chief of Engineers in which he recommends a dam at Gilbert. As you know, I did not request the issuance of this report.

The concept of a National River in the Park System was making headway until the Corps of Engineers reversed its position of the past. In an effort to resolve the issue, I met with representatives of the Executive Branch of the Government this spring and was assured that some accommodation would be reached between the Corps of Engineers and the National Park Service. As recently as July 22 I wrote the Secretary of the Interior urging him to expedite the consideration of this subject and expressing the hope that some compromise could be reached which would result in a preservation of the water in the River Basin while at the same time preserving the scenic and aesthetic values so much desired by people all over the State.

The recommendations of the Chief of Engineers will now be commented upon by the State of Arkansas, the Departments of the Interior and Commerce, the Federal Power Commission, and other interested agencies. I still hope for some satisfactory compromise.

Whatever were the senator's intentions, time and tide would, before long, remove the matter of such a decision from his jurisdiction.

Several weeks after that exchange of communications, we had word from contacts in Washington that a "compromise" was in the process of being worked out and that we had better speak up.

I thereupon called Senator Fulbright on the phone to plead for the uncompromised national park proposal. His words were disheartening. He recounted the fact that Congressman Trimble was well thought of in Congress, that he enjoyed much prestige in the government, and that the Gilbert dam project was his favorite undertaking. Senator Fulbright felt that there was no point in opposing Trimble further. The senator was sorry that nothing more could be done to stop the dams on the Buffalo. That was the low point for us in the prolonged controversy. I hung up the phone feeling that we had lost the battle and that now we would have to witness the wipeout of the whole middle section of that delightful clearwater stream. We would have to rationalize as best we could that loss and all the effort we had put into the contest to prevent it.

The Basketball Game

As summer faded into fall, repercussions of Memorial Day and the canoe race were still felt in the hills. That such could cause a major disruption in the basketball season in Searcy County is hard to believe, but it did (Yellville *Mountain Echo,* November 1965):

St. Joe Defendants Jump Gun; Have Own Trial Held Earlier

Four of the five St. Joe men who were charged with disturbing the peace following an altercation at a basketball game between Marshall and St. Joe at St. Joe a few weeks ago went into Justice of the Peace L. A. Potter's court in Marshall October 30 as summoned and told the court the case had already been tried in Justice of the Peace R. B. Glenn's court in St. Joe. …

Background of the whole affair is unofficially reported to be the Gilbert Dam issue. The Marshall newspaper strongly supports construction of the dam and many St. Joe area residents oppose it.

The Ozark Society sponsored a canoe race on an 11-mile stretch of the Buffalo River last Memorial Day and barb wire was strung across the river in an effort to halt or impede the race. Two men stood on the bank and shook the wire when the canoes squeezed between the strands, cursing and threatening occupants of the canoes as they did so, all participants of the race reported when they arrived at the finish. Some participants were badly cut by the barb wire. …

Feeling has increased since that time, and there have been numerous reports of friction in and around Marshall and St. Joe when members of the two opposing factions get near each other.

One of the Memorial Day canoe race winners was involved in the St. Joe basketball game incident, and one of the referees of the game was said at the time to have been one of the two men who stood on the bank and shook the wire.

It is reported that an argument relative to the dam preceded the game, then trouble erupted during the game which apparently had nothing to do with the game.

In addition, the four men involved in the trouble at the October 19 game will not be allowed to attend any basketball games at St. Joe between January 1, 1966 and January 1, 1967.

The school has been under temporary suspension since October 19, and all games scheduled between that time and January 1, 1966 may not be rescheduled during the current season. The school has not played since then.

The suspensions were arrived at after the executive committee heard reports from the school superintendent and principal at St. Joe, the principal at Marshall, St. Joe P.T.A. and school board, president of the student body, captain of the basketball team, the sheriff of Searcy County, game officials and others knowing about the incident.

Official Tudor was reportedly attacked by Bill Turney, a resident of St. Joe during the game, then after the game the trouble was continued both inside and outside the gym.

Warrants were issued as a result of the trouble for Emmet Slay, Jackie Hensley, Bud Gregory, Bill Turney and Edward McCutcheon.

Charges against Slay were later dropped, and trial was held Saturday, November 27 in Marshall for the other four who were charged with disturbing the peace. Hensley and Gregory were acquitted, and McCutcheon and Turney were found guilty and fined $20 each plus costs.

Tudor's suspension came about because he admitted going across the floor and challenging Turney immediately after the game, whereas association rules stipulate that all officials shall go immediately to their dressing room following a game.

Supt. D. Blackwell of St. Joe said it was just an unfortunate incident which the school could not very well foresee or prevent.

Crisis on the Eleven Point

Anti-dam activity was not limited to the Buffalo in 1965. The intense struggle over the Eleven Point ran concurrently, just as bitter and just as hard fought by the Upper Eleven Point Association.

We should take note here of their trials and tribulations as stated by their leader, John Pickett, in a press release issued by them and printed in the *Pine Bluff Commercial* for September 16, 1965:

The Fight Still Continues for the Eleven Point River

Thirty-eight Eleven Point River Valley farmers from the Dalton, Arkansas area returned home from Washington, D.C. last Friday after two days and two nights, over 53 hours, of riding a chartered bus to talk with their Arkansas Congressional Delegation. The Arkansas Congressmen were asked to reconsider their unqualified support of the controversial Water Valley Dam project on the Eleven Point River near Pocahontas, Arkansas. ...

In 1953, General Sturgis, Chief of Army Engineers, turned the project down because of its marginal economic value, but Congressman Wilbur Mills revived the project again in 1958 by getting a Congressional appropriation of $55,000.00 for review and restudy. Over $500,000.00 has already been spent for study and planning.

John Pickett, who was the leader and spokesman for the Dalton, Arkansas farmers group, reports of the trip to Washington. "We were treated cordially, but cool, by the Arkansas Congressional Delegation. They seemed completely indifferent and unsympathetic to the plight of 1,300 of their constituents who would be flooded from their homes and farms by this ridiculous project which is without precedence in manipulated and conspiratorial maneuver between Congressmen and U.S. Engineers."

... Congressman Gathings seemed nervous, irritated and uncomfortable at our presence. He admitted that he was the author of the letter to General Cassidy, signed by the entire Arkansas Congressional Delegation.

Senator McClellan stated in a recent letter his signing of the letter was a difficult and hard decision which he made reluctantly after pondering the matter for two days.

Representative Mills confided that for many years he had prevented Water Valley Dam being built as a flood control project. Representative Trimble was mostly noncommittal. Mr. Cash listened. Most of our questions went unanswered. ...

The public and the people of Eleven Point Valley are being "framed" by outright deceit and misrepresentation. The public image of professional integrity and competence developed by many generations of honest and honorable Federal employees is being used as a screen to hide the operations of a new generation of professional Federal outlaws that ignore the true National interest and prey upon public indifference. We will fight this thing through high water and hell, if necessary.

In spite of strong evidence of a compromise on the Buffalo being arranged in Washington, encouraging developments took place. One was the failure of Trimble's bill to obtain funding in the current session of Congress (*Arkansas Gazette*, October 7, 1965):

Ozark Society Hails Failure of Buffalo Act

BENTONVILLE (UPI)—Failure of Congress to act on an appropriation to dam the Buffalo River was hailed Wednesday by the president of the Ozark Society, but he added the Society will have to work harder next year to prevent an allocation.

"We're grateful for the fact that important people both of the state and national level have thought the Buffalo was worth saving," Dr. Neil Compton told the *Pine Bluff Commercial.* ...

He predicted "another assault next year by the Corps of Engineers and those representatives who may still want the dam."

Representative J. W. Trimble (Dem., Ark.), an advocate of the dam, said Congress would not act this year until reports were in from various agencies. He said as far as he knew Governor Faubus had not sent in his recommendation.

We wondered if this failure was due to executive intervention of some kind. But, however it came about, the Corps of Engineers' steamroller was definitely slowed down by it.

A Home Run for Our Side

A key figure in the Buffalo River ballgame was now coming off the bench, and he was not from Little Rock or Washington, D.C., but originally from Greasy Creek up in the Boston Mountains.

He had first come to our attention back in 1949 when he had been appointed head of the Arkansas Highway Department by Sid McMath, then the governor. That was no recommendation to those of us who had been turned off by Sid McMath's political conduct in times past. We had expected to see the last of Orval Faubus when McMath departed the capital, but not so.

Faubus had the gall to challenge Governor Francis Cherry, a good and honest administrator, for his second term. The whole world knows the result. The upstart politician from Greasy Creek defeated Governor Cherry, and twelve years later would still be in office. By then he was the whipping boy of the news media all over the land, having been billed since 1957 as the demon of the deseg-regation crisis in Little Rock. That designation he did not deserve. He was from the hill country of north Arkansas where the black population was zero, and he was without particular bias one way or the other. He was no Ku Kluxer and was not out to deprive any Negro of his due. But he did sense the danger in the pre-cipitous social change then in progress. Whether or not bodily harm would have come to either blacks or whites had Faubus not called out the Guard, we cannot now say. Perhaps he should have called for federal troops first, but time was short. The end result was that there was no real riot in Little Rock, not one per-son black or white died there or was even injured seriously.

The eastern media descended upon that unfortunate city with their minds made up, looking for Simon Legree. In Faubus they found their man, and I shall never forget that episode as it was presented to us on the TV. Six or seven of these news harpies set upon the cornered governor, lacerating him with the most arrogant commentary and rude questions that one could imagine. Faubus conducted himself with easy grace, answering every barb with unfluttered lan-guage and good reason. My opinion of him, along with thousands of other Arkansas citizens, was changed by that inquisition on the tube. It did not change as the years went on, although we were not personally acquainted in those days. But sensing that we had a friend in Orval Faubus, we did keep in contact with him through correspondence as conservation problems increased in number and magnitude here in Arkansas. That was especially true as 1965 drew to a close. In a letter dated December 2, 1965, I proposed a meeting with him:

> ... We would not want to impose on you for your time, but we would like to remind you that the Ozark Society plans a business session to be held in Little Rock on Sunday December the twelfth at 2:00 P.M. ... We would be greatly honored if you would find the time to visit with us for a short period that afternoon. We believe that it might prove interesting to you to meet some of the people who have been working for the cause of conservation in our state and to learn something more of our program. We in turn would certainly be most honored if you could spare a few moments to give us your views on this all important subject which is becoming increasingly more significant to all of the citizens of Arkansas.

Governor Faubus did not attend our meeting, but on December 10 he stepped up to the batter's box and hit a home run for the Buffalo National River. That day he made available to the news media, the citizenry, and the proper public officials the contents of a long and eloquent letter from him to General William F. Cassidy, Chief of Engineers, opposing the dams on the Buffalo. It was the most decisive development in the long, drawn-out contest up to that time.

The situation was best summarized by Harry Pearson in the *Pine Bluff Commercial*, December 12, 1965:

Battle Not Over

Faubus, interviewed by telephone yesterday, said, "The fact the dam's been stopped doesn't mean the battle's over. We have to work harder than ever now to get the National Park plan approved by Congress."

James Tudor, the editor of the *Marshall Mountain Wave* and the most ardent advocate of damming the Buffalo, was asked for comment. He said he had "not a damn thing" to say to The Commercial.

Dr. Neil Compton of Bentonville, president of the Ozark Society, ... said, "The governor's letter constitutes a termination of the proposal to build a dam on the Buffalo for as long as he is in office ..."

"Without the foresight of the governor in taking this action," Compton said, "we would have been lost, and we owe everything to him for taking this progressive step. ..."

—The governor wrote a letter to Interior Secretary Stuart L. Udall informing Udall of the action he had taken and urging Udall to move ahead with the National River plan.

—H. Charles Johnston Jr., the outspoken president of the Central Arkansas

Neil Compton, *left,* and Orval Faubus, at the Faubus house in Huntsville, Arkansas, 1982. Dr. Compton spoke to an Ozark Society group about the Faubus letter (in hand) that stopped the construction of the Gilbert dam.

Chapter of the Ozark Society, had a comment: "I think this (the letter) means the ultimate preservation of the river. It's only a matter of time now."

The *Pine Bluff Commercial* was the only state newspaper to publish the governor's lengthy letter in full at the time. For the sake of the record, it is submitted here in full:

December 10, 1965

Dear General Cassidy:

RE:: ENGGW-PD

Sometime ago you provided to the Executive Director, Arkansas Soil and Water Commission, a copy of the proposed report of the Chief of Engineers, together with the reports of the Board of Engineers for Rivers and Harbors, and the District and Division Engineers, on an interim report on Buffalo River Basin, Arkansas (Gilbert Reservoir). This was done for my review and comment, in accordance with Section 1 of Public Law 534, 78th Congress, and Public Law 85-624.

I am also aware of a proposal of the Department of the Interior, National Park Service, to create what would become known as a National River in the very same area as the proposed Gilbert Dam. Your agency is also aware of this proposal, because it is discussed in your report, and the proposed Gilbert Dam is recommended as a compromise proposal.

I have studied closely both proposals, and my comments are as follows:

1. The building of a dam (or dams) on the Buffalo River is not essential for flood control in the White River Valley area, and the creation of hydroelectric power is not essential.

2. As an attraction for tourists, or use as a recreational area, the dam and lake would be only one more attraction, of which there are already five in the White River system, five more in the state (one more under construction), and a half-dozen or more now finished or under construction on the Arkansas River. The drawing power of the dam and lake would be limited. A properly developed National River would be a national and international attraction, drawing additional tourists that would number into the tens of thousands annually.

3. Tentative plans for a National River call for the establishment of three major visitor centers. The first would be at Silver Hill on Highway 65 in Searcy County (near the site of proposed Gilbert Dam). Here would be located the National River headquarters, the maintenance area headquarters, and ranger station headquarters. Also a major camping area, a major picnic area, a district ranger station, and boat access points would be established at this point, and last but not least, a museum.

The second visitor center would be at Pruitt in Newton County on Highway 7. Besides the camping and picnic areas, boat access, maintenance, and district ranger headquarters, there would be a residence area. The third center would be at Buffalo River State Park in Marion County on Highway 14, and would be much the same as the second. (A fourth center could be located at Mt. Judea in Newton County on Highway No. 123.)

Three other ranger stations are proposed: the first located at the mouth of the Buffalo River, the second at Woolem, and the third at Ponca.

The proposal calls for nine (9) primitive camps on the river, which would be accessible only by boat. Six (6) others would be accessible by boat and by road, making a total of fifteen (15) primitive camps. There would be six (6) other boat access, or crossings of the river, making a total of twelve (12) boat launching areas in addition to the major visitor centers.

A pioneer farm is proposed for Richland Valley, with barns, log cabins, sorghum mills, and water mills.

Nature trails will lead to such areas as Bat Cave, Lost Valley, Big Bluff, Hemmed-In Hollow, Peter Point, and others. Camp Orr for Boy Scouts would be retained and assisted.

4. There would be twice as many permanent employees to maintain and operate the National River, as would be required for the dam and lake. In addition, large numbers of temporary employees would be required during the summer season for the National River (as is now the case in all National Parks).

5. With a dam and lake, the land is inundated. With a National River, the land remains, to grow beautiful trees of many kinds, dozens of varieties of wild flowers, and some crops. Many of the present residents would be permitted to remain on the land. The same fields and woods would continue to provide a home for thousands of wild birds, including quail and wild turkey, and continue to produce deer, fox, squir-

rel, rabbit, raccoon, opossum, mink, and other game. Frogs of every size and kind join with unnumbered katydids to make the summer night musical for the tired camper seeking rest and relief from social and political problems, and the fevered market place.

6. A dam and lake would cover, forever, miles and miles of tree-lined, flower-bedecked river banks; hundreds of the most beautiful holes (pools) of water that have ever been created; numbers of rock-strewn, rippling shoals; the finest sand-bar camp sites to be found anywhere; and dozens of magnificent towering cliffs. All of these are worth retaining as a part of a National River, because of a unique, inspirational, soul-resting beauty which cannot be found in comparable expanse anywhere else.

7. Already created dams and lakes are to be found on every side of the beautiful Buffalo River area within a distance of 30 to 100 miles. The creation of another such facility would add little to the attraction of the area as a whole. On the other hand, the creation of a properly developed National River would complement the attractiveness of the area. It would create a balanced recreational area unlike any to be found in any other region of the United States.

Fishing and skiing are the main, and almost only sports, on the large lakes. There is little, if any, pleasure in boating. The National River would attract the fishermen and the hunters, the boatmen, canoeists, camera bugs, campers, bird watchers, swimmers, and wild life lovers of all kinds. The area would have accommodations and unusual appeal for family groups.

Were there not already dams and lakes for flood control, generation of electric power, and recreation, this would be a different proposition. However, with the present situation, the National River can add far more to the region, and be of far greater benefit in *every way* than can another dam and lake.

It is well to point out also that by a conservative estimate, 90% of the thousands of visitors to Buffalo River State Park favor the National River over the dam and lake. Also, a college-trained businessman, operating a business in the very heart of the area of greatest controversy, became sufficiently interested to make a poll of his visiting customers. He was amazed to find that 95% of the visitors to the area favored the creation of the National River. An awareness of this sentiment probably led to the change in attitude of the members of both the Chambers of Commerce of Mountain Home and Yellville, the county seats of two counties through which the Buffalo River flows. Both groups now support the National River proposal, and have withdrawn their support of the dam.

There is no question that both aesthetically and economically, the approval and proper construction of a National River will be far better for the area, the State of Arkansas, and the nation, than would the construction of the proposed Gilbert Dam and lake.

Of course, there are other considerations. We cannot place a material value upon the soul, the spirit, and the mind of man. The mind of man must constantly be refreshed, his spirit periodically renewed, and his soul ultimately saved. Next to God's promise to man of the salvation of his soul, the greatest force for good is man's capacity to enjoy and be inspired by the unspoiled beauty of God's creation.

The Buffalo River area is one of the greatest examples of the majesty of God's creation. The beauty of the region cannot be adequately described in any of the many languages of man.

The heavens declare the Glory of God, and the firmament sheweth His handiwork.

Standing in the Buffalo River State Park, on a point overlooking a stretch of this beautiful river, is a plaque erected in memory of a little boy. The plaque bears the following inscription:

There are little corners of this earth put aside by nature to be discovered by and to bring joy to little boys. The lands over which you look here, across this beautiful river, are such a corner; and the arrowheads to be found there, the tiny box canyon with its waterfall and the springs above, are set aside forever for all little boys in memory of another little boy who did discover freedom and joy here.

Warren Mallory Johnston

In so many places, the giant power-driven machines of man are flattening the hedges, fence rows, and nooks, where the song birds nested, and the timid rabbits reared their young; draining the swamp where the wild ducks and raccoons once found refuge; leveling the forests where once roamed the wild deer; scarring the mountains and pushing down the lofty crags where perched the eagles; filling up the beautiful pools which furnished a home for the wary bass and the brilliant golden-hued sun fish.

A conscious effort on the part of society must be made to preserve a part of our God-given beauty, or very soon there will no longer be left a sufficient number of these "little corners of this earth put aside by nature to be discovered" by little boys, to bring pleasure to their pure fresh minds, and joy to their innocent hearts.

Unless this effort is made, under the leadership of the people's government, soon there will no longer be a sufficient number of accessible places where families can have wholesome pleasure and adventure together. This will constitute a loss to society, for which all the material wealth cannot compensate.

For these and other reasons, your proposed construction of Gilbert Dam is unacceptable. I praise the Corps of Engineers for its many fine accomplishments. I have always been in the forefront in supporting your program of the construction of dams as a proper means of conservation, and the building of a nation. For the very first time in my life, I must disapprove one of your proposals. However, it is a unique and exceptional situation, as the facts I have set out prove beyond any doubt.

I support the National River proposal.

Sincerely,
Orval E. Faubus, Governor

Most of those who subscribed to the national news media teaching that Faubus was the Arkansas Satan refused to believe that he wrote this letter, and many still do to this day. They insisted that it was done by us anti-dammers or

(Courtesy of George Fisher)

by some sensitive soul in his office. But it is pure Orval Faubus and deserves to be remembered in that context.

In that vein, I sent him a letter dated December 14, 1965, expressing the sentiments of those of us who had been so concerned:

Dear Governor Faubus:

… Your letter is one of the most splendid statements ever made in the behalf of the preservation and the proper use of our delicately balanced environment. I want to assure you that all of us who have been so deeply involved in this controversy will now redouble our efforts to bring to reality this most significant development in the preservation of Arkansas' God-given natural beauty. I for one shall strive with every effort to see that these words of yours will be inscribed at the entryway of the Buffalo National River as an inspiration to all visitors to this lovely bit of America and as a testimony to the fact that we had a Governor who would not let it die.

Yours most sincerely,
Neil Compton, M.D.

Congressman Trimble appeared not to be fazed by the governor's action, as was reported by the *Pine Bluff Commercial*, December 15, 1965:

Trimble Still Backing Dam on Buffalo

WASHINGTON—Representative James W. Trimble, Democrat of Arkansas, said yesterday that he was committed in judgment and in conscience to do all he can to get the Gilbert Dam constructed on the Buffalo River in Arkansas. …

Trimble said he would continue his push for authorization of the dam. He said he was committed to it as much as he was to the Beaver, Table Rock and Bull Shoals Dams, which are part of the White River flood control and water supply program.

Trimble said he would urge approval of the proposal by the House Public Works Committee. The House committee and its counterpart in the Senate will consider the proposal for possible inclusion in the omnibus water projects authorization bill. ...

In-House Problems Multiply

Meanwhile, we of the big dam opposition took advantage of this turn of events to relax through the holidays. But there was plenty of homework piled up on our agenda. The Ozark Society as an organization had never run smoothly. Our people were scattered over four states, and we had trouble maintaining contact. When we did get together, argument and prolonged bickering usually ensued, much like a session of the Arkansas legislature. There were, however, no overt quarrels. Somehow we were able to maintain a friendly base to operate from, and no real enemies were made. We had plenty of them without the fold.

Some of the in-house problems now facing us were:

How to pro-rate our limited income between the Fayetteville and Little Rock groups?

Should we put up billboards along the highways and, if so, how would we pay for them?

We desperately needed a regularly appearing bulletin to report on our activity and to keep our people informed. Who would volunteer to edit and publish it?

We had not settled the question of tax exemption for contributors. Would it be possible to arrange that with the IRS?

We needed someone to manage our film library—movies, slides, and black-and-white and color prints. Who would volunteer?

What do to about telephone bills? Most of ours were long distance, and it was becoming a strain on personal budgets.

Ken Smith's book was, by now, well along toward completion. How were we going to help him in the complicated business of getting it out to the public?

To handle all that and more, we held a two-day special meeting in the Holiday Inn in Russellville at which a number of new faces made their appearance.

Most of these were from Little Rock, coming with the idea that we should hire a paid secretary.

Charley Johnston, in a written communication to me, put the quietus on that, saying:

... I intend at the meeting to speak for a short time but to make abundantly clear that any thought whatsoever of a permanent, semi-permanent or any other paid secretary is OUT, OUT, OUT now and for the foreseeable future. I have no objection to our being ambitious, but let's not kid each other. She (Ros, that is) has invited a bunch of our mutual friends from the so-called "liberal" community, all of whom are nice, creative people who have what she calls an "interest" in the Buffalo River. They're basically the same people who ran the various organizations here when we were having all the school troubles—they're just displaced persons without an organization to hang on to and run. I'm sure Evangeline knows many of them. They'll be good people to have in the organization, but our problem will be how to make use of them.

This is why I want you, me, Evangeline, to retain a firm hand at this meeting—these people are all well meaning, but they have very definite ideas about what everybody else is doing wrong—and sometimes they're right.

The Mystery of the Missing Ultimatum

In early February 1966, there began to unfold the most puzzling happenstance of the contest for the Buffalo River. With the governor's stand on the matter, now well known, we felt that we were on firm ground for the time being—but that we were not.

Charley Johnston and John Heuston had written Senator Fulbright requesting that he introduce a bill for the Buffalo National River, stating that Governor Faubus's decision should now make such a bill feasible. In reply they received one of the senator's puzzling "compromise" letters suggesting that the Corps of Engineers and the National Park Service share the river.

After receiving that letter, Johnston and Heuston wrote a still stronger letter including some positive quotes from the Faubus letter. This time they heard from one of the senator's aides who stated that General Cassidy had never received any such letter from Governor Faubus and that what had appeared in the newspapers in December was merely a trial balloon to see what its political effect might be. (There were strong rumors at the time that Governor Faubus would run against Senator Fulbright that summer, and the senator's people were sensitive on the subject.)

As soon as that reply was obtained, John Heuston took it to his boss, Bob Evans, the head of the Arkansas Publicity and Parks Commission, which organization was then on record as supporting the national river proposal. They and other members of the commission then took that information to the governor's office where it was received with considerable surprise. John stated that no fuss was made by the governor or his aides and that a second copy was sent to General Cassidy, *this time by registered mail*!

Orval Faubus leaving the Ozark Society spring meeting at Arkadelphia, Arkansas, April 1972; Dr. Joe Nix is pictured in the light coat, standing left of Orval Faubus.

That development set all of our communication channels abuzz. On February 11, 1966, I called upon our Washington, D.C., bird dog, Spencer Smith, to find out, if he could, what was going on:

Dear Spencer:

Yesterday morning I had a telephone conversation with H. Charles Johnston Jr., the president of the Little Rock Chapter of Ozark Society, who had received disturbing communications from the office of our Senator J. W. Fulbright. I have not seen copies of these communications but understand that the Senator strongly indicated that he was in favor of a compromise between the National Park Service and the Corps of Engineers on the Buffalo River question. …

The disturbing factor in these communications from the Senator's office is … that one of his aides maintained that the letter written by Governor Orval Faubus to General Cassidy had never been delivered to the General's office. … I do wonder … if it was possible for this letter to have somehow been detoured and to never have arrived at General Cassidy's desk, this being accomplished by someone interested in furthering the big dam program. …

Smith contacted Governor Faubus, who already knew of the hiatus, by telegram and quickly received a reply verifying the second mailing.

On February 15, Smith took the matter to headquarters as well:

Dear General Cassidy:

On December 12, 1965, the Pine Bluff Commercial newspaper in Pine Bluff, Arkansas, printed the text of a letter from Governor Faubus to you regarding the Buffalo River.

In a number of our discussions, it has been reported to us that you were not in receipt of any letter from the Governor in regard to the Gilbert Dam proposal on the Buffalo River. Since this letter was made public and since it was therefore generally assumed that it was a bona fide commentary from the Governor to you, we are at somewhat of a loss to understand the persistence of these rumors. We would, therefore, appreciate it if you could advise us as to whether such a letter was received by you.

Yours very truly,
Spencer M. Smith, Jr.

The final result was stated in a P.S. in a letter to Evangeline Archer on February 17:

Dear Mrs. Archer:

I indicated to you previously that I had wired Governor Faubus as to the letter he addressed to Lt. Gen. William F. Cassidy, Chief of the Corps of Engineers.

Enclosed herewith is a copy of the letter sent me by the Governor's Executive Secretary which would seem to lay to rest the question as to whether the Governor's letter to the Corps had been received by them.

Cordially,
Spencer M. Smith, Jr.

P.S. A Mr. Olson from the Corps of Engineers' Washington office and in the office of Gen. Cassidy just called me and indicated they do have the letter of Governor Faubus and they consider this document the official position of the State of Arkansas.

A final appeal was made to learn, if possible, just what did happen, but Spencer Smith was unable to provide.

On top of the governor's NO to the Gilbert dam came another firm disapproval from an important federal department:

UNITED STATES DEPARTMENT OF THE INTERIOR
OFFICE OF THE SECRETARY
WASHINGTON, D.C. 20240

February 24, 1966

Dear General Cassidy:

This is in response to your letter of July 29, 1965, requesting comments on an interim report on Buffalo River Basin, Arkansas (Gilbert Reservoir). ...

This Department opposes the recommended construction.

We believe that the preservation of the Buffalo River in its natural state would represent the best utilization of the river's recreation and resource potentials. ...

This Department, as I have noted earlier, feels strongly that the Buffalo River should be preserved in its natural state to realize its outstanding potential for recreation. If you are unable to concur in this view, then I suggest that the problem be referred to the Water Resources Council for consideration and resolution.

We appreciate the opportunity of presenting our views.

Sincerely yours,
Kenneth Holum
Assistant Secretary of the Interior

During this flurry of confusion over the Faubus letter, the BRIA held its course, apparently undismayed by the seriousness of its potential effect on their dam. That was reflected in a report in the *Marshall Mountain Wave*, February 17, 1966:

Bria Members Optimistic;
Plan for Showdown on Gilbert Dam

Members of the Buffalo River Improvement Association met at the courthouse in Marshall last Thursday evening, and vowed their intention of continuing the fight for the proposed Gilbert Dam on the Buffalo River. ...

James Tudor told the group that the final push would be made early this year and that the association needed money to prepare all data and material which will be needed when representatives of the association go to Washington to testify before the committee on the needs for the dam.

Congressman Jim Trimble has stated that he will introduce the bill early this session, probably around the first of March, and that hearings will be held by the Public Works committee soon thereafter.

Recent floods on the lower White river caused primarily by the uncontrolled flood waters of the Buffalo river in recent weeks also lent hopes to the group that favorable action toward the dam could be obtained. The actual need of controlling the Buffalo has been proven twice in the past few weeks—Tudor told the group. ...

TWELVE

THE ELECTION OF 1966

With all of that going on, we still had time to explore some of the little-known, but truly fantastic, scenic crannies in those deep ravines along the Buffalo. Walter Lackey, the historian of Newton County, had told me about Indian Creek and its natural bridge back in 1961, and that February (1966) we had scheduled a trip into its deep recesses. On that occasion, Laurene, my wife of thirty-one years, decided to come along and afterward wrote in her own inimitable style an account of how it was during the battle for the Buffalo:

A Trip to Indian Creek
Sunday, February 20, 1966

On Sunday, the 20th, we made this great hike into the Indian Creek canyon. I hadn't planned to go because of a previous engagement, but that was cancelled at the last minute; so I ran down to the men's store before it closed on Saturday and bought me a pair of waterproof boots for walking and joined the Safari after all. And a good thing I did. Saturday night Neil delivered three babies (one set of twins) and never did get to bed at all. We bundled up against the cold and started for the far country about six A.M. Sunday morning. Neil was so sleepy he thought he couldn't make it, but I drove the old Scout and he slept some on the way over, and we joined the rest of the party at an old abandoned farm up on top of a hill near the boy scout camp, Orr.

There were eighteen in the party. Two fellows from the Rogers paper, John Heuston and a friend from Little Rock, a male technician from Harrison, Schermerhorn (the cave man), his brother-in-law and two young nephews, a tender young English professor from Harding college at Searcy (native of Tennessee), a doctor from the Veterans Hospital at Fayetteville with two more young boys in tow, Joe and Maxine Clark, and Neil and I. We finally got the cars parked and the four-wheel drive vehicles taken down the mountain and parked on an old logging road where we expected to come out, and started bravely down the mountain.

We were soon among huge boulders and descending ever deeper into the canyon, more slowly as the going got rougher, and we had to scramble down over the huge stones that lay all flung about. We came to the stream and began to see lovely

Laurene Compton at Indian Creek above the great arch.

waterfalls plunging over the rocks that were glazed by thin sheets of ice and hung with icicles in the darker, secret places. We slid and scrambled on down into the canyon, crossing back and forth the stream when needs be.

The day was pretty cold but strenuous walking kept us warm enough. Finally we came to the great arch Neil has talked so much about and tried to photograph. It was a really stunning sight. Pictures just can't convey the sculptural quality of this place. Or give a true idea of the immensity of it. I think Howard Whitlach would be very impressed with this piece of Nature's own sculpture. You come to a place where the stream seems to end abruptly, blocked in its course by a mountain of stone rising straight ahead of you. As you draw nearer, you see that the stream veers to the left and disappears through a tall narrow slit between the huge blocks of limestone. Then as you reach a vantage point and can look through this aperture or archway, you see the stone mountain curving away from you in awesome formations, while the stream falls through a V and down a solid stone sluiceway into the canyon far below. This, too, shone with the silver glint of ice and icicles hung draped in the shadows.

At this point you can no longer follow the stream but must take to a very treacherous slope of mountainside and go on over the summit and descend into the canyon from the other side. There is much loose rock underfoot, and any movement of the foot may start a small avalanche of rock descending on the climbers to the rear, so it was a bit time consuming and a tricky business to make this part of the trail. The brave, surefooted ones went ahead and secured nylon ropes to trees along this way, then the rest of us clung for dear life to the ropes and finally hauled ourselves to the top one by one, while the ones left below sought shelter behind boulders from the rain of small stones we were dislodging.

As we each reached the summit we clung trembling and perspiring to the nearest tree while we caught our breath and regained our composure, then we slid carefully on our well padded bottoms, inch by slow inch, down to the stream again. Loose rock flew all about on the descent as well. Oranges, candy bars and cigarettes flew out of my pockets on the way down but were thoughtfully retrieved by some who came behind me. When I reached the safety of the stream bed, my pockets were filled instead with bits of leaves and sticks and stones. Somewhere along the way the poor young professor lost his gloves that he had put in his pocket for safekeeping and they were never retrieved.

Finally, we were all gathered in fair shape at the foot of the great falls and cast ourselves about on the rocks to eat what lunch survived and enjoy the tremendous view. Refreshed and strengthened by this pause, we gathered up our belongings and forged on down the stream for a small distance. But what is this? The stream again falls abruptly over a steep precipice. Apparently impassable. We are deep in the canyon now, hemmed in on both sides with sheer towering cliffs, and this jumping off place dead ahead.

Great white leader says, "Fear not, to our right is a cave which we can pass through and come to a place of possible descent." Deep in the shadowy canyon sleet begins to fall from a patch of cloudy sky far overhead. With three flashlights between us, the little group of eighteen hardy explorers vanish one by one into the dark cavern in the mountain. The speleologist and small nephews are ahead. We worm our way single file along narrow ledges and descend over steep drop-offs in the dim and wavering lights.

Then disaster! One small nephew loses his footing and falls into the cold waters of a cave stream, high from recent rains. He gets thoroughly soaked and cuts a gash in his head to boot. We are almost halfway through the cave, but more and more water is in evidence. Those behind cling like bats to the ledges while the leaders make the decision. The boy is shivering and covered with blood from his scalp wound. He is

The great arch at Indian Creek from below, part of a broken-down cavern with the overhead still in contact. Ellen Shipley and Jan Dixon, *lower right*.

Wall drapes within the Tunnel Cave of Indian Creek—a passageway for experienced hikers.

256

The Tunnel Cave exit from within, forty feet above the canyon floor.

The Tunnel Cave waterfall and the cavern exit, from the western rim detour, Indian Creek Canyon.

257

mopped off and a compress put on the gash. He is stripped of his sodden shirt and jacket and dressed in extra dry clothes, and the decision is made. Turn back. Keep the boy moving so he will stay warm.

A nightmarish hush falls over the group. We stumblingly retrace our steps and pour forth from the mouth of the cave. The sleet has stopped. Neil says over the hill on the opposite side is another pass, we must now go that way. Some want to argue. John Heuston says he is going to climb over the near hill and see if he can't find a way down. "Impassable," says Neil. The speleologist and his small party do not stop to argue; they take Neil's advice and keep moving up the opposite hill and soon disappear with their wounded boy through the other pass in the mountain.

Neil starts out behind them, and finally all follow the leader. Through this hole in the rocky mountain we are greeted with tough going along narrow high ledges and steep descents, but with deadly calm and caution we go where we must and once again rely on the help of the nylon ropes to lower ourselves to the bed of the stream where we can resume our walking. The injured boy and party are now well out of sight and sound of the rest of us.

From this point on the journey goes with comparative ease. We all draw a breath of relief. The smokers smoke, the camera bugs whip out the cameras and start shooting pictures. The hazards surmounted we take almost gaily off down the stream and are able to keep to its bed most of the time and walk along its banks on one side or the other, splashing occasionally through its shallow waters and only once in a while having to take to higher ground or drop from boulder to boulder. Ferns and hepatica crush softly under our feet. We swing along faster now, pausing sometimes to examine the startling formations of lichens on gray rock or to pick up strange fossil rocks from the dry parts of the stream bed.

Finally we come to the logging road and climb up to where the vehicles are parked. The others are there waiting. They have made a fire and dried out the boy and washed him free of most of his blood. The skies are clearing. We fall into our "wheels" and rumble up the steep road to the old farmplace and the rest of the waiting cars. We say our goodbys and plan for another journey into this remote part of the world when the trees and flowers have blossomed out.

Another great adventure is over. In the warm cab of the Scout I once more take the wheel and drive Neil and myself toward home through the now gathering dusk. Immediately he falls asleep.

Laurene Compton

The Corps Eats Crow

On the Ides of March 1966, there came out of the office of the Chief of Engineers one of the most explicit examples of how to eat crow that one could imagine. In that three-page lament, there was first described in precise detail the glorious, tangible and intangible, social and economic benefits that would surely derive from the building of Gilbert dam.

All of that we had heard already, a hundred times over, but the last three paragraphs contained new and agonized language from that high office in which it originated:

10 March 1966
SUBJECT: Buffalo River Basin, Arkansas
TO: THE SECRETARY OF THE ARMY

1. I submit for transmission to Congress the report of the Board of Engineers for Rivers and Harbors, accompanied by the reports of the District and Division Engineers, in full response to a resolution of the Committee on Public Works of the United States Senate. ... the Gilbert Reservoir as proposed would be a sound economical investment and a desirable project for its intended purposes of flood control, ... I note also that the hydro-power function of the project is financially and economically feasible, and the power could be utilized on the market area load by about 1970. ... I am in full agreement with the conclusions of the Board of Engineers concerning the compatibility of plans for Gilbert Reservoir and a National River.

6. Accordingly, my proposed report on the Buffalo River which was submitted to the Governor of Arkansas and interested Federal agencies for comment, concurred in the recommendations of the Board. However, in commenting on that report, the Governor stated that construction of Gilbert Dam is unacceptable to him, and he supports the National River proposal.

7. In view of the position of the Governor and the fact that it is the general policy of the Corps of Engineers not to recommend authorization of projects which are opposed by the States directly concerned, my previous recommendation for authorization and construction of the Gilbert Reservoir project has now been withdrawn. Accordingly, I recommend that construction of a reservoir project at the Gilbert site not be authorized by the Congress. I further recommend that the Lone Rock Reservoir, authorized in the Flood Control Act approved 28 June 1938, not be constructed.

William F. Cassidy
Chief of Engineers

A month later it became official and was announced in all of the regional newspapers with the *Pine Bluff Commercial* taking the lead (April 15, 1966):

Engineers Drop Plans for Dams on the Buffalo
They Yield to Faubus Opposition

... Faubus said it still was possible for Congress to override the recommendation of the Army Engineers and vote to construct the dams, but he said "it would be unusual to bypass the recommendation of the chief of the Engineers, and judging from past actions in cases such as this, the matter probably is concluded."

... Congressman James W. Trimble has been a longtime advocate of the Gilbert Dam, and it is possible that he could continue to fight for the dam, despite the opposition of the Engineers and despite Faubus's stand against damming the river. He could not be reached for comment today.

The governor's many detractors, from then on, sought to besmirch his motives in stopping the Corps of Engineers in their tracks, charging that his reasons were purely political. But July 21, 1966, in a letter from him, are words which should leave no doubt of his sincerity in the mind of any fair-minded person:

> The points which you make in your letter of July 8th are, I believe, substantially correct. The battle to save the Buffalo River is not yet over by any means, ... True, the proposed dam is effectively blocked so long as I am Governor, but my term of office will end in less than six months. Those who favor the dam may be waiting for that time to make another all-out effort. I think it will be more difficult to reinstate the project after my stand, but, certainly, it is not impossible.
>
> I stand ready to do anything at any time to further the efforts to save this magnificent stream.
>
> Most sincerely,
> Orval E. Faubus

Shortly thereafter, at our suggestion, Governor Faubus would create the Arkansas Stream Preservation Committee. After his departure from office, it would not function well, but from it would come a related agency, the Natural Heritage Commission, which would in 1976 purchase from the owner, Arlis Coger, the Kings River Falls, affording it protection by the state in perpetuity.

Soon thereafter Governor Faubus would assume a decisive role in the now-stalled effort to "save Lost Valley." How that was initiated was related in an exchange of correspondence.

> August 20, 1966
>
> Dear Governor Faubus:
>
> There is a ... matter which I wish to bring to your attention. Recently a Doctor Edward Barron, Jr., a physician of Little Rock, informed me that he intended to purchase Lost Valley. He stated that Mrs. Primrose, the present owner, is now disposed to sell this property and that he wishes to purchase it to prevent developers from obtaining it and mishandling it. He also stated that he would, in turn, like to see the Arkansas Publicity and Parks Commission take option on the property after he has purchased it. He requested that I call this matter to your attention, and I am doing so at his suggestion. ... I would hope that you would have some of your staff who would be capable of investigating the matter.
>
> In closing, I wish to commend you again for the splendid stand that you have taken for the salvation of the scenery of our native state.
>
> Yours sincerely,
> Neil Compton, M.D.

From the governor came a reply that would set in motion a series of events that would finally see Lost Valley officially protected by the state of Arkansas:

August 30, 1966

Dear Neil:

... Thank you for the information about the intentions of Dr. Edward Barron, Jr., to purchase Lost Valley. This was brought to my attention by members of the Publicity and Parks Commission, and I gave my approval to them to purchase the property immediately from Dr. Barron, once he has obtained it. Perhaps this should be kept confidential, because I do not know what the attitude of the owners would be if they knew it was to be acquired by the state.

Certainly, I would count this as a significant accomplishment of my administration if we could secure the Lost Valley area and convert it into a state park.

Thank you very much for your interest. Please send me any information at any time, or call on me whenever I can be of assistance.

Most sincerely,
Orval E. Faubus

The Heuston Incident

In late March, the BRIA had been able to put another hard-working, anti-dam Ozark Society member out of business, at least temporarily. The situation was clearly related in the *Pine Bluff Commercial,* March 29, 1966:

Writer Quits State Job over Buffalo River Feud

LITTLE ROCK—John Heuston, widely known travel writer for the state Publicity and Parks Commission, resigned today because of controversy that boiled up after he wrote a column last week favoring preserving the Buffalo River as an undammed stream.

The column ired James R. Tudor, president of the pro-dam Buffalo River Improvement Association, and Tudor called for an apology from the Commission.

Tudor said yesterday he had received a telegram, over the signatures of Bob Evans, director of the Commission, and Bryan Stearns, parks director which said:

"The John Heuston story on the Buffalo River was released without my approval or consent. Holding up further distribution of column. Apologies."

This telegram prompted Heuston to resign.

Heuston said the column was approved for release by Lou Oberste, publicity director of the commission, while Evans was on a trip to California.

Oberste this morning confirmed this.

"Acting as publicity director while Mr. Evans is in California," Oberste said, "I approved and released the article."

Heuston also said the governor's office saw the article and approved it.

John Heuston, important journalist
and conservationist throughout the
contest.

He maintained that the commission, as a state agency, should pursue the official state position on the Buffalo River. This position, he said, was made clear late last year when Governor Faubus gave unqualified support to a plan to preserve the river in its natural state—without dams.

Heuston said he thought it improper for the commission to apologize to Tudor for further endorsing this official state position.

Furthermore, he said, neither Stearns nor Evans consulted with him before sending the telegram, and the first he heard of it was on the radio news this morning.

Heuston said today: "I consider the hasty action of Mr. Evans and Mr. Stearns in sending the telegram a slap at the professional competence of myself, Mr. Oberste, the staff of the commission, the governor of Arkansas and his administrative aides, and the people of Arkansas."

The commission last year endorsed a plan that would include the Buffalo in a wild rivers program of the National Parks System.

Tudor complained then that the commission shouldn't get involved in the controversy, and the commission consequently said it would have no further comment on the issue until the state's official position was clarified. ...

Heuston said he considered the governor's endorsement the state's position and therefore the promise no longer binding. Oberste agreed, and so did the governor's office, Heuston said.

Stearns, Evans and Tudor could not be reached for comment this morning.

262 *The Election of 1966*

Heuston, 29, has been writing for the commission for about three years.

He is a graduate of the Washington University of St. Louis with a degree in journalism. After graduation, he went to work for the *Fordyce News Advocate* and then for the *Arkansas Democrat,* where he worked four years as outdoor editor. ...

Upon learning of this, Governor Faubus tried to rectify the matter, but John Heuston had had enough of such craven deportment in the Publicity and Parks Commission (*Arkansas Gazette,* March 31, 1966):

Come Back, Faubus Tells Dam Writer

John Heuston was offered his job back Wednesday as travel writer for the state Publicity and Parks Commission by Governor Faubus. Heuston declined, saying that he had other plans. ...

When newsmen pointed out that Heuston had said that Evans wasn't around the office long enough to discuss the operation, the governor said that if he had anything to say to Evans it would be out of the hearing of the press.

Will he? inquired reporters.

"I can't, he's not here," the governor said, breaking into laughter.

Evans is attending a travel show at Los Angeles.

The governor's final comment in regard to Evans had to do with the fact that his parks commission director was addicted to judging beauty contests, accounting for his absence much of the time.

Those of us who were put out by such wishy-washy doings submitted comment of our own to Mr. Evans, along with a full copy of the Faubus letter to General Cassidy, which was no doubt discarded a second time.

Ups and Downs in the Ozark Society

One would suppose that, having come thus far, and having gained an undeniable advantage against great odds, the Ozark Society would, by now, be one big happy family. That we might be a group of random personalities, bickering, squabbling, and arguing endlessly over every move should have been unthinkable—but that is what we were. In dealing with that fact, no ill will was ever directed by me toward any one of those people who were so sincerely concerned over the prospective loss of the beautiful Buffalo River. By that we were united, and my principal task was to preserve that union—if that was possible. We were happy to have almost any volunteer who would step forth to help in whatever way he or she wished. Consequently, there was much coming and going of actors on our stage. But having done their bit, regardless of quirks of personality, they were by 1966 brothers and sisters, members of a family, if you will, although not necessarily a happy one. By that time, each knew very well his coconspirators' shortcomings, of which we all had our fair share. There was that familiarity which, if it did not breed contempt, paved the way for unabashed criticism and sticky disagreement. Like brothers and sisters, we did argue to the point of exasperation but not to full-fledged anger or fisticuffs. There was a feeling of kinship that precluded that.

All of this is best described in a series of communications between some of us characters in the increasingly intense pro-park, anti-dam melodrama. That was especially true of the argument over the proper way to get Ken Smith's book published and into the hands of the interested citizenry.

In March of 1966 Ken had sent me a long formal agreement written in legal style, describing the contract between the Ozark Society and the author (Ken Smith). Since the entire undertaking, the writing, funding, and publishing of the book was to be carried out by Ken himself, the formal contract seemed unnecessary, but if he was to be so generous, we would have been uncommonly rude to turn him down.

In that contract Ken wished to set up an Ozark Society publishing fund to be derived from profits from the book, which fund might be used to publish other works in the future. Ken would "loan" ten thousand dollars to the publishing fund to finance the effort.

He would waive all royalties on the first edition of the book with those proceeds going into the publishing fund. We should note that all of this worked out as Ken had planned with the publishing fund eventually metamorphosing into the Ozark Society Foundation.

I, for one, was strongly in favor of accepting Ken's remarkable generosity, but it wasn't to be that easy. The need for restructuring was stated in a letter to Ken dated January 25, 1966:

> First, I will say a little bit about the meeting of December the 12th. It turned out that this was more of a round table discussion than a meeting. ... One reason that you haven't heard anything about what decisions were made in reference to the book at that time is that no decisions were made about anything much. Mainly we discussed what we could do to improve the organization of the Ozark Society, which you surely know is a rather disjointed group. Unfortunately, we have never been able to get any of those who have been designated certain jobs to perform the work assigned to them. By this I mean that secretaries do not behave as secretaries or that committee heads do not function as the leaders in the various fields assigned to them. It is a sad fact that we have never even had a membership list available to the officials of the organization. Before she departed for Europe [in the middle of it all Evangeline went to Greece, remaining for almost a year], Evangeline did get together a list of sorts. It did have names, in many instances no date for the payment of dues or the amount paid, and we have never had any accounting at all from the Little Rock

group. This includes names, addresses, dues, etc. The only good organizational person that I have been able to depend upon is George Kinter, and he was arbitrarily relieved of some of his duties a good while ago and has only functioned in respect to the depositing of money in the bank. He does have at the present time a good financial report available in case we need it. The unhappy situation in regard to it is that we owe more than we actually have in the bank at the present time.

In your letter you have inquired about the matter of the Ozark Society bulletin. This is something that I have always regarded as an absolute necessity and have tried time after time to get started. We have had various ones who have volunteered to assume the task of editing and publishing a bulletin, but absolutely nothing has ever been done by any of them. ...

As far as the book is concerned, you know that everyone of us is interested in what you are doing and want to help in every way possible to see that it is a success. You will recollect, however, at the time of our annual meeting this year and also last year, the matter of financing and publishing the book was discussed. In each instance everyone agreed that we did not have the money. We also have always regarded the establishment of a loan fund for such a project as well nigh impossible. We just don't have people worth enough money to do that sort of thing. Ken, if you have the money to publish the book on your own, as you have implied, we feel that it would be the best for you to go right ahead and get it out. Even though we don't have a good organization, we can do a lot to sell the volume after it appears. That is about all that we can promise you right now, and I am wondering if it is hardly fair for the Ozark Society to be known as the publisher of such a volume.

During this forthcoming year, I am going to divert my energies toward trying to make the Ozark Society a better organization. ...

One of the first things along this line that we have done was to have a planning session for outings for 1966 at the Lost Valley Lodge a week ago Saturday. Out of this has come a real exciting program for this year, and you will soon receive a copy. I firmly believe that the very first thing that we must do to have a well functioning organization is to offer a stimulating and interesting program for outdoor activities. ...

We have also been very busy keeping our program going to service clubs, garden clubs and the like. All of this involves our own time and our own expenses.

When you realize that everything that we have done so far has been accomplished by people who have had to make a living and who have businesses and professions to attend to, it is probably remarkable that we have accomplished as much as we have. I certainly regret that we can't offer a well-knit and well-oriented organization to attend to the publication and sale of your book ... Then, of course, there is the matter of this summer's political campaign on which the whole program will hinge. That above all other things is the chief matter of concern right now.

Further comment on the impasse over Ken Smith's book was made in April and May, after Ken had written to me with suggestions about how the Ozark Society could help to promote and sell the book:

April 13, 1966

Due to circumstances which I do not have time to explain in this letter, it was not possible to go into any particular details at our last meeting concerning your book. However, by now all of the officials of the Ozark Society have had time to study it carefully. To be quite frank with you, I am going to have to admit that we don't know exactly what to do. ... Although you are acquainted with all of us, it is not possible for you to know the personalities of each well enough to make a judgment as to just who will do what in the Ozark Society.

I will discuss each person separately to illustrate what I mean.

George Kinter is a very efficient and intelligent person. His services as treasurer have been invaluable to us, but like many people who have come here to retire he has become involved in real estate and is now much more concerned with that and with his social activities in and around Fayetteville than he is with the Ozark Society. He has threatened to resign as treasurer on several occasions, and I have only been able to persuade him to remain in the Society by begging and pleading. To be honest with you, George sincerely does not want to be responsible for all the money that will be handled in connection with the book.

You probably know Evangeline better than any of the rest, but by now you should know that we cannot depend upon her to pursue any one course for any great length of time. This should be obvious to you from the character of the letters that she writes. She actually does not wish to be a secretary but chooses to be more of an advisor or to perform some sort of executive function. Evangeline is willing to work very hard in order to put the book over, but she is sincerely opposed to have the Ozark Society publish the book. ...

In your letter you have designated many responsibilities to H. Charles Johnston, Jr., who is without doubt the most undependable member of our organization. I do not say this in criticism of him, for he has done splendid work from time to time and is genuinely brilliant and very effective whenever he chooses to be. We have found out, however, that there is no use in trying to inveigle him into doing any one specific task. ... Someone else will have to be found to take care of those duties that you have assigned to him, and we will make efforts to do this. Other members who have been working closely with us but who are not officials may be able to help some. These are people such as Douglas James, John Heuston, Charles Stewart, Dr. Abernathy and others.

As for myself, I certainly feel that your plans are of the highest order and that they need to be followed out in detail if the book is to be distributed as it should. ... You must realize that I do practice medicine and that there are many times when this activity requires one hundred percent of my time for days on end. As a result of this, many, many things in reference to the Ozark Society are never accomplished. I had hoped that by this time that we would be an organization large enough to afford an executive secretary to work out such details, but unfortunately we are not that large as yet. ...

As the year wore on, Evangeline Archer's opposition to the Ozark Society being designated "publisher" of Ken Smith's *Buffalo River Country* remained relentlessly firm.

By then, I was ready to throw in the sponge, dispatching this lachrymose lament to John Heuston, September 2, 1966:

I have just received the two sheets that Ozark Society has issued concerning the annual meeting; and after reading the narrative sheet, I feel that the time is past due to make an effort to solve some of the internal problems that beset the Ozark Society.

I hate to encumber you with these things, but at the present moment you are the only member that I care to turn to for advice.

I have already mentioned this a few times, but with each new development it becomes more and more apparent that the Ozark Society absolutely must have new officers as soon as possible.

As far as I am concerned, my personal feelings are that I have not acted with enough firmness when faced with various problems that have plagued us from the beginning. By this I have in mind the disorganized manner in which our meetings have been carried on. I am thinking of the continuous interruptions that have characterized them and that have effectively prevented us from transacting needed business. Had I been more insistent on some sort of order at these meetings, I am sure that many of the things that we should have done but have failed to do would have been accomplished. ... We need someone who could and would conduct our affairs with at least the efficiency of a local Rotary Club. There are undoubtedly a lot of other things; but since even my best friends won't tell me, I am going to let it go at that and request again that you be thinking about who would be a good successor for me in this society. I believe that a change from the top down would be easier probably than to change a few.

As for the vice-president Craig Rosborough, I would want to do nothing to hurt his feelings. He has certainly been a hard worker, but from the beginning he has been confused and disoriented and since his recent illness he has been much more so. It is a sad fact that he is now no longer able to do many of the things that he did a few months ago; and, because of this, we do definitely need a reasonably effective vice-president.

As for the matter of a secretary, I probably should not say a lot since you already know most of the details. I might offer the suggestion that we have never really had anyone to function in this capacity, no one to keep records, minutes or membership lists. These are things that we have needed the worst and never have had. ... Should we be able to elect a new president and vice-president, surely a new secretary could be found. The changes that [Evangeline] has taken upon herself to make in the statements and decisions of others have been many times done by her without consultation with anyone else. If something cannot be done about this particular problem, I do not see how the Ozark Society can continue to function for any great period of time.

In discussing the Little Rock situation once again, there is no need for elaboration. We all know that Charley can do outstanding work when he so desires, and I for one have continued to hope that he would settle down and function effectively; but during the last few weeks, I have given up all expectations. You are familiar with his actions in reference to the organization and publicity in reference to the Governor's Wild Rivers Study Team. ... At the same time he promised faithfully to make arrangements for our meeting to be held in Little Rock, but instead he went to Texas for about a week, returned to Arkansas and spent another week at the Buffalo River State Park and at the present time is somewhere in the mountains of Tennessee, so I understand. The arrangements for our meeting have been made by Everett Bowman, and had it not been for him we would have been in an impossible situation as far as the '66 meeting is concerned. ...

You realize that this is a confidential letter, and I would recommend that you not file it. At the time of the meeting perhaps some of us can get together to discuss in a frank manner all of these things that need to be done. ... I for one am convinced more than ever that its aims and objectives are something that our part of the country needs in the worst way. Disorganized though we have been, we have still made some accomplishments; and with good organization, there is no limit to the good that we might be able to do.

Whatever the ruction in the Wild Rivers Study Team was about I do not now recall, but they voted Charley out and put Clayton Little in charge. After that, Charley Johnston, like an old soldier, sort of faded away, except for on a few special occasions when he reappeared as the major domo of the "Gallinule Society."

Whatever has been said here concerning the personality and behavior of any one of our people is not intended to detract in any way from their value to the long and agonizing campaign to stop the Gilbert dam. Each one played a part, critical in retrospect.

The Federal Power Commission Runs a Play

There had appeared a squib in the *Tulsa World*, July 6, 1966, which caught my eye:

White River Works Eyed
Arkansas Projects Justified, Says FPC

Five out of 10 potential water resources developments in the Upper White River Basin of Missouri and Arkansas would be economically justified, a Federal Power Commission report indicates.

The Upper White drains 9,782 square miles of Ozark Plateau of southwestern Missouri and northwestern Arkansas.

The FPC staff report says there is a need for additional flood control in the basin and for continuing expansion of recreation facilities.

The five economic developments could produce 2,337,000 kilowatts at a cost of about $321.5 million. All would be in Arkansas.

Three new projects, all of which would be located on the Buffalo River, are Gilbert Dam—87,000 kilowatts, $60 million; Compton—1,000,000 kilowatts, $100 million; Point Peter—600,000 kilowatts, $60 million. Compton and Point Peter would be pumped-storage projects.

Federal project additions are Bull Shoals—580,000 kilowatts, $93 million; Norfolk—70,000 kilowatts, $8.5 million.

Load forecasts for the Southwest Power Pool area, which includes the Upper White indicate a need for more than 20 million kilowatts of additional electric generating capacity by 1980.

I sent the clipping on to Spencer Smith to see what it might mean.

Senator Fulbright's office was also contacted, and from them we were told that it was just a "tempest in a teapot."

But Spencer Smith set us straight as to what the Federal Power Commission really was:

CITIZENS COMMITTEE ON NATURAL RESOURCES

July 12, 1966

Thank you very much for your letter of July 8 and the enclosed clipping.

This report from the *Tulsa World* disturbs me greatly. We were in the process of congratulating ourselves as to our hammering away at the prospects to include the Gilbert Dam in the Omnibus Rivers and Harbors Bill. We agree with you that these things are never dead, but the Corps and their supporters usually retreat temporarily until the climate is favorable again.

The news article raises another problem, however, since it reports a study by the Federal Power Commission. As you know, the Federal Power Commission is a different breed of cat from the Corps and can license private corporations to build hydroelectric systems. While I am sure the Corps will use this report as serving its own interest, the report is dangerous in and of itself since it indicates that the FPC is looking over possible sites with the prospects of granting licenses if there are applications. I have not been aware of any interest evidenced by private companies in filing on the Gilbert Dam site. Are you aware of any at this time? If not, do you know of any potential applicants?

To stop the FPC from granting a license, in an area where we are convinced that the damage is greater than the benefits, has been difficult for us in the past. There is no Congress and no way for local opposition to focus on the problem. The FPC is highly independent and in more cases than I care to recount has turned its back on unbelievable volumes of public protests. Thus, we are in real trouble and should be thinking of possible intervention if private applicants are interested and there are any prospects that they will file prior to our protecting the River by some statutory means.

Any information you can give us from that end on these matters will be helpful to us both.

Spencer M. Smith, Jr.

We were then able to obtain a complete report from the FPC describing their recommendations for the Ozark area. From it we prepared a special statement for our members and for the media:

> ... A brief description of the projects described in the report is herewith submitted:
>
> Grand View Dam—about 2 miles north of Highway 62 bridge on the Kings River creating a reservoir that will inundate the remaining reaches of Kings River to beyond the Madison County line.
>
> Galena Dam—north of Galena, Missouri, on the James River obliterating the remaining portions of that river as far north as Ozark and Republic, Missouri.
>
> Bull Shoals—the addition of a pump back system and additional generators to the existing Bull Shoals installation.
>
> Cotter Dam—a low dam at Cotter which will back water up river to the present Bull Shoals dam.
>
> Buffalo City Dam—at Buffalo City backing water up river to the Cotter dam.
>
> Norfork—enlargement of the existing installation by the addition of two new generators.
>
> GILBERT DAM—The main project ($60,000,000) presently under contention by the Corps, just above the Highway 65 bridge on the Buffalo River. ...
>
> POINT PETER—The Point Peter development calls for a pumped storage or pump back installation on the top of Point Peter Mountain above the Gilbert Reservoir. ...
>
> COMPTON DAM—This, the most preposterous of all projects proposed in this report, specifies that a dam be built a short distance below the mouth of Hemmed-In Hollow in Newton County. The exact height of the dam is not specified in the report but indications are that the upper Buffalo will be drowned to a point about two miles below Ponca. In this project such outstanding scenic features of the Upper Buffalo as Hemmed-In Hollow, Big Bluff, Sneed's Creek, and other areas of natural beauty will be inundated and ruined. ...

We were struck dumb by the enormity of the FPC proposal, and had we had to contest them we would not have known where to turn. As it turned out, important events would soon create a climate which would cause the grandiose FPC proposal to die on the vine.

A Sparring Session in Little Rock

In that hiatus between the primary in August and the general election in November, the factions contesting the fate of the Buffalo River jockeyed for position.

The Ozark Society held its fall meeting in Little Rock, and the Grand Panjandrum of the Corps of Engineers, Lt. General William F. Cassidy himself, came to Little Rock at about the same time, there to deliver a lengthy benediction on the now-endangered Gilbert dam on the Buffalo and the threatened Water Valley project on the Eleven Point (*Arkansas Gazette,* September 10, 1966):

Water Resources Should Help All, General Asserts

By Jerol Garrison

The approach to water resources development should reflect an attitude that will assume "the greatest good for the greatest number of people." Lt. Gen. William F. Cassidy, chief of the Army Engineers, said at the Arkansas Basin Association annual meeting at Hotel Marion Friday.

"All too often there are some groups which refuse to accept this premise and insist on fullest satisfaction of their own special objectives, no matter what the consequences may be," General Cassidy said.

"For example, when interests concerned exclusively with aesthetic values conflict with industrial needs over the development of a given reach of river, the commercial interests are sometimes characterized as greedy and self-seeking, and the non-commercial interests as civic minded and impartial. Yet in the long run, both kinds of interest are essential to the future progress and welfare of the community. Neither can be damaged without serious cost to the public. Both should share in the common resource heritage."

New Governor Could Revive Buffalo Dam

Lt. Gen. William F. Cassidy, chief of the Army Engineers, said Friday that a new governor could reopen consideration of a proposed dam on the Buffalo River by writing a letter saying he was in favor of it. He made the remark in response to a question at a press conference. ...

General Cassidy reiterated Friday that he still thought a dam was needed at Gilbert for flood control and power generation and that it would be compatible with the wild river plan by leaving the upstream and downstream portions of the River free-flowing. The downstream part of the River would be enhanced for floating because the dam would regulate the stream flow, General Cassidy said.

Harry and the National Council of the Arts

It was now Harry Pearson's turn to become a center of controversy and to leave us, his admiring readers, nonplussed. The *Pine Bluff Commercial*, the *Arkansas Gazette*, the *Wall Street Journal*, the Newark (New Jersey) *Star-Ledger*, and the *Marshall Mountain Wave* were all to have a hand in publicizing an award bestowed on Harry (*Arkansas Gazette*, October 2, 1966):

That Federal Grant to an Arkansas Reporter

The travel-writing grants made recently by the National Council on the Arts, a new federal agency, have been taken to task by the *Wall Street Journal* in an editorial based on an interview of Harry Pearson that was published August 31 in the Newark (New Jersey) *Star-Ledger*.

Pearson, 29, is the *Commercial*'s political and investigative reporter. He won one of the travel grants. He is to receive $10,000 to do a study of the Buffalo River and the Ozark hill people [and to write a book on the survey].

... Pearson himself is puzzled by the grant.

"'The grant is for me to do research and to do as I please,' he adds. 'The council is very vague about the whole thing.'"

This led the *Wall Street Journal* to conclude in an editorial published September 6:

Indeed the historian may be led to conclude that the new program for subsidizing the arts bears a remarkable resemblance to an updated and gilt-edged WPA project.

To say that the rest of us in the Ozark Society were delighted with the prospect of a book-length publication along the lines of the Pearson series in the *Pine Bluff Commercial* is putting it mildly, but there was one who was not—the editor of the *Marshall Mountain Wave* (October 13, 1966):

Editorial

People of Searcy and Newton counties remember well the infamous series of articles written about two years ago by this same writer—Harry Pearson.

Although Pearson was furnished with authentic information relative to the need of the Gilbert Dam on the Buffalo, he chose to ignore the true situation and write biased articles in opposition to the proposed dam. ...

The use of Federal funds to further the cause of those opposing the construction of the Gilbert Dam is unjust, unwarranted, and is the most ridiculous action for a national organization to impose upon a group of people who have, and are, waging a fight for their very existence.

Not until a year or two later were we to realize the proof of this National Council of the Arts pudding. By that, we, too, would be left "puzzled," since the book in question neither then or ever made its appearance.

DECISION AT THE POLLS

Nineteen sixty-six, politically, was a key year in Arkansas. After twelve years of Orval Faubus, we would select a new governor. Big dam opponents awaited that choice anxiously, having just witnessed the importance of the power invested in that office and having heard the emphasis placed upon it by General Cassidy.

A swarm of candidates would appear on the scene, three of them to our liking. Those voicing support for the Buffalo National River and opposed to Gilbert dam were Sam Boyce of Newport, Kenneth Sulcer of Osceola, and Brooks Hayes of Little Rock. On the fence were Raymond Rebsamen and Dale Alford of Little Rock, Frank Holt of Harrison, and Jim Johnson of Conway (all Democrats).

Then there was the big elephant out on Petit Jean, Winthrop Rockefeller. His attitude was an unknown factor also, but since he was a Republican, the issue in his case was not as critical. There was no sign that the well-oiled Democrat machinery in Arkansas was not running as well as it ever had during the last hundred years.

Our task was to commend and encourage the pro–national park candidates and to sell those who might not be for that proposal.

There was another and more important public office before the voters that year, the seat in Congress for the Third District of Arkansas. To have expected a favorable outcome there would have been unrealistic. Hard-core dam builder gentlemanly Jim Trimble had brought more than a billion dollars in water projects to his district. That the people would reject him now was unthinkable, and we were prepared to hold the line for another two or four years at the most.

Congressman Trimble had initiated his campaign for reelection as early as August (15) 1965, receiving favorable publicity in the *Arkansas Gazette* at the time:

Trimble Starts "Politicking" for 1966 with Vow to Support Buffalo Dam

By Leroy Doald

MARSHALL—A hale and hearty Representative J. W. Trimble of Berryville nimble-footed it around Searcy County Saturday to show his third Congressional District constituents that:

1. He was all for a dam on the Buffalo River.

2. He was their congressman and likely to ask for the position again in 1966.

... "I think personally that we are on our way," the Berryville congressman assured his audience. "Let's keep our chins up, we'll fight and we're going to win this thing."

The rally ended with a Hootenanny by groups from nearby Leslie and Dennard.

Trimble Indicates He'll Run in 1966

Trimble indicated in every sentence and move that he was getting ready to run again in 1966. Asked if he were running, he said, "What do you think I'm doing down here?"

Congressman Trimble was confronted by two opponents in the Democrat primary on July 27. One of these was our own charter member, David Burleson of Fayetteville, and the other Jim Evans of Hot Springs, who announced his candidacy in April.

In a letter to me Evans said:

We congratulate you on your accomplishment of stymieing the Buffalo River dam. The Ozark Society in my opinion has accomplished this singlehandedly ... do not be misled or drop your guard.

You know I am in favor of a national park and free-flowing river.

We appreciated those good words, but there was little we could do for either Evans or Burleson. On July 27 they received a drubbing by Mr. Trimble. We could only hope that they would seek the office again in 1968 or 1970 with Trimble readying for retirement.

In April came an announcement in various newspapers by another candidate which gained only passing interest at the time. He was a Republican and was surely not going anywhere (April 26, 1966):

GOP Chairman Files for Congress

LITTLE ROCK—John Paul Hammerschmidt of Harrison, state chairman of the Republican Party, today filed as a candidate for Third District congressman. He will be running against Congressman James W. Trimble, State Representative David Burleson of Washington County, and Jim Evans of Hot Springs—all Democrats.

Hammerschmidt paid his $750 filing fee, contributed by friends at Harrison, at Republican headquarters here. Then he went to the secretary of state's office to file his corrupt-practices pledge.

He issued a statement which said, in part:

"I am not filing as a token candidate of the Republican Party just so the party will be represented in the Third District race. I believe the people want a change—that they want more two-party representation in Washington as well as in Arkansas."

Congressman John Paul
Hammerschmidt, circa 1966.
*(Courtesy of John Paul
Hammerschmidt)*

Prior to the primary we devoted as much time and attention to politics as possible. It was mainly an effort to influence the office seekers to see the advantage of a significant national park in Arkansas. It was not, as it turned out, an effort without reward.

In a lengthy communication to John Heuston on July 12, 1966, some of the political activity here in our northwest corner was described, as well as some rumors concerning Mr. Trimble:

> We have heard that Mr. Trimble has returned to Washington. It has been said that he did this because of the fact that he is not physically capable of conducting a strenuous campaign during the remaining weeks and that he expects to coast on his past momentum ...
>
> The Governor's race is just as important, and we are beginning to find out a little bit about who is lined up with us and who may be against us.
>
> Last night we had a collision here in Bentonville between Frank Holt and Jim Johnson. Holt had scheduled the Court House for a speech at 7:30, and during the afternoon Johnson came into town with sound trucks announcing that he was going to speak at the same time.

After that matter was straightened out and the oratory over with, I was able to interview Holt for a few minutes and leave with him some of our literature. He would only say that he was "inclined our way," not too encouraging from Faubus's chosen successor. Jim Johnson, who made the best impression and seemed the most likely winner in the primary, got away before I could talk to him—a cause for concern in view of the way things stood.

John Heuston, after his run-in with Bob Evans and the BRIA, was then employed by the *Baxter Bulletin* in Mountain Home and was a contributor to other papers, notably the *Arkansas Democrat*. I needed his good judgment especially in regard to such contestants as Jim Johnson.

The results of the primary on July 26 were reported with confidence and undeniably good reasoning by the *Marshall Mountain Wave*, August 4, 1966:

Third District Gives Trimble Overwhelming Mandate for Construction of Gilbert Dam

The vote of the citizens of the Third Congressional District of Arkansas in the Tuesday July 26th primary election gave Congressman Jim Trimble an overwhelming mandate for the construction of the proposed Gilbert Dam on the Buffalo River.

Months before the campaign the controversy over the damming of the Buffalo river was injected into the Congressional campaign by members of the Ozark Society, who wished to defeat the veteran congressman because of his stand for the proposed Gilbert Dam. ...

With the issue firmly understood by all the people of the Third District, Congressman Trimble soundly walloped his two opponents in the first primary by a lead of approximately 30,000 votes over his closest challenger—Burleson. ...

The evidence thus shown by the citizens of the Third Congressional District CONFIRMS THE CONTENTION OF THE PEOPLE IN THE BUFFALO RIVER AREA—Who have repeatedly claimed that the Ozark Society was a minority group composed chiefly of about 60 active members and without the influence or strength to sway any votes—even when their pet project is an issue in a political race.

In the governor's race Jim Johnson came out on top as had been expected. He faced Holt in the runoff, giving two political seers in the Ozark Society an opportunity to exchange some educated guesses; from me to H. Charles, July 28, 1966:

Your comments in reference to the late election fit in very well with the way I feel about it. The only reservation that I would have would be in trying to get Hammerschmidt to make the Buffalo a campaign issue if he doesn't want to do so. As you have implied, it would look mighty bad to have a third man strike out for us up here. It is only natural to expect that he would do so. A Republican has never been elected from this congressional district, and I would say that conditions are no more favorable now than they have been in the past. It is true, however, that a tremendous number of people did not vote in this primary and that it is most likely that most of them were Republicans. It is also true that Mr. Trimble is vulnerable and that many of his friends admit this and that this vulnerability was not exploited by Dave Burleson. Dave was too much of a gentleman to take out after the "Judge" as he should have. But as politics becomes curiouser and curiouser, one really has no way of knowing who might not be the governor or congressman next January.

I would say that off hand it looks as if Winthrop Rockefeller may have the votes coming his direction both ways in the general election. It is obvious that the liberal Democrats are never going to support Johnson and that they will support Rockefeller almost one hundred percent. It is also true that if Holt gains the nomination he will not carry much weight with the conservative Democrats, and many of these will in turn vote for Rockefeller. If Win doesn't pull a series of boo boos, I don't see how he can miss being elected in November.

As I have said before, I think it is necessary that we see and talk to all of the individuals involved and that we attempt to get them to support conservation measures but not necessarily to declare themselves in the election. As you know I have already talked to Holt about this and find him to be friendly but evasive. … I am going to get an information kit ready for each of the candidates in the governor's race and for Hammerschmidt and have ready a series of slides to show if and when an audience can be had with any of them. If we are successful in gaining their attention in any way at all, we then need to follow this up by inviting them to see the river firsthand after the election is over. You must remember that none of them are familiar with it and that of all the important politicians who have had anything to do with it only Governor Faubus had seen enough of it to have an opinion about it.

In the runoff Johnson was the choice of the Democrats, leaving him to face Winthrop Rockefeller, who had been nominated by a Republican convention. There weren't enough of them to have a primary in those days. In the congressional race it was Trimble versus Hammerschmidt, thus narrowing the decision on the Buffalo to four men. One of these was a long-time, inflexible proponent of the big dam. The other three were not talking, and it was up to us to sell them on the national river idea.

A general meeting of the Ozark Society was scheduled in Little Rock for September 16 and 17.

From a narrative account of how it was in Little Rock in the summer of '66, we offer this description of how we spun our political wheels down there:

On Friday morning September 16 we arose soon enough to permit some visiting at the political headquarters in Little Rock since it was desired to obtain interviews with the candidates for governor.

First, the Marion Hotel was visited since it was not certain where the headquarters were. It proved that the Democratic State Convention was going on at the time, and the first person to be encountered at the doorway was Will Goggins of St. Joe. A short but interesting conversation was had with Will in reference to the behavior of the democratic candidate Jim Johnson. Will stated that he thought Mr. Johnson was in with "the wrong people" around Marshall and in Searcy County. By this he implied that he thought that perhaps Mr. Johnson was taking advice from pro-dam people. He stated that this situation was due to the fact that some of his in-laws lived in the area, and he had received this information from them. It appears that Mr. Johnson had made no effort to talk to Will, who is the Democrat County Chairman for Searcy County and who should have by all rights been consulted by any such candidate. ...

After this the headquarters of Jim Johnson in the Albert Pike Hotel was visited. Most of the occupants of this office were busy watching the television which was broadcasting the doings at the state convention in the Marion Hotel. It was possible, however, to talk to the receptionist and two or three of the other individuals in the office, and it was requested that an audience be granted with Mr. Johnson in order to explain the Buffalo River situation. ...

The offices of Rockefeller for Governor were then visited in the old Frankie's Cafeteria building, and it was discovered that this office was not fully established. A short discussion was had with one of the attendants here, ... He expressed interest in the subject and was favorable towards the Ozark Society objectives.

An Evening with WR and JPH

Northwest Arkansas had always been a populous area of the state with generally conservative tendencies. Recognizing its political importance, Republican office seekers scheduled a rally there before the general election of 1966, making it easy for local conservationists to personally meet and perhaps influence the thinking of Hammerschmidt and Rockefeller.

Such an opportunity occurred on September 29, 1966, at the Republican-Hammerschmidt dinner in the Springdale High School.

John Paul Hammerschmidt spoke first. He explained that during the last several weeks he had been visiting all the various counties in this district and that he had been meeting various individuals at county fairs and at other public gatherings. He stated that he had not made any public statements or issued any plank for his campaign but that during the next few weeks he intended to do so. He was quite emphatic in denouncing extravagant government spending and high taxes and then launched an interesting discussion of the big dams. He stated that a most difficult problem that he had to face in reference to Mr. Trimble's tenure as a congressman was the fact that he had secured so many big dams for the area. Hammerschmidt admitted that these were vote-getting enterprises and that if he (Hammerschmidt) was elected he would continue to be for the construction of more of them.

Mr. Rockefeller, having arrived, proceeded to give a short speech which dealt for the most part in generalities and which carried no implications in reference to the conservation field.

After the meeting in the high school adjourned, a reception for Winthrop Rockefeller was held at the home of Dr. John W. Dorman of Springdale. There an opportunity presented to engage in private discussion with John Paul Hammerschmidt, and I was happy to discover that he had been wanting to talk to me about the Buffalo River problem. He began his conversation by stating that he knew considerable about the river and that he had "pushed" his boat up the river from Pruitt a number of times in order to float back down while fishing the stream. He stated that in the beginning he was for the construction of the dam on the Buffalo but that after having studied various reports on the subject he was beginning to change his mind.

The matter of newspaper opinion was discussed, and I informed him of the overwhelming support of Arkansas newspapers for the Buffalo National River and pointed out that only the *Marshall Mountain Wave* and the *Harrison Daily Times* were for the dams. He admitted this and also in discussing the *Marshall Mountain Wave* implied that its publisher, Jim Tudor, was against him. I did not draw him out on this subject but assumed that Tudor considered Mr. Trimble to be a much better bet in obtaining the Gilbert dam than Hammerschmidt. Thus, even though Tudor is a Republican, he and his henchmen tend to support Trimble the Democrat over Hammerschmidt the Republican. This could very well be one of the reasons why Hammerschmidt's attitude toward the dam on the Buffalo was uncertain as he had earlier suggested. In the discussion it was pointed out that we did not wish for him to declare his position at any time during the campaign but that we hoped that he would support Governor Faubus's position later. It was learned that he had not read the governor's letter. The letter, along with other literature, was left with him.

Conclusions made at the time in reference to the discussion with Hammerschmidt are that he would be a better person in Congress than Mr. Trimble. I believed that he would go along with the Buffalo National River proposal if he were the congressman.

Following this and close to the end of the reception, it was possible to have a short interview with Winthrop Rockefeller. At first it appeared that he was uncertain as to which side of the Buffalo River argument I was on. The discussion was initiated by me inviting him to attend a meeting of the National Association of Campers and Hikers which was being held on Petit Jean Mountain two nights later. In discussing the Buffalo he suggested that Laurence was better qualified in this field than he was, and I at once requested that he invite his brother, Laurence, to Arkansas especially to see the Buffalo River. The next matter of discussion was the Governor Faubus stand on the Buffalo. It was suggested by me that Mr. Rockefeller need not make any positive commitments during the campaign, but we hoped that after the campaign he would endorse the stand taken by Governor Faubus. It was pointed out that this would be a simple and forthright way to approach the problem and that it would not necessarily cast any adverse reflection on Mr. Rockefeller were he the governor. Mr. Rockefeller stated that he considered Mr. Faubus's statement to have been made for political reasons mainly.

I asked Mr. Rockefeller if he had ever had occasion to read Governor Faubus's letter and he replied in the negative. I offered to provide Mr. Rockefeller with copies of this letter along with other information, but he refused.

Mr. Rockefeller then volunteered the information that he had been "tricked" into making a positive statement about the Eleven Point River two years ago in his campaign against Mr. Faubus. He stated that some of his friends and advisors had recommended that he endorse the dam on Eleven Point and that following this he was certain that he had lost a lot of votes in the Eleven Point area. When asked if he had ever seen the Eleven Point River before making his decision in reference to it or afterwards, he replied "no."

In concluding the conversation with Mr. Rockefeller, he made the following statement, "I shall remain enthusiastic, but I won't commit myself. You help elect me governor, and we will see what we can do."

From what would subsequently transpire, it might be construed that John Paul Hammerschmidt had evolved into a full-scale conservationist, but that is not necessarily so. What did happen was that the packet of literature left with him, and perhaps what was said during our discussion at Dr. Dorman's, was sensible enough to convince him of the rightness of keeping the Buffalo River as it was. In the days and years to come, Hammerschmidt would not deviate from that conviction, but he would not personally participate with us on any of our outings or at any of our meetings—not until the twenty-fifth anniversary celebration of the Ozark Society at Buffalo Point, when I had the pleasure of recording him and his words on videotape.

As for Winthrop Rockefeller, he didn't bother to visit the meeting of the Association of Campers and Hikers and paid no further attention to the save-the-Buffalo argument.

The winds of November brought an unexpected chill to the establishment in Arkansas that year. It was difficult to believe the headlines that broke out in the newspapers or what we saw and heard on the TV and radio all across the state the day after the election. The Democrat party suffered its first reversal since the end of Reconstruction. As a spinoff, the big dams on the Buffalo received a death blow, not by the act of one man in high office, but by ordinary people in voting booths across the state.

These were the headlines in the *Arkansas Democrat,* November 9, 1966:

W R Victory Climaxes Two-Party Drive
Republicans Also Win Seat in Congress
JPH Ousts Jim Trimble

By the Associated Press:

John Paul Hammerschmidt, a mild mannered lumber dealer from Harrison, today unseated veteran Democratic Rep. J. W. Trimble in the third district.

Hammerschmidt, the Republican state chairman, filed for the office reluctantly when no other worthy GOP candidate offered himself. ...

The 44 year old Hammerschmidt beat the hustings in the district while Trimble stayed in Washington until two weeks before the election and then campaigned sparingly.

Ignoring the opposition, Trimble's major campaign tactic had worked for him as recently as the Democratic primary when he disposed of two opponents without a runoff. ...

John Paul Hammerschmidt today called it a "tremendous honor to be elected the first Republican congressman from Arkansas in modern history." He also complimented the man he defeated, Congressman J. W. Trimble.

"I think Congressman Trimble is to be complimented for his long service in public life. It is with mixed emotions that I see him leave it. I didn't agree with him, but he's a wonderful gentleman and a long time friend."

The next day, November 10, 1966, the *Arkansas Gazette* added its confirmation of the upset:

Hammerschmidt Pulls an Upset Victory in Third District

In the most surprising upset in Tuesday's general election, John Paul Hammerschmidt of Harrison unseated Representative J. W. Trimble of Berryville, who had served 11 terms from the state's Third Congressional District.

With 815 of 824 precincts reported unofficially by the Associated Press, the vote was:

Hammerschmidt	80,796
Trimble	72,635

Trimble Shaken, Promises Statement

Trimble, visibly shaken by the results, said at Berryville that he would have a statement when the final vote was in.

"If I am defeated, I will make my concessions and notify John Paul congratulating him and I am willing to help him get started any way I can," Trimble said. "I hold no ill will toward anyone. I just didn't get enough votes."

Trimble's popularity, observers said, had much to do with House acceptance through the years of considerably more than a billion dollars in expenditures on the Arkansas and White River basins.

"It is only fitting they name one of those dams for him," said an Arkansan.

Trimble's defeat also takes away Arkansas's representation on the powerful House Rules Committee.

How we anti-dam people felt about the election and what it meant to have the Plan D logjam out of the way was recorded November 25, 1966, in a letter to the Hedges:

> As to the elections, you of course are familiar with Mr. Trimble's astonishing defeat. None of us, including the successful Republican candidate, had any idea that it would turn out like this. Mr. Hammerschmidt by now has received a total of more than a 10,000 vote advantage over Mr. Trimble, which makes it most emphatic. There is little doubt that our four-year campaign to save the Buffalo River and the presentation of our argument in a sensible fashion to the public had a lot to do with the outcome. We also realize that there was a rising tide of sentiment for the republicans this time and that the two together made the difference. I still find it hard to believe and have to remind myself from time to time that this actually is true and that the one great obstacle in the way of the Buffalo National River is no longer there.
>
> As for what is brewing in reference to the Buffalo National River, I can assure you that it is all encouraging. I had a short interview last week in Little Rock with Senator Fulbright and discovered that he intends to introduce a bill for this project in the next congress. Also we have information from Bill Apple that Senator McClellan wishes to do the same. In addition to this, we have also learned that the newly elected Congressman John Paul Hammerschmidt also is entertaining the idea of introducing a bill. This sounds like there might be some sort of fight between these people to see who now gets the honors. Still later information indicates that this will not be the case and that they will all work together for the National River Project, and I feel that it is now most likely that it will be accomplished before the end of 1967.
>
> Right now we are not doing much about the situation and are trying to catch our breath. We know that we will have to prepare reports to be presented to congressional committees during the next year, but I believe that we have enough material to take care of this and that all we have to do is to organize it properly.

My hopes for the establishment of the national park by the end of 1967 were to prove most unrealistic. We had come five years and had five more to go.

Evangeline Archer (beyond the subject of the Gilbert dam), like most who knew Jim Trimble, felt an empathy for him. She had written him expressing condolences but maintaining her feeling for the free-flowing river. She received a sad reply, pitiful and sobering in its revelation, as to how far the mind of a good man could wander once it has been embarked upon a course of delusion.

A year or two before in one of my discussions with "Judge" Trimble, the subject of Evangeline Archer came up. He knew her well. It turned out that they had been classmates together in the University of Arkansas back in 1917 or 1918. Of Evangeline, he said: "She was a little cutie."

Just how far the worm had turned was best revealed in a message from that old snooper and arm twister, Bill Apple, who sent us a copy of news that he had received from a high source:

December 16, 1966

Dear Bill:

I am sorry I was not in the office during your visit last week, but I have been advised of the matters you discussed with Parker Westbrook.

I intend to sponsor legislation in the next session of Congress designed to preserve the Buffalo as a part of the National Park System or National System of Wild Rivers. I hope that the Buffalo may be included among the rivers named in a new version of S. 1446. As you know, this bill passed the Senate but was not acted upon in the House of Representatives. I am in communication with Senator McClellan on this subject, and I hope that we may be able to act jointly in the interest of the Buffalo.

With best wishes, I am

Sincerely yours,
J. W. Fulbright

But the game was not over. On the congressional playing field there can be unlimited time-outs, arguments between referees, coaches, and timekeepers. We were entering upon a period of tiresome and sometimes exasperating delays. But perhaps it was for the best.

Important decisions such as the transfer of large tracts of private land into park service domain should not be made in a day. For the Corps of Engineers, such transactions were no sweat. The public had, during the last twenty-five or thirty years, come to welcome their huge projects regardless of whose lands were to be condemned. The incongruity of it all was best stated in a handwritten letter to Senator Fulbright by a citizen of Newton County.

February 1967

Dear Senator:

Just a few lines to Let you Pepol Know At Washington DC. Just How I feel about this park situation. I am against that thing from start to finish—not only me but Avery Body in this Hold country feels the seam way a bout this thing—there is no one in favor of such a thing. Pepol favored the Dam. But we don't this—it is a free flowing stream. And the water stream belongs to the government eny way—and why want to take the pepols land and Homes a way from them. Which they have got fixed for there finel homes. Pepol floats the stream eny way when they get ready—No body cares for them floatin the stream. If I understand the situation pepol from the head of this stream to the mouth of it is All Hard against it and why push something on the pepol they sure don't want—Nature they clame is the Beauty of the stream—so if it is just let it a lone as it is. Leave the pepols Land and Homes Alone all the land that pepol has in this Country that worth eny thing Lays up and down these Streams— Sincerely yours as senators

W——H——

Jasper, Arkansas

In that tortured message to Senator Fulbright was born the hard fact that opposition to the Buffalo National River lay no longer in the efficient, well-funded headquarters of the Corps of Engineers but in the minds of ordinary as well as distinguished men in various stations of life. Now it was going to be necessary to override them with a national law that would deny them forever those marvels of engineering at Lone Rock, Gilbert, and Pruitt.

THIRTEEN

ACTION IN D.C.—
REACTION AT HOME

By now our prolonged efforts to sell the beauty of the Buffalo River country as a potential national park had begun to bear strange fruit.

Entrepreneurs of every stripe began to see in the area opportunity for a fast dollar. We proponents of the unspoiled river and its environs had anticipated that. Our lyrical descriptions of that valley were bound to attract such attention, but we had to speak out to all listeners in order to generate enough interest and momentum to stop the dams. We had hoped to accomplish that end before "theme park" developers and other commercial and social adventurers arrived on the scene.

The national river plan would be able to nip some of these in the bud and to disestablish others under construction, but in some cases it was all simply too late.

From Dogpatch to Godpatch

In the beginning of the effort to save the Buffalo River there existed on one of its northern tributaries a place of undeniable charm and historic interest. In our early and overly ambitious planning for the national river, we had extended the boundary of the park up Mill Creek to include Marble Falls. It was not the highest but was the largest continuously flowing waterfall in the Ozarks. Mill Creek emerged from a spectacular cavern a short distance above the falls where it tumbled seventy-five feet over a resistant ledge of Newton sandstone. A short distance down the road from Marble Falls was the site from whence a block of the red St. Joe member of the Boone limestone was quarried before the Civil War. By ox teams and flat boat, that block was transported to Washington, D.C., where it now resides as Arkansas's contribution to the Washington Monument.

It was regarded as marble, although it was actually an especially attractive form of limestone. From that forgotten gift of the newborn state of Arkansas to the nation, the name of Marble Falls remains, but it is now obscured by a sad display of developmental gimmickry.

In the early days pioneer settlers kept a sharp eye out for falling water to turn the wheels of their all-important grist mills. Marble Falls was such a place, and for years (for a time known as Wilcoxson) it flourished as a center for the production of flour and cornmeal. But with the decline of water-powered mills, the community of Wilcoxson faded away until nothing remained of man's activity. By 1940 the primeval beauty of the place had been restored by the tireless hand of nature. We went there in pre-World War II times to admire the delightful spectacle of Marble Falls in its miniature canyon cut into the ancient sandstone. We never dreamed then what it would become one day. For a time, at least, we would be spared the unhappy knowledge that it was destined to be the principal location for the effort to perpetuate the hick tradition with which we are seemingly forever encumbered here in the Ozarks.

On January 4, 1967, we were afforded a stunning bit of news in the *Arkansas Gazette* describing what was about to happen to Marble Falls:

Real Dogpatch Planned South of Harrison; Al Capp Joins Venture

HARRISON—Dogpatch and its hillbilly inhabitants, which have existed so far only in the comic strip world of Al Capp, will come to life as a tourist attraction in the Ozark Mountains near here, backed by Capp and a group of Harrison businessmen.

O. J. Snow of Harrison announced Tuesday that Capp, the creator of Li'l Abner and other Dogpatch characters, had joined him and nine others in a corporation to develop 825 acres along scenic state Highway 7 south of here into a tourist magnet with the sights, sounds—and even some of the food—of the make-believe community. ...

Included in the property will be Marble Falls, one of the scenic beauties of North Arkansas; Mystic Cave, the trout farm operated by Raney; Mill Creek, which once ran a grist mill, as well as several residences and the Marble Falls post office and store. ...

Military Hero Is Not Ignored

Plans call for a city of Dogpatch with buildings to house craft shops, stores, a fire station, a chapel, log cabins and cabins of rough pine board, all furnished in Dogpatch decor, Snow said.

"We will work with Capp on the number, design and furnishings for the city," Snow said.

The replica city will even have a statue of Dogpatch's most famous military hero, Jubilation T. Cornpone. ...

Mammy Yokum Meal, Grist Mill Planned

Snow said that the old grist mill, on the same site where one operated in the 1800s when the community was known as Wilcoxson, will be activated to produce and sell Mammy Yokum corn meal. ...

Sadie Hawkins Day, in which the Dogpatch girls chase the Dogpatch bachelors, will be celebrated, Snow said.

But Marble Falls (Dogpatch) was outside the proposed national park boundary, and there was nothing whatever that we, or anyone else, could do to thwart the establishment of this hillbilly extravaganza. In fact, all other such sites of natural beauty or historic significance that then lay within the proposed park boundaries itself were equally liable to such speculative development as long as the bill to establish the park was not officially passed by Congress and signed by the president.

We were forced to wait, hoping that public law would intervene before other such travesties should come into being in this, as yet, unspoiled Ozark back country. Those not unreasonable fears were to be realized soon after the announcement of the Dogpatch plan.

The Wild Boar Hunters

On the south side of the Buffalo River, not far below the mouth of Mill Creek, Cave Creek enters the main stream. The hinterland of Cave Creek included some of the Ozark National Forest in the Boston Mountains to the south and such eminences of Iceledo Mountain, Horn Mountain, and North Pole Knob. Its valley was sparsely settled, with extensive, unoccupied, private lands available. It was a perfect place for a big tourist mecca in the deep Ozarks. In the spring of 1967 speculators from Branson, Missouri, were able to close a deal with landowners on lower Cave Creek and on Horn Mountain to the north and east.

Their project, the Ozark Wildlife Club, was going to make Dogpatch and all other such undertakings in the region look like peanut farms. The scope of that operation was listed in an information sheet prepared for prospective members of the club:

OZARKS WILDLIFE CLUB
OZARKS WILDLIFE CLUB MEMBERSHIP OFFERS THE FOLLOWING:

OZARKS WILDLIFE NEWS, monthly magazine for club members; 1,000 square foot lot; mineral rights (cooperative); timber rights (cooperative); hunting privileges for wild boar, goat, deer, wolf, mountain lion, turkey, bobwhite quail, pheasant; fishing privileges for trout, salmon, walleye, bass, pan fishcamp grounds—bar-b-que pits; children's playground; enclosed public shelter (sleeping bag only); recreation hall, game rooms; comfort stations; 10,000 acres of virginal hunting ground; 100 acres of stocked fishing pools; 3,000 foot landing strip, hard surfaced, tie downs, service facilities; 50-year full family membership—can be sold, bartered, or willed away; exchange privilege at all other clubs built in future; hiking and riding trails

CONCESSIONS TO BE SOLD OR LEASED

float trip outfitted, guide service, kennels; restaurant and coffee shop; cocktail lounge, private club; service station, garage; sporting goods store and bait shop; gun shop, taxidermist, general store; jeep rental; pack horse & mule rental; trailer park and trailer rental; deluxe motel; 1,000 homesites $350 to $10,000; locker & processing plant

SPONSORED SPECIAL EVENTS

wild turkey calling contest; "bird thrashing"; archery tournament; ozark mountain marksman tourney; wild horse roundup and rodeo; annual Buffalo River float trip; scout jamboree; camera club outings; arbor day festival; ozarks outdoor theatre; kids trout tourney; big hog roundup; annual bar-b-que; ox roast; turkey shooting contest; horseshoe pitching; holiday picnics; hog calling contest; coon and fox hunts; field trial; dog show (bench & obedience)

That such service and entertainment could not be delivered should have been obvious to the least observant huntsman or the most addle-headed tourist, but the club operators did manage to carry out some of the above-mentioned operations after a fashion. They did bulldoze out a rough landing strip on a long hogback ridge and at the project center erected a few boxy-frame buildings. Earth fill dams were thrown up in some of the ravines to create "fishing lakes," and they apparently did have some sort of title to ten thousand acres of "original hunting ground."

After that project got under way, the *Harrison Daily Times* was pleased to enhance the program with glowing coverage of events on Horn Mountain. Pictures of successful huntsmen from Iowa and other distant states appeared on the front page showing them holding the severed heads of "wild boars" and wild goats bagged in the virginal woodland.

Jack McCutcheon, a lifetime resident of Bass down on Cave Creek, an amateur archeologist and an old friend, enlightened us a couple of years later on the matter of those wild boars on Horn Mountain. The people from the Wildlife Club had a cattle truck in which they would patrol the back country roads. Upon detecting a farmstead with a good hog lot, they would purchase from the farmer a few "wild boars" to be transported back to Horn Mountain where they would be set free to roam the wilderness. When a hunter from up north or

Texas arrived at club headquarters, he would be provided with a guide who knew precisely where those wild hogs were to be found. With their sportsmen clients so certain of reward, the Wildlife Club should have prospered.

But that it did not. Before long it was on financial rocks, bankrupt and passing through various hands, none with the ability to revive that moribund enterprise.

Dogpatch on scenic Highway 7 would prove to be more viable. It came upon the scene with much fanfare in the local and downstate newspapers. As had been reported, Al Capp himself was involved to some extent and gave it his benediction. At that time his cartoon strip "Li'l Abner" was still of nationwide interest. Everyone knew of Mammy and Pappy Yokum, Daisy Mae, Injun Joe, Lena the Hyena, and all those Schmoos. In the eyes of some speculator, a real Dogpatch down in Arkansas should have been a sure-fire success, but time changes all things.

By the strangest coincidence, there appeared in the *Arkansas Gazette* on January 11, 1967, a week after the announcement of the establishment of Dogpatch, an uncanny account by two experts of what might be expected from the enterprise:

Visitors to State Compare "Dogpatch" to Disneyland; Conclude It's a Big Mistake

Two visitors to Arkansas were asked their opinion Tuesday of the proposed "Dogpatch, U.S.A." project near Harrison and their verdict: A financial and aesthetic mistake.

Edwin T. Haefele of the Brookings Institution at Washington compared the project—an attempt to create the atmosphere of cartoonist Al Capp's "Li'l Abner" comic strip community as a tourist attraction in Newton County—with Disneyland in California.

"I don't think the attempts in the rest of America to repeat the success of Disneyland have come off," he said, "and I doubt that this one will."

Leon N. Moses, professor of economics at Northwestern University, said the proposed attraction probably would be unsuccessful financially.

The two were in Little Rock to attend a seminar of the Central Arkansas Urban Policy Conference at the Hotel Marion.

Confirms Fears of State Officials

In this opinion, they somewhat confirmed the expressed fears of two members of the state Publicity and Parks Commission last week that the Dogpatch project might undermine the image of Arkansas as a "progressive 20th Century state." Bob Evans, director of the Commission, and Lou Oberste, associate director of Publicity, based their complaint on the possibility of such a project's reviving "a Bob Burns type image" of the state. ...

The estimates of Haefele and Moses would prove to be right on the money, which tender would go down the Dogpatch drain by the millions from that day on.

We in the Ozark Society took no great part in this monetary melodrama, being forced to watch the environmental and cultural insult with unease and great misgivings. For one thing, there was absolutely nothing we could do to thwart it. The National Park Service had drawn the line at Pruitt, and then there were people involved who were good friends and coworkers in our number-one objective, to stop Gilbert dam. Jim Schermerhorn, our long-time compatriot and cave expert, had assumed the management and development of the big cave, now to be known as Dogpatch Cavern. We would simply have to kiss Marble Falls good-bye and concentrate on that number-one feature of the landscape, the Buffalo River.

The Horse Ranch and the Home for Wayward Boys

Just as we expected, other such scenery-wrecking projects began to spring up all along the Buffalo. In the late 1950s a Kansas City financier had acquired five miles of the prime scenic part of the Buffalo below Ponca. There he would set out to establish the Valley-Y Ranch or the Horse Ranch as some called it. He didn't undertake massive alterations on the landscape until the early 1960s when heavy machinery was brought in and the river "channelized" according to the then favorite policy of some of our agricultural agencies. That was to have been but a part of an enormous development later on, but in this case most of it lay within the park boundary; and when the bill for the Buffalo National River became law, that type of "progress" came to a halt.

In that section of the upper Buffalo, noted for its breathtaking precipices and its great high hills, was another prime spot for something big. It was just above the Boy Scout headquarters at Camp Orr and had been acquired by a minister of the gospel who knew how to enlist the financial aid of important people. It was his ambition to establish there a home for the wayward boys of Houston, Texas. That he did, bulldozing out a road down the mountain and erecting a number of poorly designed quarters for the inhabitants of Kyles' Boys' Home, as it was called. It too lay within the proposed national park, and along with the horse ranch faded away in due time.

Back in the 1960s there was a popular TV program known as Bonanza whose principal stars had somehow come to know the Buffalo. They especially liked the middle section of the river. They were able to obtain property somewhere around Woolum, and there they proposed to create Bonanza Village. It was to be different, a resettlement village for old aviators. They would have, in addition to the residential part of it, hangars, landing fields, and machine shops. Principal access would be via air with its inhabitants flying in and out at all hours. But Bonanza never got off the ground. It died aborning when its backers realized that the national park bill was certain to pass.

There were rumors of other real-estate operations up and down the Buffalo, the largest being down at the mouth near White River where a big landowner laid off subdivisions with lots for sale in imitation of Cherokee Village and Bella Vista, but it too was forced to cease and desist by the oncoming legislation to save the river from the big dams—and more.

S.B. 704 and H.R. 7020

We did not have long to wait after the seating of our new congressman in January 1967 for significant action on the Washington scene.

In a status report on the Buffalo River from park service headquarters, we learned that:

—Senators Fulbright and McClellan had asked for drafting services for the preparation of a bill for the Buffalo National River.

—That Congressman John Paul Hammerschmidt was impressed with the National Park Service proposal and that he would introduce legislation for it in the ninetieth Congress.

—That Governor Rockefeller was apparently sold on the national river proposal, but that he had not officially written General William F. Cassidy to that effect.

—That the National Park Service would update their data, initiate further field work, and submit a master plan by July.

—That they had submitted estimates to the Bureau of the Budget for consideration.

—That the Department of the Interior would oppose tributary reservoirs on the Buffalo.

—That a draft bill would go to Congressman Hammerschmidt on January 24, 1967.

On January 31, 1967, the *Pine Bluff Commercial* was pleased to print an account of the introduction of S.B. 704 in the U.S. Senate:

Fulbright, McClellan, Hammerschmidt in Accord
Senate Gets Buffalo "National River" Bill

WASHINGTON—Arkansas's senators yesterday introduced a bill to keep the Buffalo River in north central Arkansas in its natural state by making it a National River.

The measure was introduced by Senator J. William Fulbright on behalf of himself and Senator John L. McClellan who previously had taken no stand on the controversial Buffalo issue.

Meanwhile, Congressman John Paul Hammerschmidt, a Republican, said he planned to introduce in the House soon a bill with the same general provision but that he wanted to first confer with Interior Secretary Stewart L. Udall on a few technicalities. ...

"I have not taken a definite position before this date," McClellan said. "However, it is my considered opinion, which I believe to be fully supported by a vast majority of the citizens of Arkansas, that this river, one of the last vestiges of natural beauty in my state, and, indeed in the country, should be allowed to run free and remain unencumbered."

... The bill also provides that hunting and fishing be allowed within the national river boundaries except that zones and periods can be established when neither is permitted.

Another provision would prohibit the Federal Power Commission from authorizing construction and operation of any new dam on the river without the consent of Congress. ...

That news got wide front-page coverage in all newspapers and on radio and TV stations across Arkansas and adjoining states.

It was of national significance as well because it was the first time that such a coup had been pulled off against the Corps of Army Engineers. Their Buffalo River plan had been no small thing in their grand "flood control" program, what with three major dams, their reregulation structures, and those mountain-top pump-back projects. Even though much of that had been only in the talking stage, had there been no opposition anywhere it surely would all have come to pass. The decision as it now stood against the Corps was a precedent-setting event, which they knew it would be throughout the contest and which explains their bitter-end determined opposition. It was also a comedown for related ancillary agencies such as the REA and FPC.

Congressman Hammerschmidt drove the final nail in the Plan D coffin in announcing his own bill, H.R. 7020, to Senator Fulbright.

The question of the private landowners in all of this planning and counter planning had been given secondary attention. It went without saying that the reservoirs would displace them all in toto and without delay. The general public had come to accept condemnation in that case as a proper procedure. With the Corps, eminent domain was a standard policy, and they had received little criticism because of it. The big dam proponents had, on the other hand, branded the National Park Service as voracious land grabbers and were now set to use that as a principal weapon in what remained of the contest. Actually, up until about 1966, the National Park Service shunned condemnation, most of the

The BRIA's response to Senator McClellan's dish. *(Courtesy of George Fisher)*

parks having been established on public land in the west. But as eastern parks became more in demand, the exercise of eminent domain, in some instances, was becoming a necessity. In the case of the Buffalo River, public use areas and visitor centers would be subject to it.

Congressman Hammerschmidt in his version of the bill described in a

Capitol Report dated March 4, 1967, special considerations that he extended to private landowners:

CAPITOL REPORT
CONGRESSMAN JOHN PAUL HAMMERSCHMIDT

I am today [3/4/67] introducing my bill to make the Buffalo River a National River—to be preserved as a free-flowing stream. This measure is similar to measures already introduced by Senators Fulbright and McClellan. It differs in some details. ...

The bill would, with rare exception, permit owners of "improved residential properties" to retain the rights to use and occupy that land for life or 25 years—AT THE OPTION OF THE OWNER.

The bill would provide for the acquisition of scenic easements to the river, where reasonable. This would probably mean no displacement, EVER, for those people living along the river. They would retain ownership of the land.

The bill would allow the landowner to sell only a PART of that land to the government—if a PART was all that was needed to establish successfully the National River.

The bill would meet the revenue problems which the affected counties might have. It provides "in lieu" payments of taxes for five years to the counties which will lose tax income from lands which become part of the National River plan. This five-year period would enable the counties to prepare for other sources of revenue.

The bill also provides for grazing on the National River lands where it does not interfere with public recreational facilities or activities.

I believe this bill, if passed, will bring long-range good to Arkansas—benefits for all the nation, but especially to those people who live close to this beautiful Buffalo River.

Congressman Hammerschmidt would later on be accused of foot dragging and outright opposition to the national park program by his detractors. But his policy in 1967 and from then on did not deviate from a statement made to the chairman of the Committee on Interior and Insular Affairs. It was part of a request for a prompt hearing on his bill and should serve to refute rumors that he (Hammerschmidt) was a cryptic enemy of the national park legislation:

It is indeed a pleasure to submit additional information about H.R. 7020, the bill I have introduced to establish the Buffalo River in Arkansas as a national river.

It is my hope this bill, if passed, will settle a long debated and important question of significance to the people of the Third Congressional District, Arkansas, and the country. Over the years the question of the Buffalo has created a great deal of discussion and controversy. Many people here, from time to time, have taken positions which they have later reversed. However, public opinion and public officials have recently taken a more unified outlook on the future of this river, and I feel this outlook is represented by the bill I have introduced. ...

Cost for creating this national river has been estimated at a maximum of around $10 million. This is opposed to $60 million for the Corps of Engineer proposal and $210 million for the Federal Power Commission recommendation. ...

Mr. Chairman, I have introduced H.R. 7020 for several reasons. One, to encourage the Congress to make a decision of policy which affects the country, and which is being debated by three segments of the executive branch. Two, public opinion appears to have settled on maintaining the river as a free-flowing illustration of a great, natural wonder of unlimited value to future generations. My role of representative has demanded this action. ...

The delay that would now ensue would be due to the cumbersome methods of Congress, and in this case the cantankerous attitude of Chairman Aspinall, who would not call up a bill until he felt certain of its passage in the full House. With rising real-estate values across the land, that delay would add tremendously to the final cost of the national park.

These welcome developments served to inspire activity of a different sort in the Ozark Society. From a defensive role we now operated from a base of clear advantage. Our job would be to smooth the way for the enactment of this protective law for the Buffalo River and to provide assistance wherever we could in its planning.

A Comedown for Harry

We went to great lengths to see to it that the National Park Service planners were aware of the several outstanding scenic areas along the Buffalo. One of these was the gorge of Indian Creek. Our plans for a trip through the canyon were relayed to Bernie Campbell, who was contact man for the park service planners. In the P.S. of a communiqué to him May 19, 1967, was revealed one of those unexpected crises with which we had to deal, even on the threshold of victory:

> P.S. Since dictating this letter disaster has befallen. Yesterday evening Harry fell off the roof of my house, landing on a concrete driveway. He suffered a fracture of the left wrist and the left foot, thus I have lost my partner for the outing and will most likely make my appearance alone. Harry is profoundly disappointed that he will be unable to come along, but his injuries are not serious and we shall have him back in action in two or three weeks.

The circumstances of this mishap were that Harry, in order to accumulate material for his proposed book, had been offered quarters in our home where I, by now, had a large file on the Buffalo River. One activity that Harry did engage in during this three months here was bird watching. Our big house in the woods had a gently sloping roof, an ideal place from whence to survey birddom in northwest Arkansas. Harry had managed to ascend the ladder to the roof safely, but in coming down he faced out instead of in, a no-no for ladder climbers. As a result, Harry became a patient of mine, his fracture properly casted and his foot bound up. His Blue Cross benefits in this case were most welcome, the only financial recompense that we were to have for his board and room that summer.

John Fleming Makes an Appearance

As soon as possible after hearing the news of the introduction of S.B. 704 and pending H.R. 7020, we issued a memorandum to our membership along with a press release urging all to commend our representatives in Congress and our new governor for their support of this legislation. They all had been subjected immediately to a barrage of harsh criticism by the BRIA, and our voices needed to be heard.

Alvis Owen, forest supervisor of the Ozark National Forest, had that winter invited the Ozark Society to Blanchard Spring Cavern for a tour of the great cave before developmental work on it had begun. On that occasion we were pleased to meet John Fleming, outdoor writer for the *Arkansas Gazette*. John was no youngster like Harry Pearson and couldn't go everywhere, but he was an accomplished and observant writer who would contribute much to the public understanding of conservation problems in Arkansas during the next several years. His reporting on progress at Blanchard Spring Cavern and observation on the Buffalo and Cossatot rivers are good examples. He was present at Petit Jean in April to report the Ozark Society spring meeting (*Arkansas Gazette*, April 9, 1967):

> **Battle for Buffalo River Not Won Yet,**
> **President of Ozark Society Says**
>
> PETIT JEAN STATE PARK—The battle for the preservation of the Buffalo River is not yet won "and a lot of hard work lies ahead," Dr. Neil Compton of Bentonville, president of the Ozark Society, said here Saturday. ...
>
> Dr. Compton urged the Arkansas Game and Fish Commission and the state Publicity and Parks Commission to take "a forthright stand" in favor of the bills now in Congress to make the Buffalo River a national scenic park under the supervision of the National Park Service.

He also asked the Buffalo River Improvement Association, a group that favors damming the river, to "bury the hatchet" and to help make the scenic park an economic and aesthetic asset to the area.

The Bentonville physician criticized the Publicity and Parks Commission for its decision to remain neutral in the now almost resolved argument over the damming of the Buffalo. He said that because of the location of the Buffalo River State Park neutrality by the Commission was impossible.

He also pointed out that the Ozark Society now would be engaged in a holding action until Congress has acted on bills by Senators J. William Fulbright and John L. McClellan and Representative John Paul Hammerschmidt of Harrison and the park becomes a reality. ...

Preliminary remarks in my address at that meeting illustrate the proliferation of various problems that confronted us then:

I need not comment on the contrast between this meeting and our last on which occasion the outlook for the preservation of any of Arkansas's prime scenic resources seemed grim. Today we may observe our first formal meeting in a climate favorable to the conservation of these scenic values, most notably, the Buffalo River which has received our almost undivided attention for these past five years.

I am sure that you all realize that this is no victory celebration, however, since our problems have perhaps multiplied in recent weeks and months, although they do not assume the overwhelming proportions that they did last September.

The situation is reminiscent of an incident involving a patient of mine several years ago which I sometimes like to relate in order to better illustrate the frustrations that sometimes befall us humans.

An elderly gentleman from just across the line in Missouri from us, Clint McCool, had been involved in a serious automobile accident and had suffered massive contusions. He was simply black and blue from head to foot, although he was not mortally injured, and was expected to recover in time. One morning while making rounds I greeted him with the usual inquiry as to how he had spent the night and whether or not he had rested well.

"Well Doc," he replied to this inquiry, "I rested mighty poorly last night. I had some of the worst dreams I think I ever had. You know I keep a few old cows up there and I got to dreamin' about the wild dogs that come around and chase them sometimes, and I was after them dogs all night long. I had my double barrel shotgun and plenty of shells and I wasn't havin' any trouble hittin' 'em either, but you know every time I would shoot one of them he would fly into pieces and every piece would make another dog."

This analogy, I am sorry to say, is no dream for the Ozark Society. We have shot some mighty big dogs in the last year, but like my restless wreck victim, we have plenty of pieces to worry about. ...

Dick Murray Joins Up

Nineteen sixty-seven was to be a busy year in all respects and especially in regard to outdoor activities. One of these had been arranged for Sunday, February 26, a hike to Lost Valley.

That day I was pleased to introduce another fellow who had called me on the phone inquiring about the trip. He had informed me that when I saw him I would recognize him. He had reminded me that he was one of the civil engineers before whom we had appeared at the Bureau of Engineers for Rivers and Harbors in Washington, D.C., in May 1965. Upon greeting him that Sunday morning, he was instantly remembered as one of those who had expressed interest in our presentation to the Bureau of Engineers. Dick Murray was one of those who had participated in the friendly discussion afterward. He had at that time expressed interest in the Ozark Society and had made mention of perhaps retiring here one day.

Dick Murray intended to become one of us and to enter into our activities from here on. It could truthfully be said that of all the victories that had come to us, few were the source of greater satisfaction than that of having with us then the Corps' own technical advisor on its Buffalo River project.

His special interest would be the establishment of hiking trails in the Ozarks, and in that effort he laid the base for their ongoing construction.

On one scouting trip relative to that, we were up on Penitentiary Mountain on a day in May. We had arrived in the midst of a grand display of nature, a natural wild azalea garden covering a hillside and filling a deep ravine. The gorgeous head-high shrubs in various shades of pink filled the air with their spicy fragrance. Hummingbirds and various bees darted about in swarms, savoring the sweet bounty offered them.

Then Dick said a few words that revealed just what that scene meant to both of us. He said, "Doc, do you think heaven will be any better than this?"

He was not a man given to maudlin sentimentality. His words were simple, direct and true. That sentiment stuck in my mind from then on, a basic statement as to the feeling that motivated many of us, now so concerned with the fate of this earthly scene.

It was inevitable that a day would arrive when I would be requested to deliver a eulogy for Dick Murray. On that occasion, that brief experience on Penitentiary Mountain was related, leaving his own words as a fitting memorial to a man who understood this marvelous world into which we are born and from which we must eventually depart.

The BRIA Reaction

Reaction to the introduction of S.B. 704 by the big dam people was as expected, verbally violent with veiled hints of possible physical involvement.

The *Marshall Mountain Wave* for February 2, 1967, carried double front-page headlines: FULBRIGHT INTRODUCES BILL FOR NATIONAL PARK ON BUFFALO and NO WATER, NO DAM, NO REPRESENTATION, BAD NEWS FOR AREA.

The text of the BRIA lament was reprinted in the one newspaper that could be depended upon to present the big dam case with sympathy—the *Harrison Daily Times.* In its editor John Newman's "Times Topics" column for February 3, 1967, delusions about the Arkansas Power and Light Company and the Arkansas Game and Fish Commission were presented as fact:

> The full extent of the anger of Marshall and Searcy county citizens over the deal they got on the Buffalo river dam at Gilbert is revealed in this week's edition of the *Marshall Mountain Wave.*
>
> Resentment over the bill introduced in Congress to kill the dam project and substitute a park is directed at Senator Fulbright, the author, Senator McClellan, co-sponsor, and at Congressman J. P. Hammerschmidt, the new Republican member. Searcy and Newton are Republican counties, and an estimated 95 percent of the population favored the dam.

The *Wave* report said in part:

> Reporters from the State News agencies were soon calling James R. Tudor and Gibson Walsh at Marshall for their reactions. Walsh laid it gently on the line for them:
>
> "We hold no grudges against Dr. Neil Compton, Charley Johnson and that ilk," he said. "Actually we don't feel that they had much to do with it. We have fought a good fight for a good cause, and 9,000 people in this area supported the dam—that's nearly all of them. We had to fight the private power money in Arkansas and we couldn't cope with that," he said.
>
> "Moreover, we had to fight our own tax dollars. For years, the Arkansas Game and Fish Commission and other state agencies paid people to run all over the state talking against the dam, making color slides and movies, and disseminating propaganda against any development for Searcy County. Our people could not raise the money to cope with that kind of brutal opposition."

To learn that they believed our Ozark Society color slide program and amateur 8mm movie production had been produced by the stand-offish Arkansas Game and Fish Commission was the unkindest cut of all in that tirade.

True to their convictions, they did not belabor me, Charley Johnston, or any other Ozark Society associates any further. That was reserved for the immediate perpetrators of this big dam disaster, Senator Fulbright and Congressman Hammerschmidt.

The senator was the recipient of a message from the secretary of the BRIA, February 12, 1967, bearing hints of "turmoil and sad moments" if his bill became law.

There was no comfort for the BRIA in the senator's reply February 24, 1967:

> This bill has been referred to the Senate Committee on Interior and Insular Affairs for consideration. Assumedly, in due course, the Committee will schedule this bill for hearings, at which time proponents and opponents of the proposal will be given an opportunity to make their views known.
>
> I am sure you know that Congressional approval and enactment of a proposal such as this is not taken lightly, and, at each stage of the procedure, views of all parties are sought and received. This means that the Department of the Interior, the Corps of Engineers, as well as private citizens will be heard. Only after all of these things are accomplished will the Senate and the House of Representatives vote on the bill.
>
> I am confident that should you and other interested citizens of the area wish to testify or submit statements, the Committee will receive them and make them a part of the record.

On March 13, 1967, the president of the BRIA dispatched a letter of his own in which its last paragraph specified conditions under which the construction of Gilbert dam, he hoped, could proceed:

> Senator Fulbright, the people of Searcy and Newton Counties would certainly appreciate your help. We can't see no possible benefit from a National river, when the river is dry in places, and too low to float on any portion of it, on an average of six months in every year. ...
>
> The Army Engineers will still recommend and construct a dam at the Proposed Gilbert site, according to the Little Rock office, if it is authorized by Congress and is acceptable to the Governor. I am sure that with just a little effort from you and Senator McClellan and John Paul Hammerschmidt this could be obtained.
>
> For the sake of all concerned, we should try.

The senator's reply was short and sweet:

> Thank you for your letter continuing our correspondence regarding the development of the Buffalo River. There is little more that I can say about the matter. I am committed to development of the Buffalo as a National River, and I regret that the proposal does not have your support.

If Plan D was ever to be revived, there would have to be:
—Two new senators from Arkansas in favor of it.
—A new congressman of similar convictions in the Third District.

—The rest of our delegation at least sympathetic.

—An eager-beaver, dam-building governor of Arkansas.

But the big dam fanatics simply could not believe that they had lost the contest. The above requirements were impossible, and they should have known it. They should have accepted it in as gentlemanly a manner as possible and at least given some consideration to the good points of a national park in their part of Arkansas. But that they would never do as long as most of them were alive.

We were going to be in for a long season of as nasty episodes as could be dreamed up by the angry, frustrated Buffalo dam pushers. The first of these events was a series of meetings in the county seat towns, Marshall in Searcy County and Jasper in Newton County, aimed to stir up enough local resentment to somehow embarrass or stop the national park development and to reinstate the Gilbert dam plan.

The *Marshall Mountain Wave*, March 2, 1967, recorded that in language reflecting the mood of those supposedly deprived of their birthright:

Property Owners Vow to Fight National River: Declare They Will Not Give up Their Land

A meeting of the Buffalo River Improvement Association held in the courthouse at Marshall, Tuesday evening attracted more than 100 interested people, the great majority of them property owners along the Buffalo in Searcy and Newton counties. Less than one-half dozen of those attending were businessmen of Marshall.

Property owners along the river throughout Newton and Searcy Counties emphatically stated their opposition to the proposed National River plan, and each of them voiced their approval of the U.S. Corps of Engineers plan for the proposed Gilbert Dam on the Buffalo river.

Landowners along the river in Newton county had held a meeting at Hasty last week and laid plans for opposition to the National River. ... They vowed to fight to the death the Department of Interior's plan to make the Buffalo a National River. They declared that the plan proposed by the Secretary of the Department of the Interior was doomed to failure, would not benefit the people of the area one cent and would be a detriment to the region throughout the years to come. ...

Many of the landowners along the river announced their intention of cutting the timber off their land in private conversations, and some of them have extended bulldozing operations on their lands along the river, it was reported.

In a further report in the *Marshall Mountain Wave*, March 16, 1967:

Newton County Peels off Kid Gloves—Organizes BRIA: Area Residents Hit Warpath; Buffalo Nat'l Park Forces Flinch

Derision, taunts, and sarcastic back-sass for government "representatives" was mixed with sentiment, honest tears, and sincere pleas for fair play in what was termed Newton County's first real protest meeting.

The event was Monday night in the Jasper courthouse, with more than 300 people out to fight the oppression brought by our Representatives in Congress to make the Buffalo a National River.

As a protest meeting, it was first class all the way, with grimly determined landowners serving public notice that if the congressional wheel grinds down upon them, it will be a fight to the finish. As the meeting started, J. E. Dunlap of the Harrison Boone County Headlight and Daily Times counted 280 people seated in the courtroom, with about 50 more still approaching from all sides of the square. ...

Bernard Campbell, Superintendent of the Hot Springs National Park, was the first speaker representing Fulbright and McClellan. He said he came to simply explain the provisions of the bills. He was immediately attacked by one individual who said he was a war veteran, and that he had been shot at before, and that if he was ever taken from his land it would be after a bloody battle.

As Campbell tried to soothe the land owners by telling them of options to buy their land, he was frequently interrupted by outraged citizens who challenged his statements. ...

Sweating, Campbell turned the session open for questions. An indignant landowner leaped to his feet to inquire how a national river had suddenly become a national park? ...

Maurice Tudor of Marshall asked Campbell how many National Parks there were in the United States? He answered, 31. Campbell did not know which was considered the most successful or least successful. He said there was no criteria for judging this, and he could not answer. He said he hoped Hot Springs National Park was considered the best. He admitted there were lakes in many National Parks, and man-made impoundments, and that these did not adversely affect the parks. ...

James Tudor of Marshall addressed the group as a messenger from the "Searcy County battalions of Buffalo river to tell you we have just begun to fight ... it's time to take off the kid gloves with these big money boys ... and we aren't anywhere nearly beaten. These pettifogging politicians have not overwhelmed us."

Newton County Speaks

George Viorel of Hasty set the tone of the meeting when he said that "We can't depend on the likes of Hammerschmidt to help out and Fulbright feels that our people are too damn dumb to know what it's all about." He got an ovation. ... He scored the Arkansas Power and Light Co. for their involvement, and also deprecated the University of Arkansas for their admittedly unprofessional study of the situation. The University of Arkansas allegedly owns a big block of stock in AP&L. ...

The evening's best tension-breaker was Hilary Norman [Jones, most likely] of Pruitt, a jovial man dressed in a red shirt, blue jeans, and who looked every inch the burly bulldozer operator he acclaimed himself to be. ...

"I've always been for the dam from an economic viewpoint, and because it would take less land," he said. "Not only that, but I always figured Searcy County would be giving up most of the land for a dam, and that is all right with me," he said with a twinkle in his eye that drew laughter. ...

Formation of the Newton county Buffalo River Improvement Ass'n followed with Norman elected president.

The Politicians Reply

Political reaction to the meeting was forthcoming. Next day Mr. Hammerschmidt called KYTV in Springfield (Channel 3) and told them that only a handful of people in the small area concerned were in opposition and that they didn't understand the benefits of the park. ...

Previously, Hammerschmidt had talked about this with James R. Tudor. Tudor told him people were getting worked up down here, and that if this park bill was forced on them somebody might get hurt. "That's just too bad," Hammerschmidt sarcastically replied. ...

From the *Marshall Mountain Wave*, March 23, 1967:

Governor, Congressman Scorn People in Stubborn Intent to Beat Gilbert Dam

In the talk before the Marshall Rotary Club here recently, Park Superintendent Bernard Campbell of Hot Springs said that a Survey team was expected to be in the area early in May to survey and set out the boundaries of the National Park. Property owners along the river who do not care to have their land surveyed have a perfect right to keep them off their property.

In the meantime Governor Rockefeller and Cong. Hammerschmidt continue to work as hard as possible to convert the Buffalo River into a National Park. Gov. Rockefeller flew to Washington last weekend and paid a cordial visit to Secretary of the Interior Stewart Udall and told him he was completely in accord with his idea of a National Park on the Buffalo River.

The unwillingness of the Governor or Hammerschmidt either to confer or discuss the fate of the Buffalo river with the citizens of the area pointed to only one conclusion: The Governor and the Congressman have aligned themselves with "outsiders" remote from the area which will be affected by the park, and they don't give a d—- for the people of Searcy and Newton Counties.

Prior to the introduction of S.B. 704 the people of Newton County had not been overly exercised about the big dams down river in Searcy County. Most of that undertaking was outside of their immediate area. They had not expected the national park proposal to be a likely thing, but now that it was on their doorstep they were ripe for word as to what they might have lost with the demise of the Gilbert reservoir, which would have extended all the way up to Pruitt in Newton County. That there should ever be a national park in Newton County was unthinkable and potentially unacceptable to those strong-minded, independent people.

Missionaries of the big dam faith were now in their midst bearing the word.

For some reason, the Rev. R. L. Winners had departed Marshall and taken up residence in Jasper where he sought to make converts. Jim Tudor had relinquished his job as editor of the *Marshall Mountain Wave,* to a family member, with the reported intention of becoming a pro-dam circuit rider through all the counties involved.

The meeting at Jasper was the result of the salesmanship of these original BRIA officials. That they had been able to stir up monumental animosity in Newton County was not to be denied. For a time their chapter in Jasper remained in operation, but for lack of effective leadership it languished and eventually metamorphosed into something called the Buffalo River Conservation and Recreation Council (BRC&RC).

Environmental Sinners Revealed

The reaction by the Corps to Senator Fulbright's Buffalo River bill took the form of an epistle delivered from the district headquarters in Little Rock by Colonel Frank P. Bane, the new district engineer. In it we preservationists were smitten by words that we heretofore had not expected to be used against us. Upon its delivery we were branded as sinners for sure, condemned by parables right out of the Holy Bible.

The colonel's attempt to align Plan D and the ongoing White River Basin study with the scriptures failed to draw much water in our Bible-belt state. Most of the media gave it scant attention, but the *Marshall Mountain Wave* was delighted to print the lengthy item in full. Here are some excerpts from the February 23, 1967, issue:

> **Engineer Says Conservation Biblical;**
> **Defines "Golden Rule" of Development**
>
> Perhaps no field of engineering endeavor better illustrates this year's Engineer Week theme than conservation of natural resources, for it is essentially "Engineering for the Human Environment."
>
> We seek to conserve our nation's natural resources so that they will continue to serve the whole range of our society's needs.
>
> For centuries there have been two contrasting approaches toward natural resources. One might be called the "preservation concept." This concept considers resources as precious rarities, to be preserved in a natural state and used only sparingly. The other approach is known as the "use concept."
>
> *Wisdom of Past*
>
> Several principles can be drawn from our wisdom of the past to serve as guides in today's conservation effort.

One is found in the story of Joseph in Egypt. He saved food during the seven fat years so that the people could survive the seven lean years that followed. This simple principle—saving in time of plenty for use in time of need—is the conservation principle underlying the big dams with which people the world over seek to regulate the seasonal flow of waters in rivers. ...

Another conservation principle is found in the "Parable of the Talents."

The man who buried his money in the ground was scolded. The ones who used their master's wealth to produce more wealth and increase the abundance of the community were praised. From this parable we learn that we must put our resources to use—not simply preserve them.

A third principle of conservation is the "Golden Rule."

"Do unto others as you would have others do unto you." We should not devote the resources of our own communities to our own comfort, and then demand that the people of some other community forego the benefits of modern civilization so that we may, when we wish, visit an unspoiled wilderness.

Lastly, there is the "Parable of the Prodigal Son" who wasted his resources and came up wanting in time of need. This parable teaches the lessons that our resources should not be overly exploited or expended prior to the time they are actually needed. ...

Ever since Congress initiated the general flood-control program some 30 years ago, the Corps of Engineers has enjoyed a key role in applying the "use concept conservation" to comprehensive planning for the development of the nation's water resources. ...

Management of our national resources continues to be a major challenge not only to the engineer, but to every civic minded citizen. These resources must not be squandered and misused as did the "prodigal son;" nor should they remain unproductive as in the case of "parable of talents."

By this inference of the unholy state of the Ozark Society, I was so outraged as to submit rebuttal, which received no more attention from the media than did Colonel Bane. Perhaps they were tired of the squabble, but for future generations, a couple of sentences from my lament:

As justification for the Corps "use concept" the Colonel quotes a series of parables from the Bible after which strange theological exercise the Corps of Army Engineers emerge as the true bearers of the word and intent of the Creator. Those of us who have opposed the policies of the Corps heretofore become, in this revelation, as Philistines smitten by the jawbone of an ass. After having been so chastened we should now repent and fall down before General Cassidy, asking forgiveness for our sins!

As for Jim Trimble, he is reported to have said:

I'm not mad at anybody and I plan to see the folks and my friends in the district within a few weeks—those who voted for me especially and I'll have a smile for those who didn't.

Groundbreaking for Ozark Dam, McClellan Kerr Seaway, on the Arkansas River, October 16, 1964. *Left to right:* Congressman Wilbur Mills, Col. Charles Maynard, Senator John L. McClellan, General Free, Congressman Jim Trimble, and Congressman Oren Harris. *(Courtesy of the University of Arkansas Library, Special Collections. Trimble Family Papers)*

Allan Gilbert, Jr., a reporter for the *Northwest Arkansas Times,* was a nephew of Senator Fulbright and naturally favored his uncle's plans for the Buffalo River. Later in the season Allan Gilbert would provide us with an important insight into the mind of our ex-congressman in his report in the above paper for May 31, 1967:

Overlooking an Oversight

The view from atop a towering bluff hard by the Arkansas River, just a mile or so downstream from Ozark, is as breathtaking as any from Catoosa to the Mississippi. There, high above the Ozark Dam and Lock No. 12 on the Arkansas River Navigational Project, former Congressman Jim Trimble officially dedicated an overlook for the facility named in his honor, just a few days ago.

Judge Trimble, offering a refreshing contrast (as always) to the hard-sell flamboyance of the Southern politician, predicted an age soon when there would be "cities in the sky, and cities under the sea." And the things that will happen in Arkansas "as a result of the Arkansas River development project will be just as wonderful," he further predicted.

Peering on into the future he predicted that man would continue to learn how to conquer and control his environment and that projects such as the great flood control dams on the White River were but a symbol of what is yet to come.

He also said: "I don't know when, and I didn't know how, but some day there will be a dam on the Buffalo River, too, as a part of that great complex of water and flood control facilities."

Those little slips of the tongue, which turn mundane occasions into legend and history, caught up with Berryville's Jim Trimble at the dedication of the Ozark Overlook Shelter two miles east of Ozark just off U.S. Hwy. 64, last Friday.

Last of a considerable list of speakers helping give "official sanction" to the job, Judge Trimble rose to his feet, said how proud he was to be on hand with so many good friends, and how glad he was "… to dedicate this Oversight …"

The Ozark "Oversight" was constructed at a cost of $138,400.

The former congressman wasn't guilty of a Freudian slip, exactly, either, because no one in Arkansas believes more sincerely that the Ozark Lock and Dam, its Overlook, and the entire Arkansas River project is anything but an oversight.

After Gilbert's report, whatever sympathy we might have felt for Trimble was tempered by our satisfaction that he stood only to receive an honorarium that day at the Ozark Overlook and not on the floor of the House to cast his vote for Gilbert dam.

The State Park Imbroglio

After giving due consideration to this very significant example of how a good man can become an advocate of pie in the sky, we were set upon by barking dogs from another direction. If the big dam faction could not kill the national river in Congress in one swoop, they would tear it to bits a little at a time.

That attack would be made upon the Buffalo River State Park, one of the most popular of the Arkansas Publicity and Parks Commission's units.

The "Other Days" column in the *Arkansas Gazette*, June 22, 1988, reran a news item describing the establishment of the Buffalo River State Park in 1936:

> YELLVILLE—First inspection of the new 5,000-acre Buffalo River State park and formal acceptance of the area by the state as sponsor to cooperate with the National Park Service in development work featured an all-day inspection trip in the southern part of Marion county today by members of the state Park Commission, Governor Bailey and representative of the National Park Service.
>
> The party, escorted by a committee of Yellville citizens and Marion county officials, visited several of the scenic points and natural wonders which influenced National Park officials in selecting and improving the site several months ago. All development work in the park will be done by the federal government with CCC labor under supervision of a technical staff provided by the National Park Service.
>
> The entire project, which will require three years to complete, will conform to

standards of engineering, architecture and technical development used by the National Park Service throughout the country in national and state parks.

The involvement of the National Park Service in the undertaking is now of satisfying significance to those of us who had brought about what was happening on the river in 1967. Considering its origin and design, it seems only proper that the Buffalo River State Park should now be part of the National Park Service system.

But not so to certain people in Marshall in 1967. We had heard then vague rumors that the Tudors coveted the job of superintendent of the Buffalo River State Park. Delos Dodd, the efficient and durable superintendent in 1967, was the target of those people, who sought by various subterfuge to get him fired. With one of their number in charge there they would be able to mete out unending difficulty to the national river plan. The first step would be to infiltrate the Publicity and Parks Commission, and that they were able to do. Russell Horne of Russellville, already on the state Parks Commission, was a strong advocate of the BRIA objectives, and to cap it off Jim's son Eddie Tudor had been able to gain an appointment to the commission in 1965. With Horne as spokesman they were able to sway the rest of the commission and Bob Evans, the director, as well.

It had long been the policy of the National Park Service to request donations of state land involved in any state where a new park was to be established. Until now, those states where such parks were contemplated were more than glad to render such donation. To have a national park within their boundaries was a mark of distinction; note developments at the same time on the Current and Jacks Fork rivers in Missouri. Stipulations for such donations were included in Senator Fulbright's S.B. 704, and here was opportunity for Tudor and Horne to introduce a stumbling block into the national park plan. Our complacency following the introduction of the Fulbright bill was shortly rattled by the doings of those two.

An AP release published in the *Northwest Arkansas Times* for March 15, 1967, gave the details:

Rockefeller Irritated by Park Action

LITTLE ROCK—Gov. Winthrop Rockefeller said Tuesday that the state Publicity and Parks Commission has adopted a resolution regarding the Buffalo River issue that "is diametrically opposed to the position I have taken."

Rockefeller expressed surprise when it was learned through an examination of minutes of a Feb. 10 commission meeting that it adopted a resolution saying it would not give the Buffalo River State Park to the National Park Service for a national river park.

Rockefeller, saying he had had no discussions with commission members about the resolution, said, "I consider the national river project a great triumph for our recreational development."

The commission went on record two years ago to not take sides on the controversial issue of whether or not to dam the river.

Minutes of the meeting show that C. R. Horne of Russellville introduced the resolution which was seconded by C. E. Tudor of Mountain View. It was adopted unanimously. …

At Mountain View, C. E. Tudor said the commission saw no reason to give up control of the park since persons in the area were satisfied with state control.

"They don't see any reason for the state to donate this multi-million dollar facility to the national government," he said. "This is what the resolution is about."

Tudor also said the people of the area had been trying to talk with Rockefeller to give him their views but that so far they have "been limited to negotiating with an aide of the governor."

Tudor also said he did not believe the commissioners had to talk policy with Rockefeller.

Commentary in the media statewide was generally critical of the Publicity and Parks Commission with the *Arkansas Gazette* taking the lead with a long editorial making these principal points (March 20, 1967):

Buffalo Park Lands

Despite its stated policy against taking sides in the Buffalo River controversy the state Publicity and Parks Commission has permitted itself to be dragged into the fray by those who support a dam and oppose a national park on the stream. …

In adopting the resolution the commission certainly earns no credit. Indeed, the resolution, despite the chairman's disclaimer, violates the commission's own long-standing nonpartisan policy in the river controversy, and, worse, places a state agency in the position of hindering official state policy.

The Ozark Society immediately sent out letters of protest to L. C. Dial, chairman of the commission, and to its other members as well: R. L. McKinley, Horace Fisher, Ovid Switzer, Loyd Fisk, and Eddie Tudor. These letters were respectful but firm, pleading for cooperation from the Parks Commission and pointing out the attitude of the Missouri Department of Parks in a similar situation in that state.

From none of these did we receive a reply. They were likewise invited to be present at our spring meeting, April 8 and 9, at Petit Jean State Park for mutual discussion of the question, but none were in attendance.

Bob Evans, then director of the Arkansas Publicity and Parks Commission, I would never meet, but he was advised of the attitude of the Ozark Society via the mail. He also chose not to reply. Their decision to withhold the Buffalo

River State Park on February 10 was made in secret and became public via investigative reporters. We also had heard that some of the commissioners claimed that the Ozark Society favored their withholding policy.

I deemed it wise to send a second and stronger letter to Bob Evans April 19, 1967, saying in part:

> This is to advise you that the Ozark Society does not intend in any way to support a permanent or perpetual withholding of the Buffalo River State Park from the proposed Buffalo National River. Any statements that may be made by you or any of your commission attempting to show that we are of a different opinion will result in a much more serious disruption than has occurred heretofore. I sincerely hope that such will not be the case.

To this, again, no reply, and Charley Johnston was requested to make a personal appearance and give us a report.

Again Bernie Campbell was on the hot seat as an official National Park Service representative on April 20, the next meeting of the commission.

It was a court of inquiry as to how the national river was to be managed and what the requirements were in regard to the Buffalo River State Park. Superintendent Campbell told them that the National Park Service would require the donation of these lands, as had already been stated in the news reports. Charley Johnston wasn't there, but he did state in a letter to us in Fayetteville that Bernie's flat-out rendition of that fact had the effect of making the commissioners irate and that now they would certainly hold the Buffalo River State Park out of the Buffalo National River plan if they could.

The immediate fallout of this uncooperative stand of the Parks and Publicity Commission was to create a first-class ruction within the sanctum of the Ozark Society itself.

That volatile, explosive soul, Charley Johnston, had gone as an emissary to the office of Bob Evans, the Publicity and Parks Director, to get the whole truth about the state park. Evans produced a prepared resolution to be introduced at the forthcoming April 20 meeting of the commission, praising the Buffalo National River idea but declaring the neutrality of Publicity and Parks in that matter and refusing donation of the state park. That resolution was obviously a ploy to dampen further criticism from us true-blue, hot-under-the-collar conservationists. Charley thought that we should accept the resolution under the circumstances. When he received a copy of my no-deal letter to Bob Evans, he went into a Johnstonian eruption:

April 21, 1967

Dear Neil, Evangeline, etc.

We blowed it. After talking with you and Bernie the other afternoon, I went to Bob

Evans' office to ask him what's happening. ... Bob felt that he had not been consulted by us, that we should have at least have asked him what was going to happen and why things had happened the way they did, he felt he was personally in a corner between us and the Commissioners.

After receiving your second letter, and hearing Bernie yesterday, the commission became adamant, the resolution was not even introduced, and we're DAMNED lucky that they came out in a neutral position.

This was a stupid, and needless blunder, and I personally am contacting Fulbright's office in an effort to shut Bernie up. But it was OUR BLUNDER AS MUCH AS BERNIE'S.

... But Bernie was making purported policy statements all day yesterday, particularly as relates to the State Park about which he could quite easily have been vague, evasive, or plead ignorance. I had told the haughty bastard that the resolution was prepared and to try to evade the State Park issue, and it could be handled later in private negotiations. But NO he had to rub salt in the wounds, and play right into Eddy Tudor's hands.

DAMMIT—YOUR SECOND LETTER TO BOB EVANS SOUNDS LIKE OUR POWER IS INFINITE AND DIVINE. If we came out of this thing looking as conservationists the way Jim Tudor has always been picturing us, then this little episode will be a perfect illustration of the WHY's of it.

We shot before we even had a target—much worse than half-cocked.

But we came out of all that on speaking terms and managed to live it all down as time went on. The situation gradually declined into a tooth-grinding standoff, finally coming to an end only when the terms of Tudor and Horne on the commission expired.

Four years later we were to have the last laugh in this state park set-to, as was reported in the *Arkansas Gazette*, August 28, 1971:

U.S. to Receive Buffalo River Park

FORREST CITY—The State Parks and Tourism Commission voted Friday to give the Buffalo River State Park to the federal government if Congress designates the Buffalo as a national river.

The action, ending years of indecision on the matter by the Commission, eliminates one barrier to passage of the Buffalo National River Bill later this year. ...

William E. Henderson, director of the Department of Parks and Tourism, recommended that the Commission offer the 2,000-acre park, which has been one of the state's most popular park facilities, to the federal government if the Buffalo becomes a national river.

Also part of the package will be the Lost Valley State Park, a 280-acre undeveloped park in Newton County that was acquired by the Department in 1966. ...

Earlier this summer, the state Game and Fish Commission voted to give land that it owns along the Buffalo to the Park Service if the National River bill becomes law.

The vote to offer the parks was 7 to 0. Edward Tudor of Marshall, who has persistently opposed the national river concept, argued against it but abstained on the vote, along with three other commissioners.

Tudor's opposition to the concept of maintaining the Buffalo in its wild state also seemed to have prevented his election as chairman of the Commission.

Jimmie Driftwood (James Morris) of Timbo was reelected to another term as chairman. Conservationists and wildlife groups had opposed Tudor for the position. Driftwood has supported the national river project.

Rockefeller Turns Tail

Along with the Buffalo River State Park controversy, the BRIA and VIPs in the district Corps headquarters were able to demonstrate their ability to manipulate a vacillating and indecisive, newly elected official. They were able to treat us to an eye-popping headline in the *Marshall Mountain Wave*, August 31, 1967:

WR Recants on Buffalo; Says Gilbert Dam Reasonable

New life for a prospective dam on the Buffalo at Gilbert was breathed into state politics this week by a pronouncement from Governor Winthrop Rockefeller, that he now favored construction of the Gilbert Dam.

Some area residents were electrified by the first indication that the Governor had recanted from his original opposition which came during a public appearance at Harrison Tuesday night. In response to questions from his audience, Rockefeller told the group he was beginning to see that the Buffalo dam at Gilbert was properly conceived and planned. He said it would be a good thing for this area.

"I have had a long and informative talk with the Corps of Engineers in recent days, and they have explained many things to me of which I was not aware." Rockefeller explained that heretofore he had not received the complete picture and story on the Buffalo. He told the group that the compromises offered by Searcy-Newton county interests seemed reasonable, and he felt that float enthusiasts and all types of river sportsmen could be accommodated with plans to dam the Buffalo.

On Wednesday, the Governor flew back into Harrison to dedicate the Twin Lakes Vocational School. There was a press conference at the Holiday Inn Motel, and again newsmen queried him about the Buffalo.

"Yes, I am for the Dam on the Buffalo River at Gilbert," he said forthrightly. A member of the Harrison Daily Times asked him if this was not backing down from his previous commitments. "Yes, I think you could describe it as backing down—I have changed my views to a marked degree," he said. He then enlarged upon the matter, saying that he had opposed the dam in his 1964 campaign and that his opposition had been based largely on the advice of his brother, Lawrence Rockefeller, who has a great interest in the preservation of the natural beauty of this nation. Also, he said he listened to other sincere conservationists around the state, and he indicated his

WR'S STAND ON
BUFFALO RIVER CONTROVERSY

(Courtesy of George Fisher)

personal feelings were that some other type development might be made on the Buffalo that would prove to be as advantageous or more so than the dam.

Hundreds of people came to see the Governor on Wednesday in his bus, and he gave audience to as many as possible. The governor offered his regrets to Searcy County visitors, saying that he had intended to visit in Marshall and Searcy County, but had not been able to do so the previous day.

George Fisher, cartoonist for the *North Little Rock Times*, came up with the most devastating depiction of Rockefeller's uncertain and ridiculous position with one of his best pen-and-ink bombshells.

Newspapers around the state spread the news of the governor's so-called decision. Within a few days his office was flooded with mail and telegrams berating his confused maneuvering. He had to do something. At first he gave out with garbled mumblings, which only added up to that chicken-out word, "compromise." *Pine Bluff Commercial,* September 1, 1967:

Transcript of Governor's View on the Buffalo Dam Released

LITTLE ROCK—The governor's office today issued a transcript of the remarks Governor Rockefeller made at Harrison about damming the Buffalo River.

Rockefeller was speaking at Harrison High School when he was asked the following question and made the following reply:

Question—"... Governor, how do you feel about the Buffalo River project?"

Answer—"I think first and foremost I would say to you that I am one of I'm sure many of you that would like to preserve the Buffalo River and all that it represents. (Applause.)

"It sounds a little bit like we are in the minority, but I am going to assume that some of you did not clap because you were just embarrassed; but I have really gone a long way out of my way to study this particular problem with the Corps of Engineers, and I shall study it some more.

"They have demonstrated to me that they feel that a dam should be built at Gilbert and so maintain the 'wild river,' if you will, below the dam; and in two hours, I cannot say that they've sold me, neither can I say that I have not had my eyes on it. So I have to say to you this, that I want to study it some more.

"If we can combine the things that the Engineers have told me, then maybe we can put the dam at Gilbert and still preserve that, but (applause) you can see it is obviously a controversial subject." ...

A leading spokesman for the anti-dam forces is Dr. Neil Compton of Bentonville, president of the Ozark Society. Here is the text of a telegram he sent yesterday to the governor:

"The Gilbert dam is not, repeat not, compatible with preservation of the Buffalo River in its natural state nor with the National Park proposal. Those of us who have worked so hard to save it (the river) respectfully request equal opportunity to explain why no compromise with the Corps of Engineers is possible."

Shortly thereafter, Rockefeller came up with a real revelation as to how to solve the messy situation. He would have a firsthand look at the Buffalo (*Pine Bluff Commercial,* September 20, 1967):

Governor Plans to View Buffalo

By Tom Parsons

MONTICELLO—Governor Rockefeller said at a news conference here today that he was planning a trip to view the entire length of the Buffalo River "by air or otherwise."

Rockefeller said he intended to make the trip "just as soon as I can work it in." The trip, he said, would be in the company of representatives of both sides of the current controversy over a dam on the river.

He indicated that he would prefer these representatives to be Bernard Campbell of Hot Springs, director of the National Park Service in Arkansas, and Colonel Charles L. Steel Jr., the chief of the Army Engineers Little Rock district.

In updating Harold and Margaret Hedges, in Lake Quivira, Kansas, as to what was going on, I had this to say September 22, 1967:

If you are keeping up with the Rockefeller situation you will have noted that he is hopping about quite a bit lately. ... The latest news is that he has decided to take the National Park Service in his confidence and also to see the river at the same time. To do this he is going to fly over the river in his jet plane and has requested that two men accompany him. These will be the new colonel at Little Rock, Colonel Steel, and poor old Bernie Campbell from the Park Service at Hot Springs. This it seems to me will bring about a decision from on high for sure. It will be as if the Father, the Son and the Holy Ghost will give the word from UP There. In this case it looks like Bernie may be the Holy Ghost.

The Ozark Society had sent out an emergency announcement to all members to contact the governor, by whatever means, protesting any compromise that he might have in mind. Some examples of citizens comments are these:

September 7, 1967

Dear Governor Rockefeller:

It was very alarming to the Bull Shoals Lake and White River Association (378 members) to hear your comments on the Buffalo River. We can not understand why this change "of face," when for the past years you have supported the National Park.

I am sure you are aware of the fact that there is no compromise on this issue. The "Park Service" will not support the Buffalo River National Park, with a Dam in the heart of the Buffalo River.

The facts should speak for the National Park vs. Corps dam project. If Arkansas loses the Buffalo River National Park, it will mean the loss of a much needed asset to our travel industry. I hope you will sincerely look into this issue a "little closer."

Thank you for your time. If we may ever be of any service, please do not hesitate to write or call.

Sincerely,
Jim Gaston, President
Bull Shoals Lake and
White River Association

September 4, 1967

Dear Governor Rockefeller:

Your recent statements regarding the future of the Buffalo River have caused a great deal of unfavorable comment in the Fayetteville area. A number of people, many of whom strongly supported your candidacy, have made it a point to inform me of their displeasure. Quite frankly, they feel that you have retreated from an earlier position and they regard this as something of a betrayal.

I have no doubt but that this development has seriously curtailed your strength in this area which supported you so heavily in the last election. It is my belief that the only way to offset this reaction would be for you to make a very strong and unequivocable statement that you favor retention of the Buffalo River in its natural state with *no dams at any point.*

The seriousness of this problem recommends it to your immediate attention.

Sincerely,
Lester C. Howick
County Committeeman
W-3, Fayetteville

After numerous letters from people like Lester Howick and Jim Gaston, Governor Rockefeller was strongly "impelled" in the other direction, from which course he did not deviate thereafter. The Buffalo River dammer uppers on the Publicity and Parks Commission had to eat crow as was briefly acknowledged in the *Marshall Mountain Wave,* September 7, 1967:

Rockefeller Dam Remarks Made Clear

Following publication last week of some of Gov. Rockefeller's remarks at Harrison and our interpretation of them, the Governor stipulated that he had not taken a firm position supporting the dam. The following is his explanation:

"I have in the past taken a definite stand opposing a dam on the Buffalo River. Those favoring the Gilbert Dam, however, urged me to review the situation with the Corps of Engineers, and this I have done.

"I commented at Harrison recently that their presentation was impelling, but in no way, either by word or implication, did I intend to indicate any change whatever in my often-stated position."

All of this should serve as a basic lesson to students of our system of governance. Constant attention to the opinions and performance of our elected officials is the bottom line if our belief as to what is best is to prevail. Had we gone about our daily tasks and neglected to write, WR could well have gone for that "compelling" argument to have the dam and river too.

LOST VALLEY STATE PARK

The Buffalo River State Park was not the only tract that the Arkansas Publicity and Parks Commission would be expected to donate to the National Park Service.

In a series of complicated deals going on simultaneously, the state park system would wind up in possession of Lost Valley where the Buffalo River struggle began.

Of Governor Faubus's personal interest in Lost Valley there can be no doubt. At his insistence, the State Parks Commission had been trying for two years to buy Mrs. Primrose's property, through which lay the only access to the scenic core of Lost Valley. Finally, just before Faubus left office, we were pleased to read an announcement reporting the accomplishment of that effort (*Arkansas Gazette*, December 7, 1966).

"Lost Valley" and a Mammoth Spring
Are Added to the State Park System

About 200 acres of rugged scenic land in a "lost valley" in northwest Newton County and a spring that feeds about 200 million gallons of cold water a day into a lake at Mammoth Spring in Fulton County have been added to the state park system.

Governor Faubus received the deed Monday to the "lost valley" property from Dr. Ed Barron Jr. of Little Rock, who purchased the property in June for $23,784 from Mrs. [Ursa] Primrose. He turned it over to the state Publicity and Parks Commission for what it cost him.

Dr. Barron presented the deed in a ceremony arranged by Bob Evans, Publicity and Parks director, William Stearns, associate director of Parks, and Lou Oberste Jr., associate director of publicity. ...

"I'm glad that it's over with before I leave office," the governor said. "I couldn't be happier about this."

Dr. Edwin M. Barron, a Little Rock physician, had been able to persuade Mrs. Primrose to sell her property for an inflated price. He had been inspired to make the deal by Governor Faubus; but in spite of the successful outcome, there would be tragedy in it for him. Shortly thereafter, his son was to fall from one of the tall cliffs in Lost Valley, suffering fatal injuries.

After these events, the Lost Valley problem came to a rapid but devious end in the summer of 1967. As we already knew, the scenic center of Lost Valley was owned by the Pendergrass family of Little Rock. They were friends of the governor and could be expected to cooperate, but they did not wish to donate their eighty acres. The State Park Commission, with its big dam increment, was not eager to pay the purchase price, and for the time Lost Valley State Park was left dangling. Thereupon the now ex-governor Faubus chose to take personal action. To the surprise of all, in an AP release, we were greeted with front-page news across the state (*Northwest Arkansas Times,* July 11, 1967):

Lost Valley in Newton County
Park's Main Features Said Owned by Faubus

LITTLE ROCK—Former Gov. Orval Faubus owns 80 acres of land in the Lost Valley State Park in northwest Newton County, the Arkansas Gazette reported today.

And unless Gov. Winthrop Rockefeller releases funds with which the state Publicity and Parks Commission can buy the property, Faubus intends to seal it off from the public, the newspaper said.

Reporter Ernest Valachovic said Faubus had mailed letters June 26 in which he explained the situation to the commissioners whom he had appointed.

To keep the property out of "commercial hands," Faubus was said to have written, he purchased the acreage containing the main attractions of the park— "the arch passway through a stone bridge from which gushes the main spring, and then a little further up the valley the towering cliffs, the waterfall and the cave from which many Indian artifacts have been taken."

Without these attractions, Faubus was quoted as saying, the park would be "practically valueless."

"Now, if the commission and the administration do not wish to purchase the land and create the park, please let me know," Valachovic said the letter continued.

"Then I am going up to Lost Valley and stretch a fence across the valley, shutting off the tourists from the main attractions of the area and put up a suitable sign to explain why it has not been made a state park, and who is responsible."

Faubus said he paid $80 an acre for the land, or a total of $6,400, Valachovic reported, and is willing to sell it to the commission for the same amount, plus $100 for "expenses."

The commission paid $23,784 for the 200-acre state park last December, when Faubus still was in office. No mention was made then of the fact that the most scenic attractions in the area were not included. ...

Faubus' letter said that two commission employees—Director Bob Evans and Associate Director of Parks Bryan Stearns—knew then that the attractions were not included in the purchase, Valachovic reported.

"As I told Mr. Evans and Mr. Stearns at that time, the Primrose tract of about 200 acres, purchased by Dr. Barron and in turn sold to the state, did not contain the salient features of the Lost Valley area," he quoted the letter as saying.

"I then arranged, because of my acquaintance with the owners, for the purchase of an additional 80 acres of land on which are found the main attractions of the Lost Valley area."

The commission "procrastinated" and didn't complete the purchase before his term expired, Faubus was said to have written, and "the new governor froze funds and so blocked the additional purchase."

Faubus said that after he returned to his home in Huntsville when he left office, the owners of the land came to him and told him the transaction had not been completed, Valachovic reported.

"Then, in order to prevent the land from falling into commercial hands, I purchased the tract from them at the original stipulated price, in order to have this property to create a state park ... ," he said the letter continued.

Valachovic said the commission met last Friday and voted to advise Faubus that it lacked the money to buy more land and that the federal government probably would acquire it as part of the Buffalo River National Park.

Valachovic said the letter from Faubus was mentioned at the meeting, but that no copies were released. He said Commission Chairman L. C. Dial of Brinkley promised to make it public at the commission's August meeting.

Faubus's suggestion that he might fence off his portion of the valley and put up a sign blaming the Parks Commission for the stalled state park was, of course, a measure to urge the commission to finish the work. It was not his intention to make a personal donation in the amount involved.

As for the new governor, Mr. Rockefeller said neither Faubus nor Publicity and Parks officials had written him about the matter or discussed it with him. He was displeased that the commission had not approached him.

The governor said it would be "not only pertinent but polite" for the commission to notify him about the problem. He suggested that if the commission did, a solution to the financing might be found.

The involvement of Governor Rockefeller in the affair brought the pot to a boil. All parties involved were now exchanging barbs with each other, and it appeared that Lost Valley was about to be lost for sure.

At that point the whole squabble was resolved by an unexpected benefactor (*Arkansas Gazette*, July 13, 1967):

Because State Was "Good to Me," Heyden to Buy Faubus Land for Park

By Bill Lewis

Mr. and Mrs. Albert J. Heyden, of 1409 Louisiana Street, offered Thursday to buy from former Governor Orval E. Faubus the 80 acres of Lost Valley in Newton County that contain the most scenic features of a proposed state park and then give it to the state.

Faubus was reached by telephone at his office at the Madison County Record at Huntsville. He accepted the offer, telephoned Heyden and closed the deal. He said his son, Farrell Faubus, a lawyer, would draw up the necessary papers.

Heyden, 78, a retired real estate broker, agreed to pay Faubus' asking price of $6,400 plus the $100 for the expenses Faubus said he had incurred in buying the property.

"The state has been good to me, I have the means so I feel like I want to do something for the state," Heyden said.

He added that Faubus said he intended to have a bronze plaque with an inscription of the gift placed at an entrance to the park. ...

Heyden, recently released from three weeks in the hospital, had read about the Faubus offer and said he had tried to call Faubus at Huntsville Tuesday evening, without success.

"I would be willing to go ahead and buy the property and donate it to the state," he said. "I think the state can use it to an advantage. It will add to the park—in fact, it should have been included in the first purchase. It seems a shame that that particular property was not included."

Bryan Stearns, associate director of state parks, checked with Commission Chairman L. C. Dial of Brinkley, who said the Commission would be happy to accept the Heydens' gift.

Stearns said he had looked into the Commission's records and that as far as he could determine, this was the first substantial gift of land—with no strings attached—ever offered to the Commission.

Mr. and Mrs. Albert J. Heyden were the kind of people that we in the conservation field had searched for diligently and never found. We knew that there were those who had done great things along that line elsewhere. The first important national park in the east, Acadia on the coast of Maine, had been established by grants from prosperous eastern businessmen. The Great Smoky Mountains National Park and the Grand Tetons owed their existence to generous donations from the Rockefeller family. With one of them as governor, we should have done better here in Arkansas, but the Albert J. Heydens of Little Rock were, and remain to this day, the only ones to make such an outright gift toward the establishment of park land along the Buffalo. They and Dr. Barron should be memorialized for that.

Lost Valley was accepted by the Publicity and Parks Commission without further controversy and was efficiently managed by them for a few years. It was eventually turned over to the National Park Service without a murmur, which is as it should have been.

Ozark Society
Spring 1967 Bulletin

FOURTEEN

THE BOOK
AND THE BULLETIN

Perhaps the most important factor in any effort to correct environmental problems is the matter of information, its processing and distribution. People must know what we are talking about, and our coworkers must be kept in contact. Personal contact and the telephone are important, but the printed word, if well delivered, takes precedence. In the 1960s we did have radio and television, but we were not familiar with and didn't have access to those still developing mercurial methods of spreading the word.

As has been pointed out, the Ozark Society sorely needed a good book about the Buffalo and a regularly appearing official organ in which to deliver the news and state our case.

The matter of the book by early 1967 was well along toward realization. Ken Smith had been given several duty stations by the National Park Service and now was in Ann Arbor, Michigan, working on a master's degree in natural resources administration. All the while he had applied every spare moment to writing *The Buffalo River Country* and now was nearly ready to submit it for printing.

Ken, as were most of the rest of us, was desirous that it be designated an Ozark Society book in spite of the fact that he had financed it all himself. The somewhat complicated papers of agreement had been submitted by Ken for our signature, but as usual a drawn-out argument in the Ozark Society board had to come first.

That situation was revealed to Ken in letters from me, January 12, 1967:

> … We had a special meeting in Fayetteville last week to discuss the publication of your book and the vote in favor of the Ozark Society being designated as publisher was 4 to 1 in favor. Besides myself those in favor were Craig Rosborough, Charley Stewart and George Kinter. Evangeline in her true form stated that she was just against it, that she didn't know for sure why, that perhaps it was because it was "contrived." …

I think the thing for you to do is to go ahead with your preparations as planned and to include Ozark Society as publishers or in any other way that you wish. ...

By February final details had been worked out, and *The Buffalo River Country* was ready for the press. At first its sale and distribution was to be handled by Ken's mother, who lived in Hot Springs. When that had been arranged, the Ozark Society issued an announcement of publication and *The Buffalo River Country* was on its way. Ken, with help from Harry Pearson, wrote the following press release:

OZARK SOCIETY ANNOUNCES PUBLICATION OF BUFFALO RIVER BOOK
For Release July 16, 1967

The Ozark Society announced today that it will publish Kenneth L. Smith's book, *The Buffalo River Country*, on July 17.

It is the first book to be written about the controversial Buffalo River and its surrounding watershed.

Dr. Neil Compton of Bentonville, Society president, said, "This book will stand as the basic reference on the Buffalo River for years to come."

"The book represents a first in another way," Dr. Compton said, "I know of no book that contains more comprehensive photographic coverage of the Ozark back country than this one."

The book contains 112 black and white photographs and 28 color photographs, plus eight maps and three drawings.

Dr. Compton said the Society has set up a nonprofit fund to finance publication of works of lasting value in describing the Ozark and Ouachita mountain region. He said the fund would remain separate from the operating funds of the Society and would be dependent on loans, gifts and net receipts from sale of publications. *The Buffalo River Country* is the first book to be published under the auspices of the Society. ...

Smith traveled 6,000 miles, took nearly 3,000 photographs and interviewed many residents of the watershed. When he finished his research, he returned to the Park Service and wrote the book during his free time. ...

The book will be available in paperback for $3.95 and in a deluxe hardcover edition for $5.95.

When the people in Marshall saw it, they no doubt thought for sure that the Arkansas Power and Light Company had paid the bill. That one person with a modest income could have managed it out of his own pocket was beyond belief. But it was true, an example of what devotion to a cause, in the heart of one man, and determination can do.

The Buffalo River Country received wide acclaim throughout the area, and sales were good from the beginning.

Ray Heady, in the *Kansas City Star*, gave a lengthy, favorable review, July 23, 1967:

At Long Last, A Beautiful Book
on the Buffalo River of Arkansas

Of all the people who have fought to save the Buffalo river of north-central Arkansas from becoming another impoundment—and they have included a governor, outdoor writers, bird watchers, canoeists, float fishermen, naturalists, biologists, conservationists, old and young, rich and poor—it remains for a young mechanical engineer by the name of Kenneth Smith to perhaps strike the most important blow in the prolonged battle.

Smith, whose fine book on the Buffalo was released Monday, puts together a word and picture story of the famed river that is relevant and tangible. The paperback book can (and we hope will) be sent to hundreds of key people in the U.S. who will respond to the final effort to seal the river's future as a National Riverway.

This is the Ozark Society's first venture into publishing. As far as we know, this is Smith's first book. He is not a great writer, but the book gains strength from the fact that he isn't fancy with words. He is down-to-earth and factual. He tramped the entire 125-mile long basin, and many of the side creeks that flow into the Buffalo, and he describes what he saw. ...

The book includes a good overall map of the basin; also six sectional maps that show much more detail than the overall map can contain. These excellent maps are worth the price of the book. ...

Acknowledgments for much-needed assistance in the production of this important volume should go to those who rendered service without reimbursement: to Don Winfrey, who drew the maps; and to Harry Pearson, who offered his expertise in journalism as critic and proofreader.

The Buffalo River Country served not only as a document to bolster the national river legislation then pending but as a base upon which the present Ozark Society Foundation rests. Ken Smith was discerning enough to see such a need down the line and insisted upon the establishment of what was then called the Ozark Society Publishing Fund from proceeds from the sale of the book. There were healthy profits from that source, enough to reimburse Ken for much of his expenditure and to contribute to the publishing of other books or booklets as well. From this, under good management, the tax exempt Ozark Society Foundation emerged, a successful enterprise to the present time.

The Bulletin

Just as urgent as the need for a book on the Buffalo was the need for a regularly appearing bulletin or journal to keep our people up-to-date and the public informed. The national big-time conservation groups were coming to that. The Sierra Club, the Audubon Society, and the National Wildlife Federation all had recently established elegant magazines.

We couldn't put out a sizable magazine or offer color, but a small quarterly publication with text, black-and-white photographs, and with the necessary maps and charts seemed possible.

Again I sought an editor among our specially talented writers of proven ability: Harry Pearson, John Heuston, C. C. Lambert, and others. None could, or would, volunteer. We would have to do without.

But one day in discussing the problem with Joe Clark, he looked at me and said, "Well, I might be able to edit something like that. You know I was editor of the Tulsa Rotary Club newsletter when we lived over there."

Geologist or not, we hung the editor of the Ozark Society Bulletin tag on Joe Clark, and it stuck, eventually bringing national honors to Joe and his wife, Maxine.

Joe Marsh Clark, then in his mid-sixties, tall and balding, always with a grin to go with his dimensions, had been born up near Salem, Missouri, and spoke with an easy midwestern drawl. Years of experience as a geologist had endowed him with an appreciation of the earth and its inner resources. Now he was employed by the Arkansas Western Gas Company of Fayetteville, roaming the southern slopes of the Boston Mountains. He soon knew the country, valuing it for its incomparable scenery as well as for its hidden fossil fuel.

Maxine Clark, a smallish, energetic lady from her husband's part of Missouri, had with Joe raised two boys, and then while living in Tulsa finished her master's degree in botany. Joe and Maxine provided the Ozark Society with an unexcelled team to supervise its bulletin, what with their general knowledge of the region and its natural science.

Knowing the value of easy-to-reproduce black-and-white photographs for conservation work, I had accumulated a backlog of prints, some of which were acceptable, and these I turned over to Joe for use as he saw fit along with new ones when available. We made an appeal to any photographers, writers, or commentators to furnish material for the bulletin and, after a slow start, that was forthcoming.

The first issue of the Ozark Society Bulletin, the spring issue, came out in April 1967, and we were delighted with it. Joe stood by in the print shop watching every reproduction and calling for a rerun if not satisfactory.

Some of the printed matter in that issue dealt with the concurrent Publicity and Parks Commission imbroglio, and our sometimes overly concerned secretary felt that it was out of place and too strongly stated.

It was my task to advise her that it was not, April 3, 1967:

> ... I had a talk with Lou Oberste yesterday in reference to this matter, and I can assure you that we are not being too firm about this. Extreme pressure is being put upon the Publicity and Parks Commission by Tudor and his group. They bend with whatever force affects them most. If we were to not be firm, we would be doing

exactly what Tudor wishes of us. It seems to me that both you and Charles have fluctuated considerably in reference to this matter, and I hope that we can straighten it out at the meeting next Sunday. The position of our various members must be uniform.

As for the Bulletin, I believe that it is a splendid accomplishment and that Joe Clark is entitled all the praise that we can give him. Your implication that Harry generated the statement in the Bulletin in reference to the Publicity and Parks Commission is unfair and untrue. I wrote this item myself, and Joe edited it to proper size in reference to space available. It's true that Harry did read it over and that he did offer suggestions, which I did not take. If you are not pleased with the results, it would be much better to take the matter up with me than to approach it in such a roundabout way. That could only cause a quarrel between you and Harry, which is not really necessary in this case.

In the final analysis, this bulletin is edited by Joe Clark, and he is the one responsible for whatever appears in it. He should be the first one to talk to in reference to any dissatisfaction that might arise. In this first issue it just so happens that material was not available from any other source and that he did use a considerable amount from reports that I had prepared. Since I am at the present time president of the organization, statements involving policy such as the one that was added will have to be left up to me as the occasion indicates. If everything that occurs in this bulletin has to pass through a board meeting, its continued publication will be impossible. If any of the members are dissatisfied with decisions that I might make, you know that I will give every consideration possible to those opinions.

If you consider this a serious enough infraction of reasonable policy, then a board meeting should be held and a vote should be made upon the matter.

After that Joe Clark was given free reign on the Bulletin, and it would become an unexcelled example of a regional conservation publication.

National recognition would come to Joe and Maxine two years later as was noted in the spring 1969 issue of the Ozark Society Bulletin. That announcement was modest and on the back page—reprinted here in full:

Bulletin Receives National Recognition

After receiving the Arkansas Wildlife Federation award for communications, we were extremely surprised to be notified that we were to receive the 1968 National Conservation Achievement Award—Communications.

This was presented March 1 during the National Wildlife Federation annual meeting at a banquet in the Presidential Ballroom of the Statler-Hilton Hotel, Washington, D.C. Presentation of awards of which there were 15 were by Secretary of Agriculture, Clifford M. Hardin and Secretary of the Interior, Walter J. Hickel.

Printed here is a statement given in the program:

NATIONAL CONSERVATION ACHIEVEMENT AWARD—COMMUNICATIONS. As editors of the quarterly Ozark Society Bulletin, Joe Marsh Clark and his wife Maxine have devoted many long hours practicing the percept that communication is the lifeline of any conservation effort. The Ozark Society, a conservation-education-recreation organization, was formed in Fayetteville, Arkansas in 1962. Through the Society's bulletin, Mr. and Mrs. Clark work to inform the citizens of Arkansas about the vital need to maintain a quality natural environment while enjoying the fruits of an increasingly industrialized economy. The Clarks were nominated for the national award by the Arkansas Wildlife Federation.

But about this recognition ceremony for the Clarks there was an unfortunate circumstance.

Earlier in the winter of '69 our attention had been called to a most interesting place south of Batesville, Arkansas, by Dr. Litt Craig, a dentist, who owned the land. It was Grassy Creek Canyon and the Bailey Creek Pouroff, a beautiful three-tiered waterfall, tributary of Salado Creek. Dick Murray was with me on the first scouting trip and volunteered to take the Clarks to see it later.

The day that Dick, Joe, and Maxine made the trip was cold; there was a thin skim of ice on the ground and on the trees in places. When they came to the first bluff about twenty or thirty feet high, they failed to notice a crevice through which one could walk down. Instead they would "shinny" down a small tree growing close to the bluff. Its trunk was icy and slick, and Maxine could not hold on. She fell, suffering a fracture of her right knee. She could not walk, and Dick and Joe had to carry her back to the car.

After the long and painful ride home, her recovery extended past the date of the National Wildlife Federation award. She was unable to go to Washington, and Joe was obliged to accept the effigy whooping crane trophy, symbolizing their accomplishment, by himself.

For that we were all deeply sorry, but Dick Murray, who was a carver of wood and stone, commemorated it in a strange way. He carved in sandstone the outline of a human foot about twenty inches long and one day carried it to the bluff on Bailey Creek and planted it in the place where Maxine fell, there to remain for future generations to ponder. On it there was no inscription to tell them what or why. Who could ever know that it was only an expression of Dick's wry humor? Dick Murray would present us with further such sculptured enigmas.

Upon the termination of the battle for the Buffalo River, he carved for me a commemorative object—an impossible animal. It was a buffalo, beautifully carved in walnut, but with a head on each end, a creature with an alimentary tract dilemma for sure. To top it all off, he carved its twin in anthracite coal for Col. Charles Maynard, the resident engineer in Little Rock, who most surely must have consigned it to the grate at once.

Dick Murray and Maxine Clark, RIP ...

FIFTEEN

ADVENTURES
ON THE RIVER
AND IN THE HILLS

For some time we had been aware of a worthy activity being carried on by the Ozark Wilderness Waterways Club. Once a year they would conduct a "clean-up float," usually on the often-littered Current River. That their can-pickup program was a good thing was without question. The river banks were left free of human mess so offensive to discerning eyes. It was a reminder to litterers to mend their ways and created a good image of avowed conservationists in the public eye, and it afforded positive publicity in the news media. We knew that it was something that we ought to do here in Arkansas, and in the summer of '67 we undertook the task.

Many of us had become convinced that an annual cleanup trip on the Buffalo River or perhaps some of our other streams should be made a part of the program for Ozark Society outdoor activities. The final decision came about when John Fleming, the outdoor writer for the *Arkansas Gazette,* strongly urged early in the year that we do this. Arrangements were begun to bring it off on August 12 and 13, 1967.

Not having ever undertaken such an activity before, I made a special trip to visit with Harold and Margaret Hedges at their place in Newton County in order to find out how they conducted their can-pickup trip on the Current. Harold informed me they originally started off collecting all the junk they could find and then burying it at convenient places along the river. This obviously was the wrong thing to do since the river during high water simply turned the trash up again for all to see. They then adopted the practice of carrying away all non-combustible material. This was assembled in tow sacks at some place accessible to trucks and was hauled off to the nearest city dump. Harold stated it had long been their policy to burn old automobile tires.

A news release was prepared and mailed to all regional newspapers and to our membership. With it went an info sheet listing the rules for the event:

Clean-up by Canoe
Anti-Litterbugs to Compete on Floats on the Buffalo River

Gravel bars, river banks, and river beds will be searched for their accumulation of debris left by careless visitors to the Buffalo River in an Ozark Society sponsored can pickup float on Saturday and Sunday, August 12 and 13.

… prizes consisting of pieces of equipment useful in outdoor recreational activities will be awarded when contestants gather with their loot at the base camp at Buffalo River State Park on Sunday. …

The contest is open to anyone with reasonable experience in handling canoes and small boats. All entries must register with one of the float leaders listed before being allowed to participate. Dr. Neil Compton, Bentonville, is in overall charge of the Buffalo River cleanup. Clayton Little, Bentonville, is in charge of the Maumee-State Park section and William Saunders, Little Rock, of the Gilbert-Maumee section.

Participants will meet at the base camp on Friday evening, August 11, at Buffalo River State Park camp ground. They will be divided into two groups for the Saturday morning departure. One group will be dispatched to Gilbert to float from Gilbert to Maumee. Another group will go to Maumee to go from there to State Park.

The Hurst Boat Line of Cotter will provide a service boat carrying special equipment for digging and snagging embedded junk and the transport of large objects.

On Sunday the entire group will float from State Park to Rush. On return from Rush to State Park the judging will be held and the winners named. The State Publicity and Parks Commission and the Al Gaston Boat Line of Lakeview will provide means for transportation for participants, equipment, and collected junk. …

Any sort of legitimately found litter such as plastic of all kinds, bottles, boxes, paper, cans and other items of human manufacture are eligible.

Method of scoring—One point will be allowed for every full tow sack and one point for every two automobile tires. One point will be allowed also for any other object of similar size and weight. This might include old bedsteads, pieces of old automobiles and other junk. Mattocks, picks, pry-bars and rope will be carried by service boats.

Gear—In addition to items already mentioned, it will be necessary to carry only lunch and drinking water on each date, plus one spare paddle.

Membership—Ozark Society membership is not necessary to enter.

In response to that, the *Marshall Mountain Wave* printed a very short but honest summary of our schedule. It appeared under a terse headline: "NATURE BOYS TO CLEAN RIVER," which is enough said considering.

An enormous amount of time and effort went into the planning of this event, but the enthusiasm of various participants saw it through.

One problem was the matter of reward for diligent junk pickers. The

O.W.W.C. had simply paid them twenty-five cents for one tow sack full and twenty-five cents for one automobile tire. But we wanted to make a contest of it. We solicited various firms and businesses for products to offer as prizes. The results were surprising.

Down in Arkadelphia Dr. D. D. Clark persuaded a friend of his, Mr. T. V. Sharp, president of the Ouachita Marine Company, to donate a handsome yellow aluminum canoe as first prize. Sam Walton of Bentonville, founder of Wal-Mart and on his way to fame and fortune, donated a nice umbrella tent. Stanley Kahn, of Kahn Jewelers in Pine Bluff and an avid supporter of the Ozark Society, gave an expensive scuba diver's underwater watch. The thermos company sent us one of their lanterns; Pfeiffer's of Little Rock, a Coleman lantern; and Al Gaston's Resort, an expensive rod and reel. With all of that and a few sleeping bags and boat paddles thrown in, competition was keen.

On the evening of Friday, August 11, sixty or seventy of us nature boys and girls repaired to the campground at the Buffalo River State Park where we pitched our tents and enjoyed an evening meal and a big bonfire prepared by Clayton Little on that surprisingly cool August night. Plans for the next day were recounted, and all made ready for the great trash gathering the next morning. Clayton Little, the general expediter, was to shuttle cars for both divisions on Saturday.

I was to accompany the Gilbert section, and by 9:30 A.M. we were in the water on our way to Maumee. We had originally supposed that the most litter would be discovered in the area of the State Park since it is most frequented by fishermen and campers; however, we were not long embarked from Gilbert before we realized that we may have been seriously mistaken. Within an hour or two most of our canoes were heavily laden, not so much with beer cans and bottles but with old automobile tires. These were turning up on the gravel bars along the river and in the riverbed itself in unbelievable numbers, and soon most of our canoes were loaded to the gunwales and some of our contestants looked like nothing except a couple of people riding a stack of tires down the river.

After being out about two hours and upon coming to a large gravel bar, it was decided to stop in order to discuss the situation and to determine what to do with our already overloaded flotilla. It had been my opinion that it would require entirely too much time to stop and burn the tires as Harold Hedges had said the Ozark Wilderness Waterways Club did on their cleanup trips. We were seriously considering trying to cache the tires on the bank somewhere and to send a party down later with a larger boat with motor to pick them up and carry them off. We were faced with the fact that this was not a practical idea.

The matter of burning the tires was brought up for serious discussion, and soon some of our members produced an inner tube while others started a small fire with sticks and twigs. The inner tube was added and was soon blazing brightly, and the entire crew, seeing this, began unloading their tires and throwing them on the blaze. Within moments, a huge column of black smoke rose over the Buffalo River looking for all the world like pictures that we had seen of bombing along the Mekong River in Vietnam.

Altogether thirty-four tires were consigned to the flames, and without a doubt the plume of black smoke could have been seen for many miles in all directions. Rising high into the air the smoke was caught by down drafts and was blown back down river again, thus giving it the appearance of an ash fall from a volcanic eruption. There was no danger of spreading the fire from this sort of blaze, however, since no sparks are produced from burning rubber. The blaze was entirely confined to the pile of tires, but it was considerable, and for a while it was truly frightening as the red flame rose higher and higher.

At this point my wife came up to me and said, "Neil Compton, I just can't stand this. It looks like Hell to me, and I'm leaving here!" With that she turned and walked down the gravel bar out of sight of the awful blaze. I said not a word but could not help but wonder (as perhaps most husbands of over thirty years would) as to whether or not she really knew what Hell looked like. An hour later, by hard paddling, I caught up with her hoofing down river on the gravel bar toward Maumee.

No sooner had I been subjected to this disheartening reprimand than Leonard Mizell approached with anxiety clearly visible upon his countenance.

"Dr. Compton," he said, "I don't think we ought to do this. You know the rangers are going to see this smoke, and they will be down here after us. Isn't there some way we could put it out?"

"Leonard," I said, "I am afraid it's too late. There is nothing we can do. You just as well get in there with Everett Bowman, who set it, and holler burn baby burn."

After a few minutes it was obvious that the holocaust would indeed be confined in the small section of the gravel bar where it was situated. Dick Murray and Jack Diggs agreed to stay for a short while to make sure that the fire was extinguished while the rest of the party got underway. This we did in company with another group of canoeists who were on a fishing trip on the Buffalo and who had come down the river through the smoke looking for all the world like refugees in war time. They were much relieved to find not another one of those riots in the hot summer of 1967 but only an anti-litter demonstration put on by innocent Ozark Societyans.

From this point onward our journey to Maumee was much the same as it had been before. We continued to collect large numbers of old tires along with all

Tired tire collectors on the Ozark Society cleanup float, August 1967.

sorts of bric-a-brac. By the middle of the afternoon, all involved were definitely tired, and much of what we had come to collect was left lying where it was. Our boats were loaded, and the contestants were just too tired to do more than steer their heavily laden craft to the take-out point at Maumee.

On Sunday's float all of us were considerably surprised to find that the section of the river below Buffalo River State Park was relatively free of trash and debris. This was in marked contrast to the situation just below Gilbert; and upon discussing it with some of the natives in the area, the conclusion was drawn that perhaps certain individuals living in the area, either local garage men or farmers, had disposed of their old tires by simply throwing them off of the Highway 65 bridge.

Within a short time we were back at base camp walking about among the various stacks of rubbish and trash, visiting with friends, and awaiting the arrival of the rest of our party from Rush.

When all was ready, the three who were to serve as judges (these being myself, John Fleming, and Everett Bowman) made the rounds and inspected the results of the labors of each team. In addition to the expected large numbers of tires, bottles, and cans, there were other such items as old bed springs, bedsteads,

The great tire fire, below Gilbert,
Ozark Society Buffalo River cleanup
float, 1967.

The Leviathan. Dr. Neil Compton (in the panama hat) kicks the side of the ugliest object dragged
up from the bottom of the river during the cleanup float. This homemade metal boat was constructed from two old automobile hoods welded together, with ribs added and a mount for a
motor on the back. Ozark Society members referred to it as the leviathan, since it was a monster
up from the deep. Everett Bowman, chairman of the Little Rock chapter of the Pulaski County
Ozark Society, is the gentleman in the cap standing left of Dr. Compton.

The first-place winners, Dick Byrd and son, with their Ouachita Marine Company canoe at Buffalo River State Park. The judges on the back of the truck are, *left to right*, Everett Bowman, John Fleming, and Clayton (Curley) Little. (*Courtesy of Bea Devlin*, Baxter Bulletin)

Maxine Clark with her booby prize, designed and made by Dick Murray. (*Courtesy of Bea Devlin*, Baxter Bulletin)

horse collars, washtubs, assorted pots and pans, and there was even part of an old ore bucket from the mines at Rush, along with the wheels of the trolley that it rode upon; there was also a homemade boat, which had been created by the welding together of the hoods of two old Ford cars. When all of the points were finally counted, winners were announced.

In first place were Richard Byrd, Jr., and Sr., of Little Rock, a father and son team who had entered with a small homemade canoe and whom most of us had thought to be hardly in the running until we observed enormous quantities of old tires they were able to load into their small but wide-beamed craft.

In second place by only a fraction of a point was the Naval Reserve team from the Fayetteville unit of the USNR consisting of Chief John Mader, USN, and Hospitalman Jim Turner, USNR, who were awarded the rod and reel; the Ouachita canoe, of course, having been awarded to the first-place winners.

In third place were Rob Walton of Bentonville and his partner who received a sixty-dollar scuba diver's underwater watch.

When the final winners had been announced, the big truck was loaded with as much of the debris as it would hold, and following this the first cleanup float of the Ozark Society was adjourned.

News coverage of the Buffalo River cleanup was especially gratifying. Present were John Fleming of the *Arkansas Gazette,* an initiator and judge of the affair; Bea Devlin of the *Baxter Bulletin* in Mountain Home; Allan Gilbert of the *Northwest Arkansas Times* in Fayetteville; and Bill Pharis of the *Southwest American* in Fort Smith.

These reporters obtained numerous photographs of the amazing objects produced by the hand of man and cast thereby into this, our last and best Ozark scenic river. Upon seeing these in the paper, the public must surely have been puzzled by what they saw, that especially so in the case of Maxine Clark, who won the booby prize. She had found an old, long-lost straw hat out of which grew a pretty daisy. That she displayed along with her reward, a brand-new canoe paddle into which had been neatly drilled a dozen one-inch holes by that wry woodworker, Dick Murray.

EXPLORING THE WILDERNESS

With the advent of 1967 we entered a phase of the battle for the Buffalo River strange to our view of it heretofore. We were in command of the field. The BRIA and the Corps of Engineers had withdrawn their troops to regroup, if they could. Head-on encounters with the big dam battalions were no longer in the offing, and our job was one of mopping up, so to speak. We needed to guard against overconfidence and to put our own forces in order.

We urgently needed to know more about the object of our crusade, the remote high hills, their forests, tall precipices, and the rushing river that drained that country. Frequent visitors though we had been, there was almost always some new and stunning scenic wonder awaiting us on every trip to the Buffalo River country.

By now there was important legislation, besides the bills for the Buffalo National River, in Congress which would have a direct bearing on the region.

It was the Wilderness Act of 1964. It was being altered to permit inclusion of areas east of the one-hundredth meridian (that included us) and not just purely virgin lands out west. Man had done something to just about every acre here in the east, but in the Ozarks nature is capable of restoring man's abuse rapidly, if given a chance. There were many such places that we had heard of, and we would seek them out.

A Trip to the Devil's Fork

One of nature's wonders was a southern tributary of the Buffalo, Richland Creek, starting up near Ben Hur in the Boston Mountains and entering the Buffalo River at Woolum. My introduction to that remote valley had come from a patient of mine, Mr. A. A. Lee, who with his family had moved to Bentonville in the 1950s and who originally lived on the Richland where he operated a sawmill. In 1961 he and his brother-in-law, Wesley Robb, had taken me on an all too brief tour of Falling Water Creek, the Richland, and the Point Peter Overlook. I was intrigued by their account of the main falls up the Richland and of the Twin Falls of the Devil's Fork.

Thus, on March 17, 1967, in company with my wife and a number of Ozark Society friends we set forth for the Falling Water Campground on the Richland to see what had been missed previously. Arthur Lee, Wesley Robb, and Julius Williams would meet us there and serve as guides.

That evening we lost no time in building a fire in the big stone fire pit which had been prepared by the National Forest Service and in setting up our tents. Darkness was soon upon us, and the temperature descended steadily until it was below freezing. This proved to be no terrible inconvenience since we had brought plenty of warm clothing; and with a hot supper and plenty of coffee on hand and with a big fire roaring in the pit, we were relatively comfortable.

Dawn broke with relatively clear skies, and by 9:00 A.M. the hike began with

Mr. Julius Williams, age seventy-one, leading the way. He first showed us an excellent spring that flows from beneath a bluff about a hundred yards up the Falling Water Creek from the campground. We then retraced our steps to the point where the Falling Water Creek and the Richland come together in an enormous jumble of rocks and alder bushes. Here we crossed the Falling Water Creek by moving from one boulder to the next and after passing through heavy thickets of alder and willows came to a point on the main stream of the Richland where it had to be crossed. Again it was necessary to leap from one rock to another, and on this occasion the final steps were made along a large willow limb that reached out from the bank. At this point our first mishap occurred. Joe Clark, who has always been the trouble shooter on Ozark Society trips, somehow or another in the process of leaping across the stream got his walking stick between his legs and fell waist deep into the foaming torrent. Upon scrambling up the bank out of the frigid waters he again fell back but on the third try made it to the boulders on the far side, wet from the waist down. Before the morning was over, however, he was dry again from the exertions put forth in traveling up this almost impassable, boulder-strewn gorge.

We then entered a region of huge boulders. These rocks lay tumbled about in all attitudes, some of them the size of houses, and through them the stream wound its way descending in many beautiful rapids and falls through the rocky maze. In order to bypass this barrier, we ascended the side of the hill for a short distance.

Walking became more and more difficult due to the impenetrable thickets of alder and willow and the enormous boulders that lay scattered all about. We did well to average a mile an hour. Finally we entered into an area of woodland where enormous trees stood in almost undisturbed condition. Here were white oak, walnut, black oak, sweet gum, and black gum, with pine on the hills above. Julius Williams during this time gave us an account of the area through which we were passing, describing the various types of vegetation that were present and using the local names applied to them. Once while descending a slope, he had reached for a bush to hang onto to break his descent and in so doing he let forth a howl of pain. Upon being asked what had caused his discomfort, he stated that he had gotten ahold of a "pear-blanket." The offending shrub was determined to be hercules club, which has short sharp thorns. Pear-blanket is a term that none of us had ever heard applied to any sort of tree or bush. What he had really said was *tear* blanket, an Indian cognomen without doubt.

By about noon we had arrived at a spot where the Pitkin Limestone lay exposed by the action of the waters of the stream and where its surface had been broken up into segments giving the appearance of a tile floor, an ideal spot for lunch. Since inactivity led to chilling in the below-freezing temperature, John

A shoe roast on the Richland, March 1967. *Left to right:* Everett Bowman, Jeffry and Karen Kahn, and Lee and Julius Williams.

Heuston built a fire which was most welcome and upon which he roasted wieners brought for his lunch and which little Jeff Kahn used to roast his wet shoes.

After the light lunch, we proceeded on upstream, our objective being the Falls of the Richland which we had heard of so many times and which had never been seen by any of us except our guide Julius Williams. After about another forty-five minutes, the stream bent rather sharply to the south, and here again it was necessary to cross and again boulder hopping was in order. Most of us had made it safely across, but Stanley Kahn elected to perch too long on the top of a pyramid-shaped rock, and in a moment he too had gone into the stream. He scrambled out again quickly enough and seemed to be not the least disturbed.

In a short while we came in sight of the main falls of the Richland. Here we tarried for some time photographing and viewing this waterfall. It is almost level all the way across from one side to the other and thus gives the appearance of a miniature Niagara. Below this waterfall, we found an old stone buhr or millstone testifying to the fact that a hundred years ago this area had been populated and that a mill had existed on the falls of the Richland.

Twin Falls of the Devil's Fork, Richland Creek Wilderness Area, April 1973.

Julius Williams elected to return to camp by an old road which ascended the mountain and which could be followed around its rim. We thereupon followed this overgrown trail, which seemed to proceed almost straight up for a goodly distance. We were soon on the first bench of the mountain and were afforded a view of the magnificent rugged wilderness in which we stood. The sight of these great trees alone is worth the effort that must be put forth to come to see them. We had known that there were also waterfalls in the Devil's Fork, which joins the Richland near its main falls. As we proceeded onward, our eyes searched the bottom of the gorge for the stream that flowed below, and soon we heard the roar of its falls. We descended to the bottom and soon were standing on the edge of the beautiful clear waters of the Devil's Fork. Traveling up it a distance of another hundred yards, we soon saw the Twin Falls of the Devil's Fork in the distance. These two falls are formed at the exact spot where the Big Devil's Fork and the Long Devil's Fork come together. The two streams flow over a ledge twenty feet high and fall precipitously into a pool, thus creating a truly beautiful scene. By now the sky had again become overcast, and we were concerned as to what sort of a night we were in for.

Searching for the road out, Julius Williams led us again up the side of the hill which was most steep and tiring to all of us. While resting on one of our frequent stops in order to catch our breath, we again saw flakes of white descending from the sky. It was unmistakably snow. We descended to the stream bank, picking our way along as we had done during the morning, but certain that were we to continue downstream we would soon come to the camp. By this time all of us had become thoroughly fatigued. Harry Pearson found his reserves taxed to the limit. On one occasion I by chance stepped on one end of a log at the same moment that Harry stepped on the other end. By the leverage created, he was thrown to the ground where he lay upon his back looking up at the sky unable to rise. He had not been injured but was so tired that further efforts were almost impossible.

Finally we approached again the familiar area of the great rock pile with its enormous boulders and beautiful pool of clear green water. Here we were greeted by a sudden increase in the snowfall. Great white flakes of wet snow descended from the sky so thickly that we could hardly see the opposite bank. The scene was one of incredible beauty. To be here in the midst of this true wilderness and to witness this most impressive and beautiful act of nature was to bring to us all a new awareness of the real world of which we are a part but know so little.

It had been our intention to remain another night and day in this secluded area for a hike into Bobtail Creek, but upon observing our snow-covered tents and feeling the cold wind from the north, there was no need to discuss whether or not to remain. With all haste snow was swept from the tents, they were folded as quickly as possible, all gear and equipment were thrown into our vehicles, and with darkness falling again the entire group departed for home.

Reflecting upon this brief experience in the Richland, it was obvious to us all that here was the most significant wilderness area remaining today in Arkansas. We hoped that the National Forest Service would consider declaring it a wilderness area to be left as it is without further cutting of timber or development of any kind. We believed that the Richland Creek area was every bit as significant from a wilderness standpoint as any of those regions now being set aside under the Wilderness Bill in other parts of the United States.

Efforts of the Ozark Society and kindred conservationist groups supporting the wilderness legislation would eventually bear fruit. In 1984 the Richland Wilderness Area, carved out of the Ozark National Forest, would be created by an act of Congress. That, along with other similar tracts, would constitute an accomplishment in conservation second only to the creation of the national river itself.

The Arkansas Conservation Council

Motivated by dreams of a significant Ozarkian wilderness to complement the impending national river, we scheduled a more or less informal holiday in the hills for that October 1969. We would repair to Ponca and spend a week or more in the Lost Valley Lodge with our hosts, Bob and Louisa Crow. We would explore the deep recesses adjacent to Lost Valley, would float the Buffalo, and I would journey on to Jonesboro to give a program to the Arkansas Audubon Society on the subject of the pending wilderness legislation.

Harold Alexander had been urging the various conservation groups in Arkansas to get together in order to present a united front to the various problems now proliferating in our area. He had obtained a commitment from the Audubon Society, and the Ozark Society was pleased to participate. We would take some first steps toward the formation of "The Arkansas Conservation Council." I had accepted an invitation to address the Audubon Society on that subject at their annual meeting. To accomplish that we would take a day out of our proposed ten-day vacation in Boxley Valley to make the long drive to Jonesboro and back.

Thus we left our cloud-shrouded mountain retreat in the early dawn, bound for Jonesboro and a lively get-together with all those enthusiastic bird watchers. We were happy to greet once more Charley Johnston, Doug and Fran James, and Jane Stern, who offered her motel room for us to rest in for a few minutes after the long journey.

That evening after the banquet I delivered a fifteen-minute address on the subject of a conservation council, which was well received. After the meeting, Sterling Lacey, the president of the Arkansas Audubon Society, and I made out a tentative list of organizations who might be interested in forming such a council: the Garden Clubs of Arkansas, the Arkansas Audubon Society, the Ozark Society, the Arkansas Wildlife Federation, the Northwest Arkansas Archeological Society, the Arkansas Archeological Society, the various riding clubs of Arkansas, the Hikers and Campers Association, the Arkansas Historical Association, Nature Conservancy, the Arkansas Museum Associates, and the Arkansas Speleological Survey.

Sterling Lacey agreed to call a special meeting for the formation of this council shortly thereafter. Most of the organizations listed participated, and a good alliance ensued under the leadership of Dr. Howard Suzuki of the anatomy department of the University of Arkansas Medical School. However, with his departure from the state, the Conservation Council languished, to be reborn ten or twelve years later as the Arkansas Conservation Coalition under the leadership of Dr. Tom McClure, D.D.S., of the Sierra Club. It would be a principal factor in the successful application of the wilderness legislation in the 1980s.

After the Audubon Society meeting, we drove back through the moonlit hills, arriving at the Lost Valley Lodge at 3:00 A.M. where we sneaked in as quietly as possible so as not to disturb the sleeping multitude gathered there for the hike into Leatherwood Cove scheduled for the next day.

But in the night came the patter of rain, which kept us holed up in the Lodge for the next couple of days.

During that time we had heard rumors of scenic marvels up in Whiteley Creek, the next deep ravine south of Lost Valley. Joe Clark and I made a visit to the owner of part of it, Walter Webb, a retired postal worker from Kansas City. Walter showed us the access point, and it was all so interesting as to inspire a full-scale walk-through the next day.

A Scouting Trip to Whiteley Creek

Joe Clark and Maxine and Laurene and I arose early on Wednesday, October 17, eager to see what the lower sections of Whiteley Creek might contain. The day proved to be delightfully clear with brilliant blue sky following the long, heavy rains of the weekend. We stopped briefly at the two-story white frame house occupied by Doy Scroggins to ascertain what we might expect on the road up into Whiteley Creek. Both of the Scroggins were eager to give us such information as they could, but Doy knew little about it except that he had been as far as a place where the bluffs came together which he called "the penitentiary."

Walter Webb took us back down to the edge of the bluff that marked the upper end of the Whiteley Creek gorge. Following the course of Whiteley Creek downward from the first high falls, we came upon the first large azalea beds. From here on the entire upper gorge of the Whiteley Creek is abundantly populated with wild azaleas. Trudging along the banks of the stream on the hard sandstone we suddenly came to the place where it abruptly spilled over a second cliff sixty or seventy feet into a still deeper and narrower gorge. Below this second fall huge forest trees of various kinds grew in profusion. These were the trees that we had seen the day before and which Walter Webb had entertained thoughts of cutting. Our thoughts as we beheld them, however, were concerned with what might be done to prevent any needless denudation of this wild and lovely place. Walter Webb is a kindly soul concerned with the preservation and the maintenance of the natural beauty on his property, but he had been persuaded by neighbors and by logging operators in the area that mature trees are not desirable and that they should be cut. He had explained to us the day before that timber thieves had gotten into the lower part of his place and cut eight hundred dollars worth of timber before he knew it. He told us that due to the fact that he was convinced that some of the timber on the lower part of his property was mature and thus not desirable, he had sold still more of it to the loggers. Evidence of this we were to see later on.

Standing on the brink of this second high falls, we could determine no point at which we could gain access to the gorge below. We had not brought ropes to clamber down over vertical rock faces, and therefore Joe Clark took it upon himself to scout the south side of the ravine and before long had found a way down.

After leaving the area of the falls, we soon came to a jumble of giant boulders, which had fallen down from the sandstone cliffs above and which filled the course of the ravine here. It was necessary to crawl, scoot, and slide in order to get down. Rich deep soil covered the hillsides on both sides, and very soon we heard excited comment coming from our botanist, Maxine Clark, who had been scouting on the south hill. We all gathered about to see what she had discovered and were genuinely astonished to see the fruit of the plant the likes of which none of us had ever beheld before. Maxine explained that this was the Doll's Eye, a herbaceous plant characteristic of Appalachian habitat. The fruit consisted of ivory white berries or pods the size of small grapes with each of these bearing a jet black spot in the center which resembled the pupil, thus giving the plant its name, the Doll's Eye. The stem of this cluster of fruits was brilliant red with each fruit being born separately from a common stalk. Thus the whole thing looked exactly like some sort of plastic decoration that has come to be so common these days.

Within a short while we realized that we had made our way through the great rock pile and that we were coming out into a broader valley floor. Here we were able to observe some gigantic forest trees which had not yet been disturbed. Some of these were enormous shagbark hickories, huge beeches, sweet gum, black gum, and maple. Below this point we found ourselves entrapped in a morass of downed treetops, logs, and brush. This was the area that Walter Webb had sold to the loggers the year before, and here it was obvious that our progress along the stream bed would no longer be possible.

Knowing that a log road should exist on either one side or the other of the stream, efforts were made to locate it, and after some crashing about in the bushes the road was found. Here the walking was easier, but the scenery much less pleasing and, in fact, it was with genuine agony that we observed the ruthless manner which the timber cutters had employed and the awful mess and litter that they had left in their wake. From inspecting the stumps it was clear that perhaps only a small percentage of the trees that had been cut were actually sound. From this one could deduce that the expenditure of this operation would scarcely pay for itself and that all of this destruction had been wrought for the benefit of no one except for the manufacturers of bulldozers, chain saws, and log trucks. We now assumed that perhaps we had seen the best of Whiteley Creek and that the log road would soon lead us back down the long mountain.

We had not traveled more than half a mile when this logging road suddenly came to an end. There was no way to determine where it went or how the truckers got their vehicles out. Here we began our descent through a beautiful beech woods interspersed with hickory, gum, and maple. Thus we did obtain some idea of the beauty of the forest before the cutters had laid hand on it. It was necessary for us to traverse extremely steep mossy, ferny banks, and as we descended the shadows grew ever deeper. As we approached the bottom the sound of falling water could be heard, and soon peering between the trees we saw a beautiful waterfall lost in this woodland. By chance the rays of the sun striking just above the mountain top behind played upon this waterfall setting it off like some vibrant and living jewel in the midst of the forest. At last we obtained the banks of the stream and found that the fall, which was only six or eight feet high, spouted from a deep channel that it had carved into the limestone and that it fell clear into a small pool.

By now the shadows of the evening were growing longer, and we trudged along in the twilight zone of the creek bottom, wondering whether or not we would reach the Scout before dark. We came to long sections of the creek that had been bulldozed out by the loggers, with the waters of the stream flowing down the bulldozed channel. Here walking again was difficult because of loose rock. Soon, however, the loose gravel of the creek gave way to a solid rock bottom of limestone. Nature had provided a sort of water-covered pavement for the log trucks that utilized the floor of the gorge to haul out the timber. At this point we spied in the distance what appeared to be a tributary stream of considerable size flowing into Whiteley Creek but not on the level with it. We could see that the waters of this tributary entered the creek with considerable force almost as if it were discharged from a fire hose. This we could not understand until we had come closer. Here we saw that this water came from a huge falling spring that emerged on the side of a hill a hundred feet above the rocky bed of Whiteley Creek. It descended the hillside in two gushing torrents, and when it joined with the waters of Whiteley Creek doubled the size of that stream. We stood and sadly contemplated what undoubtedly was at one time one of the beautiful springs of the Arkansas Ozarks but now it had been rudely violated. All around and about it were the stumps of great trees that had been felled not long before. At last we had about reached our destination, and taking

The "pipe organ" bluff in the Whiteley Creek gorge. The flutings on this free-standing bluff are travertine deposits left by running water that once flowed over it. The water now emerges in three "spigot" springs near the base. The bluff stands free and is perforated in the right center.

a turn to the left we soon spied the Scout hidden in the dogwoods at the base of the hill.

The scenery of Whiteley Creek compares in scenic beauty with any of the other gorges along this section of the Buffalo. Most of these gorges and ravines

are little known either to the local inhabitants of the area or to interested con-
servationists such as ourselves. The only people who traverse them are occasional hunters and the ubiquitous timber cruisers. It was surprising to learn that
Walter Webb, who had owned the property on top of the mountain for twenty-
five years, had no idea what lay below his place. Also it was puzzling that Doy
Scroggins, who had lived his lifetime in the area, knew little about the
Penitentiary. The reason for this is obvious to those of us who have been there.
The region is too far removed from urban centers to have attracted hikers and
explorers such as ourselves heretofore. It is too inaccessible, brushy, and rugged
to tempt visits from the local citizens who have ample scenery to enjoy from
their own front porches.

In 1977 on an Ozark Society hike, in company with Ken Smith, I first visited
the Penitentiary in the north prong of Whiteley Creek. It was stunning beyond
words, with its great free-standing limestone bluff, perforated in the center, its
face decorated with vertical travertine fluting and the spigot springs coming out
of the base and all the small waterfalls on the main stream. Had we known of it,
we would have intervened strenuously for its inclusion (and the rest of Whiteley
Creek) in the proposed national river.

In 1979 or 1980, we did get the Arkansas Natural Heritage Commission to
investigate the Penitentiary. It was considered worthy of inclusion in their program, and they made an effort to purchase it. They assigned Bill Sheppard, one
of their officers, the task of making the deal. He discovered that it had been
bought in 1955 by a now-elderly gentleman, a Mr. Motta, living in Washington,
D.C. Motta, who had never seen the place, was agreeable to selling it to the state
of Arkansas, but his sons objected, and the Penitentiary wound up in limbo, as
was its neighbor Lost Valley for so long.

Peccary Cave

We had known Jack McCutcheon since the late 1950s, having met him at
meetings of Harry McPherson's Northwest Arkansas Archeological Society.
McCutcheon's home territory on Ben's Branch had been a favorite stomping
ground for the paleoindian, and his interest in them had led him to become a
more or less self-taught archeologist of no mean ability. He had once showed
me and my daughter Edra Ann the impressive Saltpeter Cave on Cave Creek
where every slab of fallen roof rock was the burial site of some of those ancient
people. That place, much to Jack McCutcheon's grief, has since been rifled by
pot hunters.

In 1961 or 1962 Jack had discovered a vertical shaft cave in the limestone a few hundred feet behind his barn. He had gone down in it thirty or forty feet on a rope ladder to see what relics of the past might be there. He found teeth of the dire wolf, a carnivore that stood five feet high at the shoulder. And there were bones of the ice-age peccary, a pig big enough to out-hog any modern Razorback.

Jack's wife, surveying the situation, suggested that he could dig a tunnel into the cave from above the county road which ran below the barn. That he did and was able to enter on foot instead of climbing down and up a rope.

Peccary Cave, as it came to be called, was not as rich a site as the Conard Fissure, but it was significant in reference to its contents of peccary and dire wolf remains. Most likely it had been the lair of that big meat eater.

Streams flowing north from the main east-west range of the Boston Mountains, such as Big Creek, Cave Creek, the Richland, and Calf Creek, had carved deep valleys in the old plateau. The high ridges in between terminated in spectacular promontories that afforded commanding views across the Buffalo Valley and the Boone Stripped Plain nine hundred to one thousand feet below. Those overlooks had been accorded the names of mountains or points: Red Rock Point, George Mountain, Lick Mountain, Horn Mountain, North Pole Knob, and Point Peter.

We particularly wanted to do a walk-about on Horn Mountain, which was reportedly a part of the Ozark Wildlife Club, mentioned earlier. Just below it to the west were Cave Creek, Ben's Branch, and Jack McCutcheon's Peccary Cave.

Thursday morning, October 18, Joe and Maxine Clark and Laurene and I departed the Lost Valley Lodge bound for Ben's Branch and Horn Mountain.

After passing through Jasper, Vendor, and the village of Mount Judea (pro-nounced "Judy" by the locals) we wound around North Pole Knob, down to Cave Creek and up Ben's Branch, finally stopping at the newly dug entryway to Peccary Cave, which opened almost on the road.

After parking the car I went to the mouth of the cave and, upon shouting, received a reply from Jack affirming his presence inside. Knowing the interior would be quite muddy after the rains, rough trousers were drawn over our regu-lar traveling pants and boots put on in the place of shoes. Walking back about fifty feet, we entered the main room of the cave which had some attractive sta-lagtites and other cave formations on its ceiling and walls.

Jack was happy to see us and pleased to have Joe comment on the geology of the cavern. Jack showed us the area he was excavating and described his meth-ods in doing it. He then removed from the red clay matrix a tooth, looking something like a human molar, but which was surely from the jaw of a peccary. It was on the top of such a bank of clay that Jack had found the original mandible of the peccary with its enormous lower canine teeth. That discovery led to the comprehensive excavation that was now going on.

We were pleased to learn that a grant of twenty-two thousand dollars had been made by the National Science Foundation through the University of Arkansas for the purpose of scientific excavation in the cave. Jack informed us that from time to time university students were assigned to him to help with the digging and that Dr. Quinn of the university Department of Geology was in charge of the overall operation. This was an endeavor which we in the Ozark Society could be justly proud since some of us had been able to direct Jack's interest in the right direction so as to meet the people who made possible the grant from the Science Foundation. In this instance there would be no haphazard digging by bug-eyed fortune hunters or pot collectors.

At the conclusion of our visit with Jack, we mentioned our intention to go on to the Wildlife Club and Horn Mountain for a picnic lunch. When he heard that, he laughed and related the methods of the Wildlife Club in obtaining those wild boars. We both shared the same low opinion of that phony enterprise, but we were going to have a look at it anyhow.

The Wild Boar Hunters

We then set forth for the Ozark Wildlife Club a few miles to the east. At Casey's store was the intersection leading east to that development. There stood a huge billboard bearing the legend of the Ozark Wildlife Club and painted thereon were the various symbols of this enterprise including the wild boar, the salmon, the bob white quail, the white-tailed deer, the wild mountain goat and even a Sioux Indian complete with war bonnet.

Taking the road on up the hill toward Horn Mountain, we were soon within the development. Various brushy avenues led off into the woods on either side, each bearing street names and numbers. These "streets," which as yet contained no houses of any kind, did, however, have marked at frequent intervals the names of the owners and the towns from which they came.

At the entrance of the reservation it had been necessary to pass a ranger station, and here we had been greeted by an individual dressed in the manner of a forest ranger. He had presented us with a pass which gave us the privilege of going through the development.

A long, narrow, relatively level ridge extended off to the west from Horn Mountain at about the one-thousand-foot level. The woodland on that ridge had been scraped away for a distance of about a mile, leaving an ugly scar that

The "Lodge" or dormitory at the Ozark Wildlife Club on Lost Mountain (Horn Mountain) east of Bass.

could be seen from miles away. Its surface of gravel, chert, and rocks was the none-too-inviting landing strip for all those boar hunters who would be flying in from across the nation.

After traversing the airstrip we could see the main buildings of the Ozark Wildlife Club across a deep ravine. It was necessary to cross this ravine before attaining these main buildings and in order to do so a road had been built across it. The road was narrow and steep beyond belief and at its bottom it crossed a thin earth-filled dam. Behind this dam were collected about six or eight acres of very muddy water in which stood a number of dead trees. The road then took a sharp turn up a steep grade before coming out on the level area where stood the buildings of this enterprise. One could well imagine what the results of a sudden rainstorm would do to the dam we had just crossed. It was obvious that the road and the dam as well had been constructed without the consultation or advice of any qualified engineer. The main buildings consisted of one large lodge of squarish outline and in the poorest of architectural style. The other buildings were those which housed the office and the cafe and another which was designated as a bunkhouse. The lines of these were even more ungracious than the main building. They were so cheap and garish as to be painful to behold. All this assemblage of crummy architecture stood forth

One of the fishing lakes on the Ozark Wildlife Club property.

bare and stark along the ridge. Its builders had bulldozed down every tree in the vicinity leaving the area as barren as if it had been on the plains of Kansas.

It was well past noontime and our party being desirous of a place to eat considered the cafe at the Wildlife Club office; however, as our impression of it grew steadily worse, this was decided against. Since it had been our objective to inquire of the officers of the club on duty as to whether or not we could conduct a hiking trip around the top of Horn Mountain in November, we went into the office for the purpose of this interview. Here we met several young men and discussed our plans for a camping trip at Iceledo followed by a driving and walking trip around the north end of the mountain. Up until a few days before, we had not been aware that Horn Mountain lay within the jurisdiction of the Ozark Wildlife Club. Today we learned that it did indeed lie within the boundaries of the organization and that were we to hold the meeting as planned, we might become involved against our will in a "wild boar hunt." Upon learning who we were, the representative of the club kindly offered to move the hunting operations to the other end of the mountain while we were there. This seemed to be a fair enough arrangement, but upon reflecting upon the situation it was decided against because our presence on Horn Mountain in the capacity of an official

Ozark Society hike would give this organization another item to use in their advertising. When we expressed some concern about the probability of being unwilling targets in a wild boar hunt on Horn Mountain, an older gentleman who happened to be present and who was dressed in camouflaged clothing, obviously a hunter just in from the chase, informed us that we had nothing to worry about since most of the guides and none of the hunters were able to hit anything anyway. This conversation seemed to indicate that this particular hunter had been unable to hit the home-grown hog provided for him by this sporting organization.

Without making any further commitments in reference to our outing, we departed from the office with the intentions of driving to one of the picnic areas which lay a mile or two south of the main buildings. We had brought along sufficient lunch for this purpose, and by now we were more than ready for a decent snack. Driving over a freshly constructed dirt road, we descended a long hill and came to the picnic area. Here one of the most dismal sights that is to be seen anywhere in the Ozarks met our eyes. Bulldozers had been at work in the deep gullies and ravines on the west side of Horn Mountain and had gouged out a series of what were supposed to be lakes. All of the dams were earth fill like the one which we had driven across. Most of the fill and the borders of these so-called lakes consisted of clay and shale. The waters of these lakes were muddy and the banks slippery, muddy, and eroded. Standing about in the water were still more dead trees. On the banks of each of these bodies of water are signs indicating the kind of fish that one might expect to catch; however, it was noted that the kokanee salmon had been omitted. The picnic areas with the tables which had been provided were just as dismal as the waters nearby. While we were beholding this woeful scene, an automobile driven by one of the salesmen of the Ozark Wildlife Club came by bearing prospective clients. The salesman was busy explaining all of the various facilities which were about to be installed, and at once it was evident that the developers of this sorry enterprise were trying to build, in addition to a hunting reserve for sportsmen, a resettlement village for retired people something after the manner of the already successful Cherokee Village near Hardy.

By this time we had seen enough, and we headed back for the crossing at Cave Creek where a small picnic area had existed for many years and where the scenery was less depressing. After the usual picnic lunch we then turned back westward toward Jasper and Harrison.

That evening we traveled up the long eastern slope of Gaither Mountain and beheld the lights of Harrison along the eastern horizon and watched the moon rise over the Buffalo Basin far below. What we had seen this day had brought home to us something which we had not fully realized before. It was that commercial developments and inflationary real-estate activities pose a threat to the natural scenic wonders of our Ozark country just as great as that of the determined and senseless governmental bureaus who wish to dam, drown, and modify it beyond recognition. The real problem revolves around the definition of progress in reference to what should be done with this once beautiful land.

The Moore Creek Gorge and Sikes Bluff

On our many trips to Boxley Valley, perhaps the most impressive scenery along the way was the descent from the top of the mountain on Highway 21 to its junction at Highway 43 at Boxley. The view to the west from that road encompassed the spectacular gorge of Moore Creek and its northern rampart, Sikes Bluff. Below that precipice lie great broken-off blocks of sandstone, and in the bottom can be seen two-hundred-foot-high limestone cliffs bordering Moore Creek as it nears its junction with the upper Buffalo.

From what we had seen from the road, and after studying the latest maps of the area, Joe Clark and I determined that our next adventure would be to walk that northern rim of Moore Creek. That necessitated a visit to Clyde Villines, whose property it was. We found this elderly native cutting firewood up on his mountain-top ranch. Speaking with an inflection and style harping back to those first coming Elizabethans, he let us know that he was glad that we wanted to see the big bluffs bordering the southern boundary of his land. He gave us directions as to how to get there.

After going through a gate on the south side of Highway 21 and crossing an open field, we parked the Scout, and Joe, Maxine, Laurene, and I trudged on through brushy overgrown areas, over open marshy spots, and down long wooded slopes. Suddenly the rocky woodland gave way to nothing. Stepping across a deep crevice, we walked out upon the flat surface of a great three-cornered rock platform from whence we were afforded a breathtaking view of the Moore Creek gorge, whose bottom lay twelve hundred or thirteen hundred feet directly below us with the wider valley of the Buffalo River and the mountains of its eastern rim filling up the distance. Looking westward along the face of Sikes Bluff, it was observed that it was not one uninterrupted cliff line. At six or seven places projections similar to the one we stood on could be seen, with the last and largest of these standing prominently like an inverted cone at the farthest visible limit of the bluff about a mile away. We would visit these overlooks with our main objective being the big rock in the distance. The first of these overlooks was covered with a grove of gnarled pines. From here we could look down almost vertically into the bottom of Moore Creek gorge and could

Joe Clark and the big black gum log, Casey Mill, Boxley, October 1967.

see the tall limestone bluffs bordering its course. We could hear the roar of waterfalls, borne upward from that abyss. One was at once reminded of the similarities between this and the Grand Canyon. There one can stand on the topmost rim and hear the roar of the Colorado River almost a mile below. Here the distance between the observer and the rushing waters of the stream is only thirteen hundred feet, but the eye and the ear find it hard to detect the difference between the two.

While exulting in the visual impact of the vast scenery, a flock of pine siskins arrived in the tops of the pines over our heads, and soon chaff from the pine cones, which is their source of food, came floating down around us. Laurene and Maxine, who are confirmed bird watchers, were enthralled with all of this avian activity.

Clyde Villines had told us about a big, free-standing column of rock with one slab piled on another, and presently we came down out of the woods again to the bluff line and beheld this immense piece of natural masonry. It stood about fifty feet away from the bluff, and its vertical height was about one hundred feet. It was composed of one flat slab of rock stacked on another, and on its very top grew a small pine tree and a clump of huckleberry bushes. So far as we knew it

The coral head, from Moore Creek, upper Buffalo River, 1967.

had no name, but it might as well be called Pancake Rock since it looked like nothing more than a gigantic stack of stony pancakes.

Shortly after noon, we were on the big overlook up in the sky above the soaring buzzards who wheeled and circled far below us. In the distance down in Boxley Valley we could see the shiny thread of the highway winding past the almost-invisible form of the Boxley Community Building. To the east stood Winding Stair Mountain, which is ascended by Highway 21 as it proceeds from Boxley to Kingston.

As we returned to the Lodge we passed by the Casey sawmill in Boxley where a brief stop was made to behold something which we had been informed of by Harold Hedges. A few days before he had observed an enormous black gum log lying in the yard at the mill. He stated that he had counted its rings and believed that it was over five hundred years old at the time it was cut. We discovered that the log was still present, and we too made an estimate of its age. An accurate count could not be made since the center six inches of it was hollow, but it was ascertainable that this big gum was easily more than five hundred years old. It had been a living tree when Columbus discovered America, standing in some lost hollow all these many years and had witnessed the rising tide of our American empire. In October 1967 in a few minutes it had been felled with a chain saw by some logger who stood to profit practically nothing from its ancient carcass.

In harvesting timber in this country no rhyme, reason, or long-term plan is exercised. Long ago the prime hardwood such as white oak, walnut, and hickory had been removed, and today the loggers range the hills and hollows in search of trees once held in disdain, such as black gum, sweet gum, sycamore, and beech. It seems hopeless that these local sawmillers can ever be educated into methods of sustained, nondestructive forestry.

On our way back to Ponca we resolved to return to Moore Creek at the earliest opportunity and to descend from the cliff top all the way to the bottom and to follow Moore Creek on down to Boxley.

Thus a few weeks later we were back on the high overlook considering the best way down into the gorge. Instead of bright sunshine that day we were to behold long curtains of dark clouds sailing in from the Gulf to be impaled on the high ridges of Cave Mountain, Shiloh Mountain, and Winding Stair. From time to time streamers of mist from the bottoms of the cloud bank swirled majestically down across the face of the cliffs, giving us an inspiring view of the Ozark mountain country as any we had ever had. But on that trip there was more in store for us than mist.

When we finally attained the bottom heavy rain began falling, and it was necessary to don our ponchos and watch our steps on the slippery boulders. While studying the creek bed for a safe place to set foot, my attention was drawn to a strange limestone rock about eighteen inches in diameter lying in the bed of the stream. Its surface was rounded and presented a pleasing pattern, obviously a fossil of some living thing. Joe Clark was called in consultation and immediately identified it as a coral head, which I should have recognized from my exposure to such on the reef-ringed isles of the South Seas. What a discovery it was to stir the imagination! Here once flowed the tropical Ozarkian Sea, and here those soft-bodied polyps built their stony abode to form reefs in that shallow ocean, just as they do today in the waters off Espiritu. Here in this lump of limestone we beheld a marvel of nature and a miracle in time, a token of aeons past to remind us of the continuity of life and the mystery of this earth upon which we pass our days.

Joe Clark was so taken with this hefty artifact that he declared his intention to carry it out with him, and this he bravely undertook to do. We could not help but be amused at his appearance, trudging along like a farmer on his way to market with a heavy cabbage, first on his shoulder, then under his arm, and sometimes resting on his stomach.

Presently we came to the upper end of the great limestone bluff that could be seen from the overlook, so far above us now. Bee Bluff, as some of the natives call it, towers 250 to 300 feet above Moore Creek and in some places overhangs it. At its upper end was an enormous boulder dam that had been laid down by nature. It consisted of a multitude of rounded sandstone blocks from the cliffs above, now piled across the course of the stream. They were covered almost entirely with a carpet of gray-green lichens, bright green mosses, and ferns. In this case there was no pool of water above this dam, but the stream in its sharp descent had churned out a big hole below it, a pool of blue-green water.

By now the rain had slackened, and we continued our way slowly down the bed of Moore Creek. The colors of the forest, the hills, and the boulders were especially pleasing under these atmospheric conditions, however. The mosses and lichens were especially brilliant in the even light from the overcast sky, and the brown leaves on the forest floor took on an especially rich and lustrous tone.

On this second visit to Moore Creek we had with us a number of Ozark Society people: Jim Ranchino and his wife, Veda, from Arkadelphia; Dick Murray and Jack Diggs from Fayetteville; my wife, Laurene; and Joe and Maxine Clark. Attending to the car shuttle were Chalmers and June Davis from Pine Bluff, and from Boxley and Ponca Louisa Crow and Mrs. Ott Young. They had moved our cars down to the property of Mrs. Young's parents, the Jim Reynoldses, who lived near the mouth of Moore Creek. Jim Reynolds, who had been at his barn at the time we appeared, returned to his house and announced to his family and to our drivers that the hikers had arrived and that one of them was carrying a big rock.

An Assemblage of Characters

On the occasion of that first trip to Moore Creek in October, we had returned to the Lost Valley Lodge as early as possible so as to prepare for a two-day float on the Buffalo from Ponca to Pruitt on Saturday and Sunday. It was to be a combined activity of the Ozark Wilderness Waterways Club and the Ozark Society, with a few wandering characters thrown in for good measure.

Upon entering the lodge, we found that Dick Phelan had come over from his home in Pea Ridge to join us. Dick was an outdoor writer for *Sports Illustrated* and other magazines. It was his intention to write a report on the Buffalo for *Sports Illustrated,* and we were happy to have him along. He had just returned from a three-hundred-mile walking trip through the Sierra Madre Mountains of old Mexico. He stated that he had been in country where the natives could not speak Spanish, let alone English, and that he was soon to return there for another three-hundred-mile hike to finish it up.

About then we heard a clomping on the stairs, and in walked a couple of fellows whose appearance was startling to say the least. The first of these was tall, broad shouldered, and rugged looking. He wore a full beard but no hat to conceal a balding dome and had on a loose blouse, shorts, and sandals. In this raiment he looked like an athletic version of Jesus or certainly one of the disciples. His blue eyes sparkled with friendliness and good humor. In a whiskery sort of way he resembled someone that we should have known somewhere. That he was indeed. He was Ron Guenther, come all the way from California for another float on the Buffalo River. His partner, Charley Penn, was a shorter, beardless individual, who in place of going bareheaded wore a large handsome digger hat. He too wore shorts and tennis shoes, and both of them were in need of a change of clothes and a bath.

They informed us that they intended to float the full length of the Buffalo from Ponca to White River, and that they would do, all in one week without accident or capsizing in their heavily laden canoe, a remarkable achievement in our book.

Another character of a different sort had arrived the day before from the opposite coast. We had known that we were going to be joined by Bill Henry, who lived in North Carolina and who did his floating in a German-made Klepper Folbot, a kayak that could be folded up and carried in a car trunk.

On Friday morning, the day before the regular float, Joe Clark helped Bill assemble his beautiful blue kayak at the low-water bridge and reported that it disappeared into the mist down river, the big paddle flapping, like a blue goose trying to become airborne. We were astonished beyond words to learn that evening that Bill Henry had completed the twenty-six-mile run from Ponca to Pruitt in eight hours or less. For us it had always been a two-day trip.

Back at Ponca that evening we had a short discussion with Bill concerning what he thought of the Buffalo River. He stated that it was not nearly as wild as some of the waters he had been on elsewhere but that it was possibly the most beautiful stream he had seen, that because of the tall cliffs and the clarity of the water on which he was borne. After finishing his solo trip, he would join us the next day, only to leave us canoe paddlers drifting in his wake.

Buffalo River Float, Ponca to Pruitt

The morning of Saturday, October 21, found the Buffalo River Valley full of heavy fog. We awoke in the cold and darkness, and it was with real effort that we managed to dress and to prepare breakfast and to get our equipment and canoes ready for the river. Harold and Margaret Hedges appeared upon the scene and graciously offered to ferry one of our canoes down to the river bank since it was perched on Dick Phelan's truck in a rather precarious way.

Both banks of the river at the low-water bridge were that morning alive with

Put-in activity at the Ponca low-water bridge.

activity. Altogether thirty-five or thirty-six canoes and kayaks had to be put into the water with all the gear and equipment necessary to take care of double that many people. We did not have an opportunity to find out just who was present, and each group departed from time to time down river disappearing into the mist and fog. As is usual, the Ozark Society group proved to be slow on the take-off, and it was around ten o'clock before we finally got our part of the expedition underway. By this time the fog had lifted and beautiful clear blue sky appeared overhead. The sun on the sparkling waters of the Buffalo was a joyous sight to behold and lifted our spirits out of the gloom. The only major problem now seemed to be overloaded canoes. In our enthusiasm not to leave behind anything that one could possibly need, it seemed that we had aboard everything but the kitchen sink, and I sincerely expected to discover this on the bottom upon unloading that evening. One likes to feel his craft quick and responsive to the stroke of the paddle, but today it seemed almost as if I was paddling a loaded truck. In spite of this and the lack of comprehension of each other's navigation technique, Laurene and I managed to make the journey from Ponca to Pruitt without capsizing. This does not mean to imply that humiliation was unknown to us. On more than one occasion the gunwales were awash, we were broached in the rapids with disaster seeming certain, we went pinwheeling down long chutes bow after stern, we crashed into willows and into rocky banks and slithered over sunken logs and limbs.

The water in the Buffalo River was falling during this float, and thus the trip was not unduly hazardous. In fact, only two capsizings occurred. One of these was in deep water where the current makes a sudden right angle bend, and the canoeists Stanley Kahn and his wife struck the bank and overturned but fortunately lost little of their essential equipment or supplies. In another long, rocky chute a member of the party struck bottom and broached with the canoe filling rapidly. Here it took prolonged effort to extract the wrecked canoe from the channel, and a considerable amount of stomping and bending was necessary in order to straighten out the metal of the hull afterward. Again, no important items were lost, and at about 4:30 P.M. we had all arrived at the gravel bar below Goat Bluff where we were to spend the night.

Here in very short order a small tent city sprang up within a few minutes, and fragrant campfires were going from one end to the other. In a little while our canoe tent was up, and one of our craft was removed and turned upside down to serve as a table. Laurene and Maxine busied themselves with preparations for supper while Joe and I rounded up enough wood for the evening and morning fires. After we were well situated, we indulged in a period of visiting others of our party. We were especially interested in Bill Henry's shelter. In a kayak very little room is available for tents and equipment. He was just as comfortable as any of the others, however, with a small nylon tarp which weighed only a few ounces serving as protection against the nighttime dew. Under this he had unrolled his sleeping bag which was also of unusually light weight, and with a minimum effort he had made himself warm and comfortable. For this narrator it turned out to be another one of those nights on the hard, hard ground. The air mattress is a fine invention. It is lightweight and when deflated fits into a small space, but experience has shown that these devices all too often leak. Within twenty minutes, I was on the rocks and there remained for the rest of the night.

We had not been through this section of the river since the famous tree-cutting incidents and other nasty doings on Memorial Day 1965. On this part of the river there were several obstacles which had been there for literally years and which we expected to see again. We were certain that the trees cut by disgruntled dam supporters would make matters all the worse; however, on arriving at the first obstacle where two large sycamore logs had lain across the channel of the stream for many years, we were pleasantly surprised to find that they had been cut and were no longer in the way. Also we well remembered Dead Man's Curve where the swift current actually makes a U-turn which is almost unsurmountable even by the best canoeists. Many, many dumpings had occurred at this point, and again we were surprised that the big tree which had stood at the base of the U and which helped to hold the treacherous bank on that side had been cut. It was one of those felled by the dam supporters, and after this the current of the river had straightened out the channel with the big log lying lengthwise in the stream bed and not representing any serious dangers to floaters. This section of the river is thus open and easily traversed, and we made much better time than had been anticipated. At about 2:30 P.M. we arrived at Pruitt, beached our canoe, and trudged up the bank to have another interesting conversation with Pearl Holland.

It was with genuine regret that we bade good-bye to Louisa Crow and all of our friends, for this had undoubtedly been one of the most pleasant experiences that we have ever had. One could not help but feel deep remorse upon realizing that in all of the years of visiting and traveling through this part of the high Ozarks that we had never spent more than a night or two in succession in this interesting country. However, it was not too late to make this old lodge our headquarters on future trips to the Upper Buffalo, and this we resolved to do as we departed on that Sunday evening in October 1967. As we drove up the steep grade to Highway 21 and over the mountain toward Kingston, the sun sank ever lower toward the western hills, and on this particular day a glorious deep bronze illumination spread over the sky and over the hills. Turning north at Old Alabam we took the cutoff to bring us out at Forum on Highway 23, and just as we approached the junction of these two country highways a line of moving objects appeared, flying low in the sunset, moving swiftly from north to south. It was a flock of wild geese on their way to some winter feeding ground far from these dry Ozark uplands. As it has always been, it was thrilling to see these great birds winging so surely the route determined for them by countless generations of other wild geese who saw below them not highways, chicken houses, villages, and towns but the limitless forests and prairies of yesterday. In those days here below were the great magnificent animals of the ice age, the mastodon, and the mammoth, the giant bison and the musk-ox and the dire wolf and the saber-toothed tiger. Here they were to be found as numerous as are cattle today on these hills, but they have perished to the very last, perhaps by the hand of man.

In considering this, one could not help but feel apprehension that even the wild geese are following the same route toward extinction and for the same causes. I could recollect that during boyhood, now many years ago, each fall and spring migrating wild fowl crossed our Ozark skies in considerable numbers. Each day the ducks and geese were to be seen winging their way across the blue, but on this day I knew that this would be the only flock of geese that we would be privileged to see or hear during this fall in 1967, and so it was. The sight of this single flock hurtling into the darkness of night brought home the thought that the problem of their existence and ours is not so far removed one from another. In contesting the senseless ruination of the American environment by engineers, agriculturists, real-estate developers, and politicians, we are fighting a battle that involves the fate and happiness of humanity as well as the marvelous creatures that inhabit the earth with us. Our task is thus clear. We must instill into the hearts of a majority of common men and women a compassion for and appreciation of the natural world. This is our basic obligation—so simple—so reasonable—and yet so difficult.

SIXTEEN

THE TRANSMOGRIFICATION OF THE BRIA

Following all of these setbacks we had reason to expect that the BRIA would go off somewhere and die, leaving us the job of attending to the birth of the Buffalo National River without further harassment—but not so.

The new year had scarcely begun when we were presented with an unsavory headline in the *Arkansas Gazette*, January 7, 1968:

> **Landowners along Buffalo May "Close" It**
> **Plan Moves in Fight to Defeat Proposal for National River**
>
> MARSHALL—Landowners along the Buffalo River are contemplating closing the river to floating enthusiasts this spring and summer.
>
> Apparently with little hope left of getting a dam on the controversial stream, this will be the way the landowners and pro-dam advocates in Newton and Searcy Counties plan to fight a proposed national river status for the Buffalo.
>
> About 150 landowners and businessmen from Newton and Searcy Counties, the principal counties through which the river runs, met at the Searcy County Courthouse here Friday night to talk strategy, according to James R. Tudor of Marshall. ...
>
> "They were all enthusiastic to keep the park from going in," Tudor said. "It's strongly opposed by all the property owners up and down the river ..."
>
> About two weeks ago, property owners along the river in the eastern part of Newton County met at Hasty and vowed to stop the national river plan. This group also proposed forming an organization of the landowners. ...
>
> "We've forgotten the dams. There's no possibility of getting that now. We're just trying now to save the Buffalo. We don't want a park on the thing."
>
> The majority of the landowners, Tudor said, are talking about closing the river "and keeping them off it. They're emphatic about trespassing—floaters, Park [National Park Service] personnel or anybody else."
>
> Tudor said that the river is not considered a navigable stream, so property owners along it have ownership of the river bed and control of the use of the river.

Compton Feels Quarrel Sought

Dr. Neil Compton of Bentonville, president of the Ozark Society, which has fought to keep the Buffalo a free-flowing stream and for the national river, felt that the landowners' efforts were an attempt to pick a quarrel. ...

However, he said, if the landowners persist in trying to keep the floaters out "I think people ought to stay off. We want to show that we aren't out to pick a quarrel with them. We believe they are wrong. I feel that if they do this, the consequences will rebound against them pretty hard."

Compton said he thought the time had come when the anti-national river people "should accept their defeat gracefully and go along with the rest of the state to get the Buffalo into the national river system. I'd rather not go over there and pick a fight with them. It's much better that we try to be reasonable and hope they'll eventually come around."

... Tudor said that one reason for the fight against the national river plan was because it would take more land off the tax books. The government already owns "no telling how many millions of acres" along the lower end of the Buffalo in Searcy and Baxter Counties, Tudor said, and several million more along the upper end of the stream.

"They've already developed camp sites and got streams all over the place," he said.

... "We want it left just like it is. We're not arguing for the dam now. We just want it [the river] left alone. The dam's already out."

The BRIA had retreated, not to some dark hollow to breathe its last, but to its lair in the Searcy County courthouse, there to become a changeling whose face and form would appear in various guises as time went by.

They now were in the process of usurping the title of those authentic landowners who were watching these contortions with amusement and with the hope that neither contestant would win out in the long run. But Will Goggin and his people would still choose the park over the dam if it had to be.

Sailing under these new colors, the BRIA crew set forth to Jasper, Pruitt, Hasty, and Western Grove to foment an anti-national park revolution. In that they would meet with limited success, and several new, strange, and troublesome characters would join the fray.

The Standoff at Hasty

Early in February 1968 a sticky proposition was delivered to me via certified mail. It was a polite invitation from J. Maurice Tudor (a son of Jim), who with his brother Eddie had taken over the management of the *Marshall Mountain Wave*, requesting that I and others from the Ozark Society attend a countywide meeting of the Newton County landowners at the hamlet at Hasty on the middle Buffalo River on March 2. At that meeting, of which he would be moderator, he stated that we would be given a chance to plead the case for the Buffalo National River and that we would not be subjected to any of the violence that had been so strongly hinted at in news reports recently.

My first reaction was that this was a cease-fire and the BRIA and its successors were ready to negotiate. By return mail I accepted the offer and sent out requests to all of our supporters to be there, if possible. That included Ed Freeman at the *Pine Bluff Commercial*; John Fleming at the *Arkansas Gazette*; Ken Smith, (now in Arlington, Virginia); John Heuston; H. Charles Johnston; and our legal advisor, Clayton Little. We soon learned that John Paul Hammerschmidt, Senator Fulbright, and Bernie Campbell had also been invited.

Much publicity of the event was soon in evidence, in the newspapers, on the radio, and even some short clips on the TV. The "Landowners Association" had even had handbills printed to boost the event.

We were taken aback a few days later to read an account in the *Tulsa World* which sounded as if it might have been written by the editor of the *Marshall Mountain Wave*. The *Tulsa World* had been for years a popular and influential newspaper in northwest Arkansas and had previously not been concerned in the Buffalo River problem. If they were to enter the fray on the side of this pro-dam Landowners Association, our position in the matter could be seriously embarrassed.

Since Ken Smith had been so pointedly selected as an invitee, I felt obliged to keep him up-to-date on the forthcoming Buffalo River summit at Hasty, February 19, 1968:

> I have further important information concerning Maurice Tudor and the meeting at Hasty on March the 2nd.
>
> I did not regard a simple letter of acceptance to Tudor as being sufficient, and therefore this morning I called him on the phone to talk it over. In my conversation with him he was somewhat guarded and tried to make it appear as though he was not deeply involved with this meeting. He tried to make it seem that his being assigned moderator of the meeting was a sort of accident and that he didn't really know very much about the people in the Hasty area over in Newton County. He said that Mr. Viorel, the leader of this group, had requested that he serve in this capacity and that it would not be unlikely that we might meet with some hostility from some of the members of that group. He reiterated his intention of conducting a peaceful meeting, however, but I am certain that this get-together will be no love-in. It does represent, however, a real opportunity to take a first step toward better relations later on. ... I feel certain that Jim and Eddie and all the other Tudors know quite well what is going on and also that Maurice knows much more than he was willing to tell me on the phone.
>
> The thing that will interest you most, however, was the fact that they would be

mighty pleased to have the fellow who wrote the book about the Buffalo River present at this meeting. He stated to me that it was a "splendid" piece of work, and in saying this his tone was sincere and I have no doubt that he meant it. My immediate thought was that if he felt like that about it he was on our side. … This is something that you have cause to be more than proud of, and I am herewith suggesting that if you can persuade your superiors to give you the time off that you should be sent to this meeting expenses paid in order that some of these unsuspected admirers be given an opportunity to meet you. …

The National Park Service, of course, had no intention of dispatching Ken Smith all the way from Washington, D.C., where he was now stationed, to Hasty, Arkansas, to make an appearance before such a questionable panel.

Reflecting upon the situation, I, by now, was seriously wondering whether or not I had committed our people to a set-up of some kind. I suggested that Dick Murray and Col. Jack Diggs go over into the combat zone as special agents and find out more about this business at Hasty. That they were glad to do, making the rounds of the entire area and contacting numerous local citizens. They found out the sum total of nothing. Nobody seemed to know much about it; although they did refrain from going directly to the Tudors or to Viorel and his delegation.

I then called John Paul Hammerschmidt on the phone. His advice was emphatic: "Don't go." His experience with some of these people firsthand had been most unpleasant. In his opinion, there was nothing to gain and a lot to lose in attending. We should have been invited to assist in the planning of the event to begin with. We should have had a say as to the selection of the place, time, and the mechanics of such a meeting in order to assure fair treatment. Congressman Hammerschmidt's advice put a stop to any plan that we had made to attend. The word was passed to our people to stay at home, and it was my obligation to advise Maurice Tudor of that decision.

It turned out that a phone call to him was all that was necessary to end this series of negotiations with the BRIA. He did not seem to be at all surprised when I stated that we did not intend to be present. He seemed to accept as valid our contention that we just hadn't had time to get ready for this and that we did not have the information that we really needed to transmit to the people over there. He still seemed to be quite vague as just how this meeting was to be brought off, and I feel certain now that he was not one of the main planners of the meeting. He said that the reason that the Hastyites asked him over there was just because they liked him so much he reckoned. My impression was that he almost seemed relieved that we weren't going to go.

I have been left to wonder to this day what might have ensued had we been present at the Hasty meeting. We might very well have made some converts, and some of the unpleasantries that ensued from then on, even after the creation of the Buffalo National River, might have been avoided or at least lessened. On the other hand, there might have been bloodshed. There certainly were some present who would have been glad to participate in the letting.

The Hasty affair was reported by three newspapers whose generally anti-park slanted reporting we read with interest and unease (*Harrison Daily Times*, March 4, 1968):

Landowners to Plead Case in Washington;
Vow to Block Park

HASTY—With a fervent plea for private prayer "in your own homes for all those people who are persecuting you," about 200 landowners and citizens along the Buffalo River departed from a three-hour meeting in the small two-room school house here Saturday night vowing to continue their fight against a proposed National Park plan. ...

Almost identical bills have been introduced by Sen. J. W. Fulbright and Cong. John Paul Hammerschmidt.

Control of the river would pass from the hands of landowners to federal administrators.

It's the consensus of the landowners that if they can't have a dam, earlier recommended by the Corps of Engineers and killed by the Bureau of the Budget during the Eisenhower administration, then they want the river left as it is.

George Viorel, chairman, set the tone for the assembled gathering in repeated applause early in the Saturday night meeting when he accused "the very people we've tried to help—those who floated the river all for free—" as now "trying to take away what we wanted them to enjoy."

... [T]here were at least two political faces in the crowd and both drew applause for their remarks.

Hardy Croxton, Rogers attorney, who is rumored as a Democratic candidate for the Congressional seat in the Third District to oppose Hammerschmidt, said he was glad to accept an invitation to "discuss a common problem that has not been adequately gone into."

... Croxton declared, "When I do talk it will be clear, strong and unequivocal."

Bobby Hayes of Calico Rock who has already filed for U.S. Senator against Sen. Fulbright said, "I didn't come over here to straddle the fence, I'm with you."

... Al Hochberger, a real estate man from Jasper, voiced the fear that citizens would have to stop letting the federal government take land away if the county is to have sufficient revenue from the taxes to operate county government. ...

Mrs. Lucile Hannon of Pruitt and Harrison jeered at the ostensible popularity of polls taken to indicate interest in the park. "Anyone can take a poll," she said. "But we're here."

Harry Willmering of St. Joe offered a resolution which was unanimously adopted expressing the desire that the Buffalo remain as it is and title of ownership remain invested in the landowners.

The plea for private prayers came from Maurice Tudor, of Marshall, moderator of the meeting.

He said, "The only right way to handle anything in this life is to do good unto those who persecute you and turn to God. That means you all go home and pray for Dr. Neil Compton and Cong. Hammerschmidt … All those other people need our prayers … And if you pray long enough, hard enough and honest enough, the Lord might just come down here and give us a little help and the Lord knows we need it," he said.

In this report we note some newcomers to the fray, most of them recent arrivals in Arkansas. They would come to be major participants in the transmogrification of the BRIA from big dam builders to oppressed landowners.

George Viorel, a short, sturdy, balding, ex-navy boxing champ, had come from Oregon a short time before. He had bought a farm on the middle Buffalo below Hasty and was an authentic landowner but one who should have known that his property, when he bought it, was located in a zone being contested by the Corps of Engineers versus the National Park Service. Federal ownership was inevitable within a few years, a fact that many gung-ho real-estate agents failed to advise their clients of in those times.

Viorel was a man of quick perception, a man of decisive action and with words to go with it. For that reason, he was a natural for the chairmanship of the 1968 version of the Buffalo River Landowners Association.

Al Hochberger and Charles Petree were two new members of the lineup from whom one would not hear much more, but Miss Lucile Hannon was one who would set our ears abuzz. A grandmotherly figure, then in late middle age, she had been born and raised in Mississippi, had lived for several years in Kansas City, Missouri, and then a year or two before the Hasty affair had landed in Pruitt on the banks of the Buffalo where she set up shop as a CPA. She owned no real estate there but leased a tract which she defended with the zeal of Queen Boadicca leading the Britons against the legions of Rome. Her principal weapon was her typewriter from which poured forth as great a torrent of words as could possibly be generated in the mind of woman. The verbal lashing from that source we would have to endure from then on, even until the Buffalo National River had become a reality and for years thereafter.

Even though she was not in a position to do real damage or to alter the outcome, her rantings appeared regularly in the *Harrison Daily Times,* whose editor must have been in sympathy with the content. Finally, realizing the embarrassment that it was to his publication, he eventually phased out the Buffalo River Report by Lucile Hannon.

With that he was also obliged to discontinue the long-winded harangues to the editor generated by another malcontent about to come on the scene, Herb Van Deven, who would become a pillar in the evolving pro-dam anti-park coalition. Eventually that group was to parade under the cumbersome cognomen of: The Buffalo River Conservation and Recreation Council. For it, Van Deven would serve as the chief verbigerator, spouting lengthy platitudes, often with religious overtones, always ending with a question which his opponent could never answer without self-incrimination. He was a master of the loaded question, in other words.

The editor of the *Harrison Daily Times* was, in those days, John Newman, a long-time friend of fellow newspaperman Jim Tudor of the *Marshall Mountain Wave.* For that reason Newman afforded the big dam plans for the Buffalo the full support of his paper. That policy would continue after his retirement, with J. E. Dunlap as editor.

But in 1968 our concern over the hostility of these two local papers was increased by the entry of the *Tulsa World* on the side of the BRIA-BRLA. We would soon learn that its editor, John Steen, was a Searcy County boy, raised over around Marsena, near Snowball, Arkansas, and that he, too, wanted to do something for his old friend and colleague at the *Marshall Mountain Wave.* For that reason, he would send a special reporter to the Hasty jamboree.

How we felt about that was best said in a letter to Ken Smith, March 7, 1968:

You may recollect that I had mentioned that the BRIA had contacted the *Tulsa World* and that that paper had assigned a reporter by the name of Tom Omstead to the meeting. This man had written me early stating that he wanted to present both sides, and in my reply to him I suggested that he stop here on his way to Hasty for a talk. Following this I sent him a copy of *The Buffalo River Country,* also a copy of Governor Faubus' letter to General Cassidy and a copy of the Pearson newspaper series. He later acknowledged receipt of these but admitted not having read them. He had informed me that his boss had requested that this meeting be covered and so it appears that Tudor had some sort of connection with some of the editors at the *Tulsa World.*

As I have indicated before, this is an important paper in our part of the country. Normally it is a conservative Republican paper and should be willing to give space to the opinions of such a capable Republican representative as John Paul Hammerschmidt. During the last few days we have become aware that Hammerschmidt is to have hardcore democratic opposition and that the person concerned is going to go along with the BRIA and the Hastyites. To have the *Tulsa World* support Hammerschmidt's opponent in this would be a serious matter indeed.

Last night I had a telephone call from Tom Omstead, the reporter who covered the meeting, and this conversation lasted very nearly an hour. I was quite shaken to discover that he had been sold on most of the propaganda put out by Tudor and his henchmen. The most disturbing was that the Ozark Society is being financially backed by the Arkansas Power and Light Company. During this conversation it became clear to me why they wanted you to be present at the meeting. They had hoped to be able to harpoon you with accusations that your book also was financed by the Arkansas Power and Light Company. They do not believe, and I am afraid that this includes Omstead, that we have been able to do all that we have done with the limited resources at our disposal. When this reporter leveled the accusation that we were so financed, I am afraid that the conversation became slightly acrimonious. It was all that I could do to retain my temper, but following that exchange the conversation went on for about another thirty minutes during which time I tried to explain to him how we actually did operate. It all wound up with the understanding that perhaps I would be given an audience with either Mr. Omstead or some of the other people on the staff of the *Tulsa World*.

Next week on Thursday I am going over there to meet with the Tulsa Canoe Club and to discuss the possibility of forming a Chapter of Ozark Society in Tulsa. I took the liberty to suggest that Omstead or some of the other reporters be present at the meeting and have since contacted our members in Tulsa requesting that they formally invite them. I am also going to talk to the executive editor of the paper asking that he give me an audience in the afternoon before the meeting so that I can go over our side of it a little bit. ...

Joe Clark and I were successful in getting an audience with editor Steen, who was polite and attentive, but it didn't change a thing. However, after moderate reporting of the plight of threatened and oppressed landowners and a few inept remarks about the Ozark Society, the *Tulsa World* dropped the subject. The Buffalo River was two hundred miles, more or less, to the east in another province and was not that big an item in Oklahoma news.

But before that Editor Steen had also assigned his reporter, Omstead, to cover the Ozark Society spring meeting at Petit Jean State Park on April 6 and 7. We attempted to show Omstead every courtesy, but he was not very interested in us or the program. In reporting that event, his main statement was: "The Ozark Society was a small but educated band of nature lovers"—not necessarily an offensive phrase, but for some reason it raised the gall of Lucile Hannon when she read it. With that phrase she would belabor us with every other breath from then on, as if us "nature lovers" were agents of the true Satan fresh out of Hell.

Dissent Spreads to Jasper

Actually there was a fourth newspaper dedicated to the defeat of the national park plan for the Buffalo, one that we didn't see much of because of its limited circulation. *The Informer* was published in Jasper, the county seat of Newton County. However, we did come by one issue (Friday, March 28, 1968) which carried an account of another meeting of the anti-park people in Jasper not long after the Hasty convocation. In that account the name of a professional troublemaker first appeared. With the influence of the Tudors on the wane, Charles Thompson, of Harpers Ferry, West Virginia, would, for a time at least, call the shots for the crumbling BRIA, the current BRLA, and the about-to-be-born BRC and RC. The following is from *The Informer,* March 28, 1968:

Many Still Concerned over Fate of Buffalo River

Approximately 100 persons attended a meeting in regards to the proposed national park on the Buffalo River, March 19 in the Courtroom in Jasper.

George Viorel of Hasty, president of the Buffalo Land Owners Association, presided.

Featured speaker of the evening was Charles Thompson, president of American Landowners Association of Washington, D.C.

The meeting was held to present the local side of the issue to Mr. Thompson.

George Viorel said, "We have a fight on our hands."

He went on to say that the real issue is self determination and you do not force an issue upon a county. ...

Thompson said that the well to do have money and leisure time and want the Buffalo for only two days a week (the weekends) whereas the landowner uses the land seven days a week.

Thompson ... congratulated the residents of this county for being so stubborn and honest and said it was the first time he had seen a county population so stubborn to fight to the finish for what they believed in, in this case against the proposed park on the Buffalo River.

But for the moment, in early 1968, the principal pacesetter and warmonger for the BRLA was Lucile Hannon.

For some time we had been preparing to attend a hearing in Washington, D.C., which was scheduled for April 18 and 19, but just before our meeting at Petit Jean we received word that it had been cancelled, and we passed the word to our members. Word of that cancellation sent Lucile Hannon into a monumental rage. She thought that it was a trick to get the BRLA to make an unnecessary and expensive trip to Washington to oppose the program. Her account of that underhanded nastiness appeared in her column in the *Harrison Daily Times* for April 19:

Buffalo River Report by Lucile Hannon, Owner Shady Grove Trailer Park, Pruitt, Ark.

This particular report is an open letter, with special attention to Dr. Neil Compton, President of the Ozark Society Bird Watcher and Canoe Floating associates; all reliable and false news media, Ark. Dem. Senators J. William Fulbright and John McClellan, Republican Congressman John Paul Hammerschmidt, Secy. of the

Dept. of Interior, Stewart Udall, and all multiple allies for a National Park or River on the Buffalo River in Arkansas: Just ten minutes after the release of our paid for Buffalo River Report in Thursday's Harrison Daily Times, true confirmation reached us of a fact given by Cong. John Paul Hammerschmidt to one of our land owners on the Buffalo River Sunday night over the phone, stating that: "THERE WOULD BE NO MEETING IN WASHINGTON, D.C. ON APRIL 18TH AND 19TH INVOLVING THE BUFFALO RIVER IN ARKANSAS."

We didn't believe him then, but now we do, though we are not apologizing, because we felt like a reliable paper such as the Tulsa World, and its Staff Reporter, Tom Omstead, was printing the true minutes "out of the horse's mouth" of the last meeting of the Ozark Society at Petit Jean State Park on April 6th and 7th, in which the following FALSE STATEMENT was made: (we don't have the society member's name now, but we'll get it for you) "ON APRIL 18TH, 19TH, A CONGRESSIONAL HEARING IS SCHEDULED IN WASHINGTON, D.C., WHICH, IF APPROVED WILL ESTABLISH A NATIONAL PARK ON THE RIVER" (BUFFALO).

All we can say, and with reliable information, that Staff Reporter, Tom Omstead of the Tulsa World is now in the same "scapegoat pen" with us people here on the Buffalo River. And so is Republican Congressman, John Paul Hammerschmidt. How does it feel to be cramped up in this "scapegoat pen?" A "Judas" of the Society informed reliable sources that it was the true intent of this particular false conspiracy when they used this reporter and his reliable paper, the Tulsa World, to nationally syndicate this news, was to bluff the Buffalo River delegates into hopping a plane, with a brand new pair of slick-sole cramping shoes, and a borrowed suit case, to convene at a hearing on the Buffalo River in Washington, D.C. which was not to be. ...

Shortly after the meetings at Hasty and Jasper the BRLA, feeling that they had started a real anti–national park revolution, decided to schedule an even larger meeting in Western Grove. Chairman Viorel, seeking a top figure, invited Senator Fulbright to be the principal speaker.

Senator Fulbright's views were already well known by everyone concerned, and he chose not to participate in an event wherein his political image could be made to suffer. He knew very well that the ALA-BRIA people were already committed to the election of his opponent, Bobby Hays of Calico Rock, in the next election.

Just what Charles Thompson's motives were was open to question. How he had made contact with the anti-park faction we didn't know, but we would discover that George Viorel, in whose home he was first received, mistrusted him from the outset. Thompson was in the business of organizing a nationwide lobbying group to oppose all new national and public parks, all wild rivers projects, and all proposed wilderness areas. We were aware that he had had success in some quarters and regarded his appearance in Arkansas with concern. He was chairman of what was called the American Landowners Association (the ALA), striving to increase its membership and influence throughout the country.

Senator Fulbright declined the invitation, and we have no record of what happened at Western Grove.

The same political alignment existed in the case of Congressman Hammerschmidt, whose opponent was Hardy Croxton, a Rogers lawyer. Thompson was astute enough to recognize the fact that Congressman Hammerschmidt was the key figure in their effort to stop the national river plan. Papers that later fell into our hands revealed that he had contacted Senator Fulbright's office to sound out what the result of Hammerschmidt's defeat by a Democrat might be, hoping perhaps that Fulbright (a Democrat) might be partisan enough to offer assistance. His report was: defeat Hammerschmidt and you have won a large-scale victory. Commentary from Viorel was: "Many here voted for John Paul. He also carried Searcy County."

In a message to Viorel, Thompson decried the appearance of other national river plans, notably the Potomac, to which Viorel inscribed a note: "How funny—Maryland passed the Scenic Rivers Bill—This is Thompson's home state—Where were you, Thompson?"

The Meeting at Petit Jean and the Pine Bluff Riot

Through all this the Ozark Society had attended to its own knitting, some of which was, as usual, tangled up. In a letter to Ken Smith dated January 24, 1968, it was necessary for me to extend some brotherly advice concerning our star media man, Harry Pearson, who had been gone from the scene to parts unknown. The gist of that message was:

> Harry is back working for the *Pine Bluff Commercial,* and his performance has improved. The truth is he was stymied by the prospect of writing the book—as it now stands, he will not produce a book on the Buffalo River within the foreseeable future. I do not know what effect this will have on his federal grant, but nevertheless it is true. The thing that concerns me is that he is in possession of all of your original papers and excellent maps that you had given him in good faith to help with his book. He is an unreliable custodian for such documents, and I would like to suggest that you retrieve this material in a diplomatic way when you are back here.
>
> You realize that these papers grow increasingly more important as time goes on and as events transpire. I would like very much to make copies of some of these papers to add to my own files, which are now very large indeed.

Whatever Harry's shortcomings, we valued his contributions to the cause whenever he was moved to make them. He had just organized the Delta Chapter of the Ozark Society in Pine Bluff, an enthusiastic and important group of coworkers, and we were not anxious to alienate him.

Fortunately, Ken already had copies of most of his papers to replace those that would go up in flames along with Harry's big house out on Long Island several years later.

In those first months of 1968 we were busy with plans for the spring meeting of the Ozark Society at Petit Jean State Park in early April. The new Delta Chapter in Pine Bluff had been delegated as host and, not to be outdone by the Hasty affair, had nerve enough to ask Jim Tudor to appear on the program. Like me, he first accepted but in the same vein finally declined, which was probably just as well.

Arriving at the park office in Mather Lodge shortly after 7:00 P.M. on April 5, Laurene and I proved to be the only members there that early, and we elected to take lodging in one of the new larger cabins. There we would have room enough to assemble easels and stick photographs in mounts for the photo display the next day. Laurene had brought her small portable radio and had kept it on most of the evening listening to a Little Rock station. As a result, we learned something which had a direct bearing on the course of our meeting. At about 12:30 A.M., a brief announcement was made stating that extensive rioting was at the moment in progress in the city of Pine Bluff. We immediately visualized the plight of our colleagues and especially of Harry Pearson and others of the *Pine Bluff Commercial* who would necessarily be involved in this catastrophic event. We felt that this might be the cause of considerable absenteeism from our meeting, and that it did; but the personal safety of our people was no less a concern.

Saturday morning we were happy to see some from the Delta Chapter on hand. Chalmers and June Davis, Alice Dickey, and Harold Franklin were there and ready to start the meeting.

From Tulsa there was Tom Omstead, the reporter for the *Tulsa World,* and his two daughters. Omstead spent most of his time out on the entryway to the lodge watching the comings and goings of the humans in attendance and the big birds that sailed gracefully in the updrafts over the bluffs of Cedar Creek. In honest ignorance he inquired of us as to what they were. Surprised that an Okie wouldn't know a buzzard when he saw one, we told him what they were. With that he was so impressed as to categorize us as "educated nature lovers" in his write-up, whether we deserved it or not. From him it was all that we got.

That evening we had a call from Harry Pearson in Pine Bluff. He told us in horrified tones that he had nearly been killed in the riot, not that he had had hands laid on him, but that he would have had he not taken supreme evasive action. He had been out riding his bike when he unexpectedly came upon a band of hoodlums who threatened to do away with him in more ways than one. At that he turned his bike about and rode it through a man-proof hedge, making good his escape except for a few scratches.

We had made special efforts to have good news coverage at this meeting in order to dispel some of the propaganda being generated by the BRIA-ALA, but all the newspapers had held their reporters at their stations, except for the *Tulsa World,* because of the rioting across the nation. But in spite of everything, at adjournment we felt that it was the first meeting at which the functioning of effective organizational machinery could be observed at an Ozark Society affair.

By 1968 we had come to realize that the scenic beauty of the Ozarks was imperiled by much more than too many big dams. Some of the most serious damage was at this time highway overbuilding. Having been slow to recognize this fact, some of us by 1968 had been awakened by a new surge of road building then in progress in Boxley Valley. Upon personally investigating it on several occasions, we were overcome with remorse at what we saw. Highways 21 to the south up to Mossville and to the west to Kingston, 43 north to Compton, and 74 east to Jasper were all in the upheaval of excessive construction. There was nothing that anyone could do, but I felt compelled to, at least, report it to the Ozark Society at the Petit Jean meeting:

> In discussing highways, I wish to emphasize that I am in no disagreement with the need for good roads. ...
>
> But I do oppose the enormity of some of the projects now underway. As an example, attention is called to the present paving of Highway 43 from the village of Compton in Newton County down to Ponca on the Buffalo River. Up until lately this was a narrow gravel country highway winding down a deep and picturesque ravine. We knew that it had been scheduled for paving but understood that the blacktop was to be applied upon the old right-of-way or relatively close to it. If this had been the case, no essential damage would have ensued, but what actually is being done is shocking beyond words. Extra money became available for this project, and an 80-foot swath has been blasted out leaving a scar worse than that of a major volcanic eruption. This road will transform forever the nature and the character of the country into which it extends. An urban thoroughfare will have been built between villages whose population numbers only 40 or 50 each. It will be not unlike thousands of other miles of modern highway throughout the country and shatters permanently the sense of remoteness and the serene beauty of the upper Buffalo Valley. Even the local citizens who worked for this project are shocked and dismayed by its magnitude.
>
> It will not be possible for an organization such as ours to continuously evaluate activities of the State Highway Department but we do want to offer suggestions as to how such mistakes can be avoided in the future. We can urge the Highway Department to establish a policy for the tailoring of construction to fit the terrain. We recommend that they hire a qualified landscape architect to supervise such activities. Damage already done might be partially rectified thereby. Certainly it could be prevented in the future.

Formerly an arcaded mountain road, Arkansas Highway 21 descending Winding Stair Mountain into Boxley Valley lies sundered by excessive new construction, 1968.

The Highway Department should also be encouraged to adapt their construction program to take advantage of scenic overlooks where possible. Furthermore entire sections of some highways could be designated as scenic highways after the example of the Blue Ridge Parkway.

Later the Arkansas Conservation Council did meet with the highway department to discuss these recommendations. We were given patient attention, and

The bed of the Buffalo River on the Lewallen place above Boxley was ripped and torn by the highway department to obtain fill material for construction on Highway 21, 1969.

Part of a two-mile-long landslide on the Mossville hill, Arkansas Highway 21, south of Boxley Valley, the result of injudicious modification of the original road.

cooperation in the matter was promised but nothing ever came of it. We are as yet without those roadside scenic overlooks, let alone an entire well-designed scenic parkway.

The summer of '68 was a time to agonize over headlong efforts of various small-time entrepreneurs to make a few dollars out of whatever natural resources they could come upon in the Buffalo Valley before the park went in.

The most regrettable damage then being done was helter-skelter logging operations. Small locally owned sawmills had existed in the area for generations, and now their operators were out to cut everything that they could find within the proposed park boundary before it became a reality. Bernard Campbell had urged that we put out the word that the timber would be worth more standing were it sold to the government than it would be cut and sawed in the mill. No one understood that, and we could only hope that Congress would act soon to stop that rape of what little was left along the Buffalo.

We had heard rumors of such goings-on in that special sanctum of Hemmed-In Hollow and as soon as possible went there to investigate. One would have thought it too inaccessible to be a target of local loggers, but there was the old wagon road, long abandoned, that we had traveled in 1961, which went by the mouth of Hemmed-In. It was the route used by the timber cutters who had been in there with their big trucks. There was the scene of despoliation that we had witnessed all too often already, a jumble of treetops littering the area, sections of unused logs lying all about, thickets of brambles everywhere, and trash heaps left by the perpetrators.

We later learned how devious that activity could be. Our friend Bob Crow, who was then in real estate, had sold the place to out-of-state speculators who, wishing to make something out of it, sold the timber to a local logger.

The BRIA Down for the Count

During the winter of '68 and '69 ALA-BRIA activity had been slack, but they were not to be taken for granted. Their new leader gave us cause for concern.

Charles Thompson was not without insight. He recognized the need to correct the unconvincing dry Buffalo dirge, the Batesville flood bugaboo, and the non-navigable stream gabbery that had been the staple of the BRIA these many years. The dam builder-newcomer-landowner-leaseholder coalition was out of a program, and Thompson was on the spot to think up one. The result was a plan of action in which he suggested that: The anti-park group first *delay* the current legislation then to modify it so as to *kill* it. To do that the BRIA should *lay low* so as to minimize fear of a dam on the river. The anti-dam people should be

Dick Murray and Veda Ranchino view logging debris in Moore Creek.

mollified by a new concept, an agricultural river, which would undermine the position of the Ozark Society. That concept would be renamed the *pastoral* river plan, a gritty impediment in the way of the national river for the next several years.

Thompson in his plan of action noted well the importance of the donation of the Buffalo River State Park by the Arkansas Parks Commission to the National Park Service, a stipulation in the current bill then in Congress. He provided the park opponents with good information on how to do that and how to bring suit against the government if the bill did pass.

Thompson's analysis of the plight of the BRIA was not illogical and his advice that they "lay low" was a bitter pill that they would have to swallow. It did, in fact, mark the end of the Buffalo River Improvement Association. But Thompson left ajar the door of compromise and perhaps a dam someday for the Buffalo. That was apparently a stratagem to enlist the Marshallites into the wide-ranging American Landowners Association and to firm up its influence in Arkansas via its newborn chapter, the so-called Buffalo River Landowners Association (BRLA).

But in spite of what might have seemed a sensible and sincere plan designed by Thompson, there was something in his method that nourished the seed of dissention in the minds of some of his disciples, notably George Viorel. Letters from Harpers Ferry to Viorel began to contain increasingly nasty marginal notes written after their receipt by the former pugilist.

In a communication from Thompson to Viorel on April 30, the former issues instructions on how to raise money, some of which was to find its way into the coffers of the American Landowners Association in Harpers Ferry. Viorel's tacked-on footnote says: *"I wish we knew about Thompson's other chapters. We are not after money. Thompson is."*

Both sides in the controversy were now gearing up to appear before a hearing of the Senate Subcommittee on Parks and Recreation in Washington, D.C., scheduled for May 27, 1969. Senate Bill 855 to establish the Buffalo National River had been introduced by Senators Fulbright and McClellan of Arkansas. The false start of the year before had put us on edge, and Lucile Hannon into a temper tantrum, but the action now impending was the real thing.

Charles Thompson was more or less in command of the national park opposition. He was thoroughly familiar with the Washington scene and the ways of Congress and turned to with energy to arrange for the appearance of five anti-national river delegates at the Senate hearing.

He dispatched a letter of instructions to Viorel upon which the latter chose to make many uncomplimentary marginal notes for the amusement of none except us when those papers fell into our hands.

A communiqué from Thompson shortly after the hearing contained the first wording of the revised title of the floundering anti-park organization. It was from henceforth to be the Buffalo River Conservation and Recreational Council (BRC&RC). With it went a more acceptable designation of what the river would be known as: a pastoral river, not an agricultural river. Details of how it was to be managed would be revealed later. Again, Thompson's advice that the visibility of the previous partisans be kept under cover was proper for their objectives.

Communiqués from Thompson to Viorel in June and July 1969 revealed his concern for Viorel's intransigence and the falling out between the latter and Lucile Hannon and others in the group.

The last memo contained Viorel's kiss-off notation in the margin.

From then on silence would be Thompson's lot from this quarter, but the Ozark Society would not be relieved of the mess cooked up by him. Local leaders Marvin Sherman, then G. L. Hutchingson, and finally Herb Van Deven would carry the colors of the Buffalo River Conservation and Recreation Council and its pastoral river plan on to its niggardly and costly end long after the Buffalo National River became a reality.

SEVENTEEN

THE RIVER WAR WIDENS

The final weeks of 1968 found the Buffalo River improvers permanently stranded on the rocks and the national park detractors floundering for identity and for a workable program.

Meanwhile, we in the Ozark Society wrestled with day-to-day problems:

Should we raise our dues to four dollars per year?

How do we breathe life into the proposed Arkansas Conservation Council?

Who would serve on our nominating committee?

If nominated, who was there then to accept?

Who would undertake the revision of our constitution?

How could we get members in Fayetteville and Little Rock to form functioning chapters?

What could we do to get S.B. 110 (State Senator Oscar Alagood's bill to create an Arkansas Stream Preservation Commission) through the state senate?

That we were now moving from provincial skirmishing to a final decision in the halls of Congress was obvious. In the months ahead we were going to need the very best people available, willing to devote their time and energy to the legislative process.

At the same time there came the distracting call for help from other quarters. There were other concerned citizens with other rivers to save—if that was possible. Among them there were some who would become champions for the cause of conservation in our part of the world. We were to have going a contest on more than one front.

Some Outstanding Leaders for Our Side

At the November 1968 meeting of the Ozark Society in Little Rock we were pleased to meet a husky young man from Muskogee, Oklahoma, David Strickland, a descendant of those original Americans, the Choctaws. David was consumed with a desire to save what was left of Oklahoma's Illinois River. The lower end of it was already impounded by the Tenkiller Dam, but the Corps wanted the rest of it, the only clear, free-flowing canoe water left in the old Indian Territory.

David Strickland had just come from a hearing conducted by the Corps of Engineers at Tahlequah, Oklahoma, for the purpose of initiating the Chewey dam project on the upper Illinois. David and his followers had been able to submit 159 statements opposing the project, but the Corps claimed that the Chewey Dam Association had turned in over 1,000 in favor.

David's work was cut out for him, but work he did, eventually becoming the principal personality in the creation of the Oklahoma Scenic Rivers Commission. That organization was eventually able to stall the Chewey dam and to discourage further studies by the Corps on the Glover River in the Oklahoma Ouachitas. As a reward for his accomplishments, he would, on March 21, 1970, be named National Conservationist of 1969 by the National Wildlife Federation at their awards banquet at the Ambassador Hotel in Chicago. He was at the time president of the Oklahoma Scenic Rivers Association and a member of the Ozark Society and had already been chosen Oklahoma State Conservationist of 1969.

During the next few years David Strickland became one of our hardest working members, never missing a meeting, a hearing, an outing, or an opportunity to speak out against the big dam program in Oklahoma or elsewhere.

His stature as a conservationist was such that he was selected by the National Park Service to serve on their Southwest Regional Advisory Committee. In attending committee meetings in Santa Fe it was my privilege to have David as a traveling companion, but not without some concern. When taking his turn at the wheel, we sailed down the road at eighty miles an hour with David in lively discourse, which he embellished with energetic flourish of hands and arms. There were moments when the wheel was free of either hand, moments when I was overcome with consternation.

David's lack of consideration for the automobile was perhaps a factor in what finally came to pass in 1976. At a meeting of the advisory committee in Big Bend National Park, which David had to miss, Joe Rumberg, the park supervisor and moderator of the meeting received a brief written message from an attendant. When he read it, his face fell in shock, and he handed it to the rest of us. In it we read that David Strickland had that morning been killed in an automobile collision in Muskogee.

We had known Bob Ferris since our first contacts with the Ozark Wilderness Waterways Club in 1962. Dependable and low key, Bob Ferris was one to stay the course.

Sometime in the mid-sixties he had moved from Kansas City to Tulsa where he soon affiliated with the Tulsa Canoe Club. The club was especially concerned with Oklahoma's last remaining canoe streams, the upper Illinois, the Barren Fork, and Lee Creek. For them the Ozark Society might be of some help, and Bob was anxious for us to organize a chapter in Tulsa. Responding to his plea, in November 1968 Joe Clark and I assembled material for a founding session, gathered up our wives, and repaired to Tulsa for the establishment of what was at first called the Northeast Oklahoma Chapter of the Ozark Society, but which was later changed to the Indian Nations Chapter. That chapter, being in such a large city, did not grow as spectacularly as might be expected, but they have persisted over the years and have been, under the encouragement of Bob Ferris, a dependable and effective voice in conservation problems in our area until this day. In 1988 along with other Oklahoma conservation organizations, Bob Ferris and the Indian Nations Chapter were successful in gaining wilderness designation of significant sections of the Ouachita National Forest in their state.

One day not long after the founding of Ozark Society I had a phone call from Mrs. Bryant Davidson, from Shreveport, Louisiana. Originally an Arkansas girl, she had dwelt for many years in Shreveport where she had been a director of health and recreation at Centenary College and had served in similar capacity in the YWCA, the Girl Scouts, and the Boy Scouts there.

She had been greatly disturbed by the Corps of Engineers' plan to obliterate her favorite recreation stream in the whole country, the Buffalo, and wanted to do whatever she could to stop that plan. From that time on, "Tip" Davidson, as we came to know her, would be with us through thick and thin on most of our outings, at many of the hearings, and at almost all of our general meetings.

During her time in Shreveport, Tip Davidson made many converts for conservation, some of whom went out and did something about it. One of these was a young lawyer who came to our spring meeting in Little Rock in 1968 with an ax to grind.

That young lawyer's name was Wellborn Jack, and he was an avid and even militant outdoorsman concerned for the integrity of those uplands, the Ouachita Mountains, that arise not far north of Shreveport. The Army Engineers in their headlong rush to flood control the continent had scheduled every mountain river in the Ouachitas for damnation.

Wellborn was especially concerned about the Gillham Dam on the Cossatot River, a wild rambunctious stream when in good water. The Gillham Dam would take out a significant midsection of it, but the dam was already nearly halfway completed. Wellborn wanted to know why the Ozark Society hadn't done something to stop that piddling pork-barrel project.

We knew all about it and the big job the Corps was doing on the equally attractive Caddo River, which project was then more than three-fourths complete.

Harold Alexander kept us up-to-date on such things, but we had been obliged to witness these earth-shattering, river-wrecking projects become a reality with helpless remorse.

I had to tell Wellborn that the Ozark Society had been put together almost too late to knock out the Corps' plans for the Buffalo, let alone those abuilding elsewhere. "We just didn't have enough soldiers to take them on down there," I said.

"Well," he declared, "I am going back to Shreveport and organize a chapter of Ozark Society to fight the damming of the Cossatot, no matter how far along they are!"

That he did, which is a story in itself, but from his resolution did come the Bayou Chapter of Ozark Society in Shreveport. That enthusiastic group, which is still in existence, spawned temporary chapters in Monroe, Louisiana, and even in Lafayette on to the south.

Wellborn and his newly created Bayou Chapter, with the full backing of the parent society, by clever legal maneuvering were able to cause the Corps of Engineers to provide an environmental impact statement and to defend themselves in federal court, an indignity which they had never before been subjected to. Construction on Gillham Dam was held up two years, not a victory for Wellborn but an accomplishment nevertheless.

After that Wellborn Jack left the stage, but the Bayou Chapter continues to the present.

The Gulf from East Hanna

While Wellborn Jack's bold venture against the Gillham Dam was fore-doomed he sparked an alternative that was an unequivocal success. In conjunction with Harry Pearson and the Delta Chapter in Pine Bluff, he would lead an overnight backpacking trip to Caney Creek, a tributary of the Cossatot. They enlisted the cooperation of Alvis Owen, by then superintendent of the Ouachita National Forest, and his staff. All of that area was national forest property and would be eligible for wilderness designation under the new law. On this trip it would be determined if it was qualified for that.

The Caney Creek backpacking expedition turned out to be a very important affair, with Rupert Cutler from the home office of the Wilderness Society being present in addition to top officials of the Ouachita National Forest. As a result, the stage was set for the inclusion of the Caney Creek back country in the wilderness program a few years later.

The date finally selected was March 22 and 23, 1969. Assembling at the Shady Lake campground in the Ouachita National Forest on the night of March 21, we arose the next morning to receive direction from our trip leader, Wellborn Jack.

So many individuals registered for this affair that the group was necessarily divided into three divisions, all with different departure points with a common campsite designated in the midst of the Caney Creek valley. It was my privilege to accompany group C on the rigorous ascent to the summit of East Hanna Mountain.

As we began our climb up Buckeye Mountain the first objects of interest were the various forms of plant life observed on the way. We had not gone far up the ridge when numerous small white trilliums were encountered blooming among the leaves on the forest floor. Also at these same elevations, which reached two thousand feet or more above sea level, were clumps of azaleas, thick colonies of buckeye, and an unknown species of witch hazel. Along the crest of the ridge more than one species of hawthorn was observed, and underfoot there grew a thick tangle of blackberry, fragrant sumac, and barbed thickets of green brier. Most of these mountains terminate in very sharp ridges with a rocky spine protruding through a mantle of fragmented particles of sharp novaculite. On the north slope fairly tall specimens of hickory, oak, wild black cherry, and walnut were noted. On the south slope loose flinty novaculite in places remained bare and uncovered, but in other places a thick growth of stunted black jack and post oak impeded our way.

Generally there was almost no clear open space for easy walking on the entire hike.

Our first good view of the Cossatot Mountains came after attaining the crest of Buckeye Mountain. Here we could look back to the east toward the jagged and irregular peaks of Raspberry and Blaylock mountains. Directly to the south of us, Tall Peak was visible with the fire tower on its top. Away on the western horizon was the blue pyramidal outline of East Hanna Mountain, our destination for the day's hike.

Shortly after our lunch stop, the long knife-like spine of Buckeye Mountain began to dip sharply as it proceeded westward. Suddenly we came upon relatively open space where the rocky backbone protruded above the stunted trees. From this vantage point we were afforded what is surely one of the grandest views in the middle west. Immediately on the left the rounded dome of Katy Mountain rose into the sky. It wore about its base and on its rocky ledges a skirt of giant pines interspersed with hardwoods with the latter growing all the way up to the top. In the near distance to the west stood East Hanna Mountain dominating the horizon and looking for all the world like a forest-covered Egyptian pyramid. Connecting us with East Hanna was an immense dipping saddle or hogback ridge. It fell away sharply for hundreds of feet and then rose again in one grand sweep clear to the crest of East Hanna. This sharp eastern edge of the pyramid could be observed all the way up through heavy under-growth looking like a gigantic, crumbling staircase. Spread out below us on the south was the wide valley of Caney Creek down which other members of our party were at this moment proceeding toward the campground. The entire northern and western horizon was filled with wave upon wave of the undulating sharp-peaked ridges of the southern Ouachitas, a motionless blue sea attesting past heavings and contortions of our not-so-solid earth. Lying far on the north-ern horizon could be seen the high blue outline of Rich Mountain north of Mena. After enjoying this inspiring vista as long as time permitted, we picked our way down the great saddleback ridge to its lowest point. Here under giant pines we indulged in a short rest stop before attempting the ascent of East Hanna. Here while enjoying the view down a deep gorge to the south some of us observed high on the spine of Hanna Mountain a small red speck moving about. It was Dick Byrd's backpack, and we realized then that Dick and Harold and Margaret Hedges had jumped the gun and were already on their way to the top. We immediately set out after them on what proved to be a jack-in-the-beanstalk experience. Our pathway led forever upwards. East Hanna Mountain, like most of these Ouachita peaks, presents no sheer vertical bluff or cliff line to the climber. With its strata standing on end, great plates and slabs of rock rose up alongside us like a wall providing shelter from the fierce wind that had now begun to blow from the south. By now the sky had become gray and overcast with high clouds obscuring the sun and darkening the landscape. Onward and upward we slowly trudged, grasping and holding to whatever firm support that might sustain us. Before long all the world lay beneath us. We could see in the far distance Buckeye Mountain from whence we had departed that morning and all of the immense ranges of the Cossatot and Ouachita mountains in the south and east.

By now the afternoon was wearing on. Some voiced despair of ever reaching the top, and upon inquiring of our leader Wellborn Jack as to whether he had ever been there the dismaying information was that he hadn't been. Jim Brewer of the Ouachita Forest staff then had to admit that neither had he. Some of our party wondered aloud if anyone had ever been to the top of Hanna. Then we heard voices in the woods on the south slope below us. It was Harold and Margaret Hedges and Dick Byrd. They had given up hope of obtaining the top and were on their way down into Caney Creek to the campground area. After a brief consultation, however, we decided that having come this far we should

Ozark Society backpackers on the spine of East Hanna during the Caney Creek trip, March 1969. Wellborn Jack and Jim Brewer, *far left*, Dick Murray and Rupert Cutler, *far right*, others unknown.

proceed onward. After a last determined effort, the knife-edge spine began to level off, and we perceived that we had attained the wooded crest of the mountain. Continuing on for two hundred yards we came to the western terminus of the ridge. Below us and to the west the mountain fell away, continuing as a long high ridge toward the Cossatot River in the distance, beyond which stood the outline of West Hanna Mountain. This combination of East Hanna, West Hanna, and the long ridge between is known as the Hanna Range. Immediately to the south of it is the broad valley of Caney Creek, which is bordered on its southern rim by the lower Porter Mountain. Beyond Porter Mountain the Athens Plateau slopes gently into the gulf coastal plain. Standing on the topmost ledges of East Hanna, the moist south wind blowing with gale force about us, all that lay beyond the ramparts of Porter Mountain so resembled an ocean that I was moved to announce to our party that it was the Gulf of Mexico. So convincing was the idea that we all sat for a long time contemplating this vast inspiring void stretched out below.

Despite the darkening skies, there was in the hearts of those of us who had attended an informal meeting at the Ouachita National Forest Headquarters the day before a certain exultation. We carried with us the knowledge that this mountain peak and nine thousand acres around it was to become the first true "wilderness area" under the management of a qualified agency in the Ozark-Ouachita uplands. To those of us in the Ozark Society it was especially satisfying to know that this had come about as a result of a resolution requesting such preservation presented to our last annual meeting by Wellborn Jack and approved by those present and voting. Following this action Alvis Owen, supervisor of the Ouachita National Forest, and his staff had drawn up plans for the maintenance of the Caney Creek area in its natural state. Its designation under their concept is to be the "Caney Creek Back Country Area." It has not been heavily logged since the 1920s. No further timber cutting will be allowed within it. Such jeep trails as now penetrate part of it will be closed. The presence of livestock will be prohibited. Footpaths for hikers will be established with only enough clearance of vegetation to facilitate passage with backpacks.

Those of us who attended the meeting at which these plans were revealed cannot commend too highly the sincerity of interest and efficiency demonstrated by the staff of the Ouachita National Forest. They have submitted a program for the Caney Creek area which is in keeping with the best principles of true conservation and which should serve as a model and good example for other agencies from now on.

For the Ozark Society it is gratifying to see so much good achieved without the usual opposition of obstructionists, promoters, and developers. From this accomplishment we must be encouraged to overlook no opportunity to aid and assist in environmental improvement no matter how difficult or how easy the problem may seem.

Our meditation was brought to an end by ranks of gray cloud marching in across the darkening sky from the south. With haste we descended from that Olympus to the banks of Caney Creek, there to establish as secure a shelter as possible before nightfall. For Dick Murray and me that consisted of a tube tent, an eighteen-foot tubular section of clear plastic, four feet in diameter and open at both ends. Through it ran a nylon line which, when tied to two stout saplings, supported the "tent" with its bottom at ground level. With one of us at each end, Dick and I settled down within that cocoon like two man-sized caterpillars after closing the ends with clothespins.

For all of that night we lay, squirming and wormlike, fearful of the tread of Thor, who ranged the length and breadth of the Ouachita Mountains until dawn. Through that clear plastic the glare of lightning came through closed eyelids and past the tortured retina to sear directly upon the brain the fact of nature's celestial violence. With that the crash of thunder literally shook both mountains and the valleys between with vibration stunning to the eardrums and even transmitted, so it seemed, from the ground up to our rigid backbones.

With dawn the storm drifted off to the north, but yet the rain came down while we rolled up our shelter and bedding and marched off to where we had left Dick's pickup. In that distance we learned that human ingenuity had not been able to devise a garment that a prolonged downpour would not penetrate. Soaked to the skin we crawled into the front seat, started the motor, turned on the heater, and set out in quest of Umpire, Arkansas, and the road home.

Mile 16 on the Mulberry

The Cossatot and the Illinois were not the only rivers then the subject of controversy. There was the Mulberry, a steep, gradient stream coming down off of the Boston Mountain crest and coursing south into the Arkansas, a more exciting whitewater canoeing stream than the Buffalo, most of it within the Ozark National Forest boundary. It had not escaped notice of the river dammers and had been offered up as a source of municipal water for the small county seat town of Ozark.

In initial arguments over its fate a year or two before, the Ozark Society and others had suggested a smaller dam much nearer to the town on White Oak Creek, which was not important as a recreational or scenic waterway. It could serve the purpose at much less expense. But the old formula for building the big one was in effect. The local public officials and newspapers would have none of White Oak. The Corps of Engineers was glad to go along with that, and as a result I received a message from an important public servant:

January 24, 1969
Reply to: 2320 Wildernesses and Primitive Areas
Subject: Caney Creek Backpack Trip (Your ltr. 1/13)
To: Dr. Neil Compton, Bentonville, Arkansas

I'm late in answering your letter of January 13, but want you to know I will do my best to be with you on the hike in the Caney Area March 22 and 23.

I see the Corps of Engineers are holding a public hearing on damming the Big Mulberry February 25 at the National Guard Armory, Ozark, Arkansas. I was hopeful some legislation could be passed by then that would be helpful in opposing this.

Alvis Z. Owen
Forest Supervisor

Alvis Owen, formerly forest supervisor of the Ozark National Forest, had been transferred to Hot Springs as supervisor of the Ouachita National Forest.

He had been in charge of the development of Blanchard Spring Cavern. He was a sympathetic but practical conservationist and didn't like the dam on the Mulberry any better than we.

Colonel Charles Steel convened the hearing on February 25 in the drafty Ozark Armory where we were obliged to listen to the same fatuous claims that we had heard for Gilbert, Lone Rock, and Water Valley. The water supply argument was lost in flood control, irrigation, recreation, and industry. Somehow hydropower wasn't mentioned. The preponderance of opinion at that hearing did not favor the Mulberry dam. A significant opponent of the dam was the new forest supervisor of the Ozark National Forest, Jim Sabin.

The most vociferous proponent was a local lawyer, Jeta Taylor, president of the Western Arkansas Development Association and a member of the Arkansas Basin Association and the Mississippi Valley Association. In other words, he was a main supporting wheel in the then-under-construction Arkansas River Seaway. Taylor wanted to see all the side streams of the Arkansas dammed. Otherwise, their rise and fall would disturb shipping in that about-to-be industrial waterway, he said.

Like a hell-fire-and-brimstone holy-roller preacher, he warned us that if we didn't let the Mulberry dam proceed we would be in the awfulest bear fight that ever took place in Arkansas.

But significant legislation was then on the way in Washington to thwart the plans of the Corps on the Buffalo. That had a dampening effect around the country on lesser jobs like Mile 16 on the Mulberry.

The bear fight on the Mulberry turned out to be a compromise on White Oak Creek. For Jeta Taylor and his followers, however, the water supply idea has survived to this day on Lee Creek and Illinois Bayou in spite of good alternatives. For true conservationists, the watchword is constant vigilance.

A River's End—The Caddo

Dr. Joe Nix had a Ph.D. degree in chemistry. He was an authority on water quality and a professor at Ouachita Baptist University in Arkadelphia. He had spent much of his boyhood in the out-of-doors along that nearby Ouachita Mountain stream, the Caddo. It was an outstanding river in that overthrust zone where it, like the Cossatot, was forced to tumble over upturned ridges of hard sandstone and novaculite. Between the ridges there would be wide basins—perfect places to hold the waters of a dammed river, and by 1968 the Caddo was in the final throes of strangulation at a place called DeGray.

It was inevitable that we would meet Joe Nix who, in his grief for the Caddo,

Dr. Joe Nix, second president of the Ozark Society, roasting chickens in the coals, Caddo River trip.

sought those who would understand and who would come and have a last look with him. That we did, on several occasions, recording as best we could on film and paper what this charming river had looked like before its day was done in the summer of '69.

There was no move by us to contest that project as was the case on the Cossatot. We came only for a last look. What we were afforded from that experience was a revelation, not so much of the beauty of the doomed river but of the enormity of the power of man over the natural order of the world.

I had seen firsthand some of that in the valley of the upper White River above Beaver a few years before, but DeGray was something else. It was no vertical slab of concrete to stop the flowing waters but an earthen barrier of mind-boggling proportions, literally a mountain moved across the course of the Caddo. Behind it the water would rise two hundred feet to bury forever Parker Mill Falls and the International Paper Company campground almost as far up river as Amity. We were witness to the army of heavy dump trucks ranging the country for miles around to obtain enough precious dirt to lay that mountain across the course of the Caddo and the frightful loaders that dug it up for them.

The DeGray earth-fill dam under construction on the Caddo River, 1970.

The Caddo River exiting from the temporary tunnel dug under the south hill, to be blown and sealed when the dam was finished.

The intake tower (glory hole) for warm top-water outlet, DeGray Dam, under construction on the Caddo River, September 1970.

By this time the Corps of Engineers had become aware of criticism by conservationists, especially the sport fishermen, for building dams with cold-water outlets that destroyed the native fish in the river below. At DeGray they would rectify that by installing a warm-water outlet at an added expense of millions of dollars. A tall silo-like intake tower had been constructed just above the dam into which warm surface water would fall to be transmitted to penstocks below to drive the generators of the hydropower phase of the project. A service ramp connected it to the hill on the south side. Below the dam there were only seven miles of the Caddo remaining before it joined the Ouachita River, which poses the question of whether the perch, goggle eye, bass, and catfish remaining in that short section were to be worth it all.

To divert the water of the Caddo while the man-made mountain across it was being built, a diversion tunnel was dug under the rock-ribbed south abutment, a temporary channel of egress. When everything else was in place and finished, the tunnel would be blown with explosives, sealing it off, after which the waters would rise to the level of the "glory hole" (as Dick Murray and his former pals in the Corps called the opening at the top of such intake towers). There the exact lake level could be manipulated to satisfy various purposes—flood control, hydropower, recreation, and water supply.

The DeGray Dam had no traditional concrete spillway. To accommodate inevitable flood waters, which might overwhelm the glory hole, a diversion channel was dug through a high hill on the north for a distance of a mile or two so as to bypass the dam. The bed of the channel was of hard regional stone and would withstand any anticipated high water. Its cost was no factor.

But this was not all. The basin of the Caddo River on the north side was lower than the proposed lake level. That defect in topography was five miles wide, and the Corps, not wanting for money to do it, was obliged to build a retaining dike seven miles long and seventy feet high in some places.

Such labor and expense should have been prohibitive. It already had been in the case of the Arkansas Power and Light Company, who surveyed the DeGray site in the late twenties or early thirties and found it impractical. They moved north and built the Carpenter Dam on the Ouachita forming Lake Hamilton.

Being aware of rising criticism of all of this, the Corps went all out to underwrite recreation and water sports on Lake DeGray. All of those facilities that the public had come to expect of them were provided in abundance, and, in addition, land on the north shore was made available for an elegant state park with all accommodations. That DeGray was a pretty lake is not to be denied, with its expanses of open water wider than those narrow serpentine impoundments up in the Ozarks and its shores embellished with embayments and scattered islets all studded with pine groves, green in summer and winter.

The principal use of Lake DeGray may lie in the future, a purpose not bargained for in 1969. The Little Rock complex of cities and towns, now striving for megalopolis status, have nominated DeGray as their future water supply. They surely will not be denied that federally provided commodity when the time comes.

Students of the new geology—plate tectonics, subduction, transverse faulting, and volcanism—should, upon visiting DeGray Dam, take into account its significance. Here lies a mass of earth as great as those shaken loose by earthquake or dumped out by volcanoes across many valleys around the world to create nature's own impoundments.

But by 1968 the DeGray Dam was all but finished. It was a favorite project of Senator John L. McClellan, the father of the Arkansas River Seaway. We chose not to butt heads with him. If we were to get anywhere with our hope for the Buffalo, we were going to need him on our side.

EIGHTEEN

SENATE BILL 855,
THE BEGINNING OF THE END

By 1969 we had good reason to expect continued support of national river legislation from Senator McClellan, but it was most reassuring to receive notice of that directly from him:

January 28, 1969

Dear Dr. Compton:

This will acknowledge receipt of your recent letter regarding legislation to establish a National Park along the Buffalo River.

As you know, Senator Fulbright and I co-sponsored legislation last session of Congress to provide for the establishment of the Buffalo National River. We shall again introduce this legislation in the 91st Congress and you may be assured that I shall do everything possible to see that this bill is passed and becomes public law.

I appreciate hearing from you, and with best wishes, I am

Sincerely yours,
John L. McClellan

In those days McClellan and Fulbright were two of the most powerful people in the U.S. Senate.

On the heels of that came a telegram bearing the message that we were on our way. The draft bill replacing S.B. 704 submitted to Senator Fulbright would be designated S.B. 855. Its wording was essentially the same, but the area of the proposed national river had been reduced from 103,000 acres to 95,730 acres.

The date set for the hearing on S.B. 855 was May 27, 1969, and that set off a flurry of preparatory activity in our ranks. An instruction sheet was mailed to all who might consider making an appearance and/or a statement by mail. In those instructions we were pleased to list as an on-the-scene expediter Ken Smith and also to recognize the hospitality of Mrs. E. C. Johnson.

The presence of Ken Smith in Washington, D.C., was not by design but

393

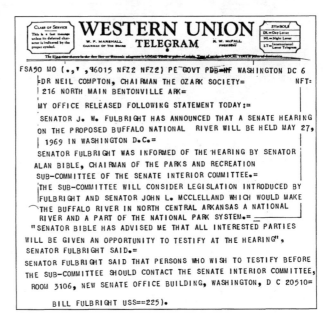

FSA50 MO (..,▼ ,96015 NFZ2 NFZ2) PE GOVT PDB=NF WASHINGTON DC 6

=DR NEIL COMPTON, CHAIRMAN THE OZARK SOCIETY=   NFT=

216 NORTH MAIN BENTONVILLE ARK=

MY OFFICE RELEASED FOLLOWING STATEMENT TODAY:=

SENATOR J. W. FULBRIGHT HAS ANNOUNCED THAT A SENATE HEARING
ON THE PROPOSED BUFFALO NATIONAL RIVER WILL BE HELD MAY 27,
1969 IN WASHINGTON D.C.=

SENATOR FULBRIGHT WAS INFORMED OF THE HEARING BY SENATOR
ALAN BIBLE, CHAIRMAN OF THE PARKS AND RECREATION
SUB-COMMITTEE OF THE SENATE INTERIOR COMMITTEE.=

THE SUB-COMMITTEE WILL CONSIDER LEGISLATION INTRODUCED BY
FULBRIGHT AND SENATOR JOHN L. MCCLELLAND WHICH WOULD MAKE
THE BUFFALO RIVER IN NORTH CENTRAL ARKANSAS A NATIONAL
RIVER AND A PART OF THE NATIONAL PARK SYSTEM.=

"SENATOR BIBLE HAS ADVISED ME THAT ALL INTERESTED PARTIES
WILL BE GIVEN AN OPPORTUNITY TO TESTIFY AT THE HEARING",
SENATOR FULBRIGHT SAID.=

SENATOR FULBRIGHT SAID THAT PERSONS WHO WISH TO TESTIFY BEFORE
THE SUB-COMMITTEE SHOULD CONTACT THE SENATE INTERIOR COMMITTEE,
ROOM 3106, NEW SENATE OFFICE BUILDING, WASHINGTON, D C 20510=

BILL FULBRIGHT USS==225).

A Western Union telegram.

because of routine assignment by the National Park Service, his employer. As was the case so many times, he happened to be in the right place to help out. This time he was able to furnish car transportation for those who needed it, to run errands, to offer guide service in the city, and to familiarize us with the legislative process of bills in Congress.

Our contact with Mrs. E. C. Johnson was no accident. She was the mother of Lois Imhoff (Mrs. John Imhoff) of Fayetteville. The Imhoffs had been close associates since 1964, and Lois would before long serve as secretary of the Ozark Society. Mrs. Johnson generously offered three spare bedrooms in her home for the use of myself and my wife, Laurene, and Orphea Duty and Margaret Hedges.

In the weeks remaining it was necessary to conduct an intensive campaign to obtain delegates to appear at the hearing on May 27. Even though we knew that the outcome was tilted in our favor with such senators as Fulbright, McClellan, and Committee Chairman Senator Alan Bible supporting S.B. 855, we wanted to appear with as impressive a delegation as possible.

The State Parks Commission Changes Its Tack

In the midst of these preparations for the senate hearing a game of musical chairs had taken place in the state capital in Little Rock, which added greatly to our advantage. Jimmy Driftwood, that modern-day troubadour and champion of the real pioneer style, would join our ranks. He was to occupy a very effective position in the hierarchy of state agencies.

Up until only a few days before the senate hearing, the question of the disposition of the Buffalo River State Park remained a thorn in our side. As has already been mentioned, the National Park Service had requested the donation of state-owned lands within the proposed national river boundary. With the presence of Gilbert dam supporter C. E. Tudor on the State Parks, Recreation and Travel Commission and his influence on Bob Evans, the director, and other members of the commission, that state agency remained a very real obstacle in the way of S.B. 855.

In early May we had made a plea to Director Evans to resolve that issue but had once more been rebuffed.

However, within a few days "the will of the people" took care of the problem to the satisfaction of river savers all across the state (*Pine Bluff Commercial,* May 24, 1969):

Director since '62 Quits Parks Agency

Bob Evans, executive director of the state Parks, Recreation and Travel Commission, turned in his resignation Friday, probably heading off a request that he resign. ...

Gov. Rockefeller's appointees gained control of the seven-member commission last month when James C. Morris (Jimmy Driftwood) of Timbo was named in Rockefeller's fourth appointment.

Evans said later, "They want one of their own people to work with. I was hoping they could work with me, but when they couldn't I said fine and told them to go ahead and put one of their people in there."

... "To the victor belongs the spoils," said Evans. ...

Lou Oberste, acting director, had from the start been a strong supporter of the Buffalo National River and would continue to be. That same day the State Parks, Recreation and Travel Commission passed a resolution favoring the passage of S.B. 855. That was not without a significant drawback, however. One of our eager beavers had appeared before the commissioners and informed them erroneously that the state of Arkansas could retain the Buffalo River State Park and Lost Valley if it wanted to. That was an honest error and not an act of trickery, but it would be ironed out in the months ahead after the appointment of William Henderson as director of the State Parks and Travel Commission.

On the heels of that endorsement came the full approval of S.B. 855 by the heretofore standoffish Arkansas Game and Fish Commission. We could now go to Washington with the belated backing of those two state agencies.

Senator Bible Hears the Case

While some of our delegates chose to fly, others elected to drive, and that included me, along with my wife, Laurene, and two passengers from Boxley, Margaret Hedges and Orphea Duty.

In addition to the out-of-state press, reporters from the Arkansas newspapers were there to give a firsthand account of the proceedings and of the sometimes surprising, if not shocking, statements made to the Senate subcommittee (*Arkansas Gazette,* May 28, 1969):

Buffalo Hearings Open; Chairman Supports Proposal

WASHINGTON—Witnesses both praised and condemned the proposal to add the Buffalo River to the national park system in four and a half hours of hearings before the Senate Parks and Recreation Subcommittee Tuesday. Thirty-one persons were heard.

Proponents got a boost from Subcommittee Chairman Alan Bible (Dem., Nev.) early in the session when he said, "I am sold on the preservation of great national river systems, and I believe we have to move quickly because the bulldozers are not far behind."

Bible, who expressed eagerness several times to see the North Central Arkansas stream, told a half dozen opponents present that he thought the National Park Service could work with them to solve any problems in an equitable manner.

Foes Say River Really Not Much

The opponents downgraded the river's scenic and recreational value, said feasibility surveys were mere propaganda and raised questions about the motives of Arkansans sponsoring the Buffalo legislation.

Fulbright, McClellan First to Testify

Lead off witnesses were Senators John L. McClellan and J. William Fulbright, co-sponsors of the bill. Both said that National park status for the river now enjoys overwhelming popular support and that controversy over it had largely been resolved. ...

Once Supported Damming Buffalo

Fulbright noted that he had once supported damming the stream, but added that the Engineers had determined that the benefits from damming the stream were "marginal at best." The senator argued that a dam could always be built later, after creation of a national river, but he held that unlikely.

Fulbright's contention was that Congress could always undo a park, but could hardly justify removing a dam. His remarks caused Senator Frank Church (Dem., Ida.) to seek assurances that the legislation would foreclose any reservoir construction.

Fulbright replied that no dam was contemplated now, and that he meant that one course was always reversible while the other was not.

He then said that he perhaps should not have raised the issue.

Interior Department Endorses Measure

National Park Service Director George B. Hartzog added the Interior Department's formal endorsement and detailed the bill's provisions—acquisition of 95,730 acres in Newton, Searcy, Marion and Baxter counties, and government spending amounting to $9,200,000 for site acquisition and $8,224,400 for development. ...

Governor Asks "Immediate Action"

Governor Rockefeller's plea for "immediate action" was read by Harold E. Alexander of Conway, a member of the Arkansas Planning Commission. ...

Bible Asks if State Would Yield Parks

Senator Bible wanted to know if the state would cede the acreage it owns along the Buffalo, largely in two state parks, to the federal government. Donation of state-owned lands in all such cases is a prerequisite to creation of a national park.

Alexander noted that the Arkansas Parks, Recreation and Travel Commission had passed a resolution favoring the national park. Bible said this did not answer his question. Alexander indicated that the Commission understood the federal requirement. ...

Buffalo Landowners Attack Proposal

Leading off the opposition, George Viorel of Hasty (Newton County) chairman of the Buffalo River Landowners Association, castigated the Park Service report for "minimizing" the river's "good, prime bottomland" and also attacked a University of Arkansas study on a national river's economic benefits as "an attempt to sell the park idea for the Interior Department."

He scoffed at the Park Service's contention that the area was full of wild game:

"No one, to my knowledge, has ever seen an otter. Fifty bears were released to starve. It is doubtful if as many as three are left, and they are in an unhealthy condition. One can float the river and not see the equivalent of one rabbit. ..."

Views Abound in Ozarks, Foe Says

James R. Tudor of Marshall said that not by the wildest stretch of the imagination could a national river be justified.

"To an old hillbilly fresh from the Ozarks there is a scenic view outside this window. But what gives it preference over another lake view in Chicago or any city? Scenic views abound in all the Ozarks. Views along the Buffalo are no different from those throughout the Ozark area. Why not take all the Ozarks into the park plan?"

Tongue in cheek, Senator Bible cautioned Tudor to be more circumspect in what he was saying, noting that Park Service personnel were inclined to make everything in sight a park, given a chance. ...

Hilary Jones of Pruitt asserted that park status would destroy the quaintness of the Ozarks and its social life—the things that tourists really go to see.

Suspicious of Backing by Two Lawmakers

Jones questioned the support of the national river proposal by Fulbright and Representative John Paul Hammerschmidt (Rep., Ark.) who has introduced a companion bill in the House.

He said they had had wild rivers in their home counties, but had worked to get them dammed.

They did not oppose development of Beaver Reservoir, Table Rock or Bull Shoals "because it enriched their home territories," Jones said. ...

The *Pine Bluff Commercial* (May 29, 1969) reported the event in even more detail, a high point of which was Margaret Hedges rebuttal to James Tudor, *et al.*:

Mrs. Harold Hedges of Boxley, formerly of Lake Quivira, Kansas, and a long time canoeist and conservation militant, put the finishing touches on the opponents:

"For many years, a newspaper at Marshall, Arkansas, namely, the Mountain Wave, has promoted the damming of the Buffalo.

"They have claimed that the river scenery has been vastly overrated, that the river is only floatable a few weeks out of the year, that no tourists come to Arkansas now to view the Buffalo and no tourists will come if it is made into a park.

"Yet in the new Arkansas Tour Guide for 1969, there is a page—sponsored by the Marshall Mountain Wave—advertising the beauties of the Buffalo River.

"The color picture used on page 90 was borrowed from Dr. Neil Compton, president and founder of the Ozark Society, an organization formed for the purpose of preserving the Buffalo. The title for their page was borrowed from Ken Smith's book *Buffalo River Country*.

"But the caption under Dr. Compton's picture is entirely their own and I will quote in part—'The wonders of the exotic Buffalo River will fascinate you beyond your wildest imagination. This river is one of America's most beautiful. The scenery, the rugged mountains and the free flowing river are a must on your trip to Arkansas' End quote.

"Thus we are in one accord in describing the wonders of the Buffalo River. Now there remains the problem of keeping this river beautiful and inviolate."

It was Mrs. Hedges and Orphea Duty, respected citizens of Boxley, who stole the show away from the men folks.

The following delegates all submitted strong endorsement of S.B. 855: Jim Gaston, Lakeview, Arkansas, president-elect of the Ozark Playgrounds Association, representing the travel industry for the Ozark area; Bob Ferris, Tulsa, Oklahoma, the Tulsa Canoe Club; Col. Jack Diggs, Fayetteville, Arkansas, representing the Arkansas section of the Sierra Club; and Mrs. Bryant (Tip) Davidson, Shreveport, Louisiana, representing the Council for Youth Groups for Natural Beauty and Conservation, the Boy Scouts and Girl Scouts, the 4-H clubs, the Red Cross Youth, and the YMCA and YWCA of the Shreveport area; George Smith, Mountain Home, Arkansas, the Mountain Home Chamber of Commerce.

Orphea Duty, Boxley, Arkansas, lifetime citizen of upper Buffalo area, submitted this testimony:

Honorable Chairman and fellow members:

My name is Orphea Duty. I have been a resident of Newton County, Arkansas all my life and have lived in the same house in the Boxley community for 59 years. ... I was appointed Postmaster in 1917 and, as clerk, still maintain Boxley Rural Station. I also operate Boxley Mercantile Company, first established in 1914 by my father, the late Ben E. McFerrin, who served county, district and state as Representative, Senator and Lt. Governor.

Since that time much progress has come to the valley. Roads have come and trucks and cars have replaced horse and buggy transportation and most of the homes are modern. Many tourists find a warm welcome at Boxley. Improved grasses make the pasture land green and lush and there is little to remind us of former labors in the row crops, once so common in the Buffalo valley. The blacksmith shop is gone and a roadside park stands in its stead, maintained by the Extension Homemakers Club. There is no evidence of the civil war battle waged here on the Buffalo except a few remains of the large iron kettles used to make gunpowder. ...

We who are older have watched the changes come. We've seen the tomato canneries thrive and prosper and vanish. We've seen the old water mill fade and fall into disuse, with the advent of store bought meal. Only the saw mill is unchanged. ...

The nearby church, founded in 1838, stands serene against a beautiful green hill. Ancient oaks shelter it from the summer heat. This little church has been the scene of worship and wedding, revivals and funerals for many years. But change has come here, too, for the old log church house is gone. In its stead now stands three structures, the community building, built in 1900 to serve as church, school and lodge, the modern church erected in 1956, and a modest parsonage, a symbol of the humble prosperity of the Boxley Baptist Church. Near the church is a cemetery where our babies, our spouses, our parents and our grandparents rest in peace.

Truly, we have a wonderful way of life along the upper Buffalo. People are happy here, prosperous and blessed in so many ways by the Creator. And now comes the question, what about a National River on the Buffalo? Well, first of all, I am proud that you have found this part of Arkansas to be so beautiful, with scenery unsurpassed, and I am happy to know that you will maintain the river as nearly as possible just as it is today. I am pleased with the area designated on the park proposal as a private use zone, for agriculture uses. Most of the people in the valley "heired" their land and their roots are deep. Like all proud farmers they wish to retain their land to pass on to their children. If you can bring a park that will not disrupt the citizens, that will keep the river free of factories and taverns—a park that will allow reasonable use of the land, then we see eye to eye. I have seen changes come and I enjoy the conveniences that modern ways have brought but I realize disruptive changes will come if we do not control our river. I want to keep my farm—the cattle, horses and hogs, the pastures and meadows, the little stores, the rodeo arena, the saw mill, the Boxley Rural Station, and, of course, the church.

Yes, I favor Senate Bill 855 that all America might enjoy the mountains, the bluffs, the free flowing river, and beyond that—the peaceful way of life. I feel sure that our representatives in Washington, the most wonderful government officials will do that which is best for all concerned and thank God for such men. Thank you.

Orphea Duty was surely born with the spirit to speak such words, but that spirit, serene, tranquil, and filled with the joy of living, has been enhanced and fulfilled by her years in Boxley Valley.

The words of another who knew her are here in order. Robert Flanders, editor and publisher of *Ozark Watch* (an organ of Southwest Missouri State University), describes a visit in her house in the spring of 1988:

> Orphea (pronounced Orphey) dwells in a fine two story house overlooking the Buffalo River in Newton County, Arkansas. Mrs. Duty is a country gentlewoman now in her late eighties. Her father was a progressive state senator, a founder of Arkansas' public schools. Her husband was a prosperous farmer, merchant, miller, broker and leader of the valley. Her children and grandchildren are pillars of society. She still "receives" in her home. Although she lives alone, her large dining table remains set with china, crystal and silver for up to twelve guests. She is thus always ready to serve her ready pie, cake, cookies and coffee. Always. The sense of abiding for a long time, of a continuing home, is palpable. Being there, if but for an hour, close to that abiding, is an emotional experience. One senses, even as a guest, the fabric of this home, knit over time of many skeins: the woman; the women before her, and their men; the children, who in their time became old themselves—all in this house in this very place, dwelling.

Also submitting testimony were Everett Bowman, Little Rock, Arkansas, president of Pulaski Chapter of Ozark Society; Dr. Joe Nix, Arkadelphia, Arkansas, associate professor of chemistry, Arkansas Baptist University; Clayton Little, Bentonville, Arkansas, representing the Arkansas Stream Preservation Committee; Thomas L. Kimball, Washington, D.C., director of the National Wildlife Federation; Ben H. Thompson, Washington, D.C., the National Recreation and Parks Association; Rupert Cutler, Washington, D.C., assistant executive director, the Wilderness Society; Barry R. Weaver, Washington, D.C., representing National Sierra Club; Dr. Spencer M. Smith, Jr., Washington, D.C., secretary, Citizens Committee on Natural Resources; Ed Stegner, Jefferson City, Missouri, executive secretary of the Conservation Federation of Missouri;

Orphea Duty, *far left,* escorting visitors into her home, October 1988.

Dwayne Kelly, Kansas City, Missouri, Ozark Wilderness Waterways Club; Maurice A. Crews, Washington, D.C., the National Audubon Society; and David R. Strickland, Muskogee, Oklahoma, president of the Scenic Rivers Association.

The Road Home

On the evening of May 27, 1969, having done our bit and the affairs of state having come to an end, we had a chore to take care of. Margaret Hedges was in a hurry to get home. Harold would meet her at the Springfield, Missouri, airport that evening, and with her would go her good friend Orphea. They would fly. Without delay we loaded their bags into the Chrysler and headed for the nearby Washington National Airport where we bade them good-bye. There as the big jet lifted off and disappeared over the darkening horizon bound for those distant hills beyond the Mississippi, one could not but marvel at the import of this great lady, Orphea Duty, born in that other land and in another time. There in 1920 Orphea McFerrin, who was not to be denied the spice of life, and Fred Duty were married—on horseback.

Her plea to the Senate Subcommittee on Parks and Recreation has today been brought to pass. Now on the approaches to Boxley Valley are modest signs erected by the National Park Service informing the visitor that he is entering THE BOXLEY VALLEY HISTORIC DISTRICT OF THE BUFFALO NATIONAL RIVER.

Following the hearing we all returned home to our appointed daily tasks and waited with confidence the decision of the full Senate Interior Committee. We were not disappointed when a telegram arrived addressed to several of our number:

August 7, 1969

Pleased to advise that Senate Interior Committee has recommended favorable action to create Buffalo National River. Hopeful of Senate consideration of bill in September. Best regards. Bill Fulbright, U.S.S. and John McClellan, U.S.S.
Book Telegram:

Dr. Neil Compton	Mr. Tom Dearmore
Mr. Harry Pearson	Dr. Joe Nix
Mrs. Laird Archer	Mrs. Orphea Duty
Mr. Charles Johnston	Mr. Jim Gaston
Mr. John Heuston	Mr. Gus Albright
Mrs. Margaret Hedges	

With that in hand, the attitude of most of our people and of casual observers around the area was that the Buffalo River war was over. We had won resoundingly and could now sit back and let the National Park Service run the show. But a few of us realized that the final verdict lay somewhere down the road. Our legislative structure is a bicameral one (oftentimes a house divided), and the House of Representatives could not be counted upon to go along with the Senate. Representative John Paul Hammerschmidt had reintroduced his bill for the national river, but those who knew cautioned us that unbridled optimism was out of order.

In reflecting on our experience in the Senate hearing, we were left with a sobering conclusion: how easily this elected body sometimes responds to whims of public outcry; how precarious the fate of the natural world, whether to change it irrevocably or to protect it as it came from the hands of creation; how tenuous the fate of this river that a few (possibly unintentional but equivocal) words from Senator Fulbright might set up a lethal compromise—if not now, then later. How surely the continued existence of the Buffalo River depended upon what passed in the minds and hearts of those gathered in room 3110 in the new Senate office building that day.

NINETEEN

DREAMERS AND SCHEMERS ON THE BUFFALO

Almost anyone would have expected that with the approval of S.B. 855, culminating almost eight years of a widely publicized program to establish national park jurisdiction along the Buffalo, private developers and real-estate dealers would limit their activity along the river, that they would exercise caution and go elsewhere. But NO—They would instead redouble their efforts amidst much hue and cry as to the injury that was about to be inflicted upon them. They chose to ignore a directive, issued by the Department of the Interior in 1967, stating that any development or construction work within the proposed national river boundary would not be eligible for reimbursement as "improved" property after that year. Such builders or investors would thereby stand to lose the difference they might claim between that and unimproved property. That did not mean that fair market value would not be paid for such lands as the park service might need for the proper management of the national river. It did mean that the government would not honor belated claims for construction undertaken within the proposed park boundary after 1967. That policy net was bound to catch a few fish, some most surprising.

How Kyles Landing Came to Be

It was never our desire or intent to have to stand up against an ordained minister, but on the old Frank Villines place in the midst of that most scenic stretch of the Buffalo between Ponca and Erbie such would come to pass.

The first inkling that something was going on there came in the spring of 1966 during a float from the low-water bridge at Ponca to the Boy Scout camp at Camp Orr. Just above Buzzard Bluff, a short distance up river from Camp Orr, we heard a disturbing sound from up on the mountainside above the mouth of

Indian Creek. It was the screech and growl of a bulldozer, and it could be up to no good in our book. Shortly thereafter, we were to learn that the dozer was gouging out a road down to a new development at the mouth of Bear Creek to be known as the Buffalo Bluff Boys' Home. Without protective legislation on the books, there was nothing that anyone could do to stop it. More disconcerting news soon made known the scope of this potentially ruinous project.

The *Ozarks Mountaineer* printed a lead article in July 1968 warmly supportive of the undertaking, saying in part:

Dream on the Buffalo

By Clay Anderson

Men can dream great dreams along the majestically beautiful Buffalo River in Northwest Arkansas. The Rev. Floyd Harris envisioned a home for homeless boys in the valley of the Buffalo in Newton County. It was no idle dream. After years of planning and work, the dream is now demanding every free minute and every ounce of energy Rev. Harris can muster. ...

Rev. Harris was pastor of Southside Baptist church in Shreveport, La. when he became deeply aware of the need for Christian homes to accommodate boys from family courts. It was Marion Underhill who suggested the Ozarks and drove with his pastor to Jasper where the two men promptly bought 40 acres. Shortly thereafter the preacher went to Venezuela as a missionary where another Arkansan was caught up in the dream. Lindsey H. Edwards added another 235 acres. The home was chartered as a non-profit organization in October, 1963 with the assistance of a number of Jasper and Harrison leaders.

Back from Venezuela, Rev. Harris continued to line up assistance for the home while pastoring a church in Houston. Whenever they could, Rev. and Mrs. Harris and their youngest son, Kyle, would get up to the Buffalo to plan and dream. ...

A drive to raise funds to start building the home was already in the works in February of 1967 when 13-year-old Kyle was killed in a motorcycle accident.

The loss might easily have shattered the dreams of a lesser man, but it only strengthened the resolve of Rev. Harris. He re-doubled his efforts. To the average of 10 letters a day he has written over the last 30 years, he added one more: A letter to Kyle telling of the progress, problems and plans for the home, which has been appropriately named "Kyle's Boys' Home."

... The plan calls for creation of a real family atmosphere. Two level living units designed to house 12 boys and their foster parents will be built. Each house will have dining room, kitchen, recreation room, fireplaces, and workshops. Sites for the first 120 houses have already been selected on the hillside facing the Buffalo River and bluffs beyond. The houses will be built one by one as funds become available. Rev. Harris says eventually as many as 500 boys could be living at the ranch.

... Kyle's Boys' Home has opened a store in Jasper selling antiques, crafts, hobby supplies, coins and the like—most of which have been donated.

Rev. Harris hopes the home will eventually be self-supporting to a large degree. A full scale farming operation will be conducted, and in the summer plans are for a sports summer camp with assistance already promised from Frank Broyles, University of Arkansas football coach and such famous athletes as Alvin Dark, Ronnie Bull, Bobby Richardson, and Lance Alworth.

Rev. Harris wants friends of the home to think of it as their personal retreat. Seven fishing lakes have been planned as well as a hunting lodge, museum, chapel, administration building and shops. Some 700 additional acres are being held by friends for future expansion.

Early construction has centered on laying water lines, building roads and facilities for the home's first big event—an old fashioned camp meeting July 4-6. Nationally known evangelists and musicians will be on hand and an abundance of campsites are available to accommodate 3,000 or more people.

The boys for whom all of this is planned will be under 10 years of age. "We will take little boys who seemingly do not have the opportunity to better themselves," said Rev. Harris. He added that the boys will be cared for right on through college, then proceeded to outline promises of support from a number of major colleges.

It's quite a dream. Of such are great things born.

Several months later we were shook up to read an account in the *Harrison Daily Times* by our old friend Jack McCutcheon, who was no mean amateur reporter. He had been gathered into the fold and related developments at the Boys' Home under an inflammatory headline: "Buffalo River Park Plan Jeopardizes Boys' Home."

Shortly thereafter, a similar account appeared in the *Arkansas Democrat,* May 26, 1969:

Threatened by a Park

Boys' Home Being Built Near Buffalo River
May Have No Future in the Wilderness

COTTER (AP)—The walls may come tumbling down at a wilderness home for boys if the Buffalo River and its shoreline are declared part of the National Park System.

Two stone homes, tons of sand and a pile of mortar and cement mark the location in Newton County where the Rev. Floyd Harris of Cotter hopes someday will be a home for delinquent and homeless youths. ...

Funds for the project came from Kyle's insurance and college savings, from donations, lectures and the sale of family antiques at the Kyle's Boys' Home store at Cotter.

Harris said he had about $100,000 tied up in the project when a government official told him the area was included in a proposed national park and that he could not be compensated for any expenses incurred since 1967.

The information came after one home had been completed and another was nearing completion. Ample supplies already had been purchased and stockpiled for further construction.

"We were told we would just be losing money if we continued to build," Harris said. "He (the government official) said he hated for us to lose money when we gave up the property.

"We've got 450 sacks of cement just sitting there—we got it the week before he came—sand and stone piled up and we plan to just go on and build it," he said. …

Financial problems also are beginning to plague the proposed settlement. "No one can lend us any money under the circumstances," Harris said.

The fate of Kyle's Boys' Home is not dependent upon whether it can be built on the Buffalo, Harris said.

"Some friends in Texas have offered to purchase the land and build the entire home if we will return there," Harris said. "But this is where we wanted to build the home—in these mountains. They're an ideal place to raise young boys.

"We're selling everything we've got to remain here," he said, "so I don't want to quit now."

The whole affair was a hot potato that we in the Ozark Society didn't want to hang onto. To confront the bureaucrats in the Corps of Engineers was one thing, but to find ourselves at odds with an ordained minister of the Lord was something else.

I had never met the Reverend Harris, but in his various fund-raising efforts he had established antique and curio stores in Cotter and Jasper. He had requested consignment of Ken Smith's book, *The Buffalo River Country,* to sell in those stores. But some of us had observed that much of the stock he offered for sale was spurious, and we respectfully declined the reverend's request. This ticklish impasse was related in a letter to Ken Smith.

To those of us committed to the idea that the Buffalo River should be left as made by the Creator, the Kyle's Boys' Home was a dream all right—a bad dream. Our policy was one of nonintervention and limited accommodation. Ken Smith permitted the sale of his book in Kyle's curio shop in Jasper. There were no further comments from Jack McCutcheon in the *Harrison Daily Times.* The name of Frank Broyles was heard no more in connection with the project. Final legislation for the Buffalo National River moved slowly through the House of Representatives, and Reverend Harris finally accepted a more than generous offer from the National Park Service for his property on the Buffalo.

As for Kyle Harris, his name will be perpetuated far better than it would have been as a title for a foundering boys' home. Kyles Landing is today on the tongue of a host of delighted visitors: a take-out place for canoe paddlers come down from Ponca; a stopping point for hikers on their way down the Buffalo River Trail; a pleasant place for overnight campers in the nearby pine grove; a great place for swimmers on a hot summer day; an unexcelled place for photographers and students of natural science; and for sport fishermen any time of the year. To get to Kyles we all use that original bulldozed road, now maintained by the National Park Service. In doing so we pass by the site where once stood those out-of-place, eyesore dormitories, a reminder that the Buffalo Valley has been saved from much more than flood control and pump-back hydroelectric power projects.

Bonanzaland-Ozarkland

The spring and summer of 1969 was a time of continued stimulation of the startle reflexes of those of us committed to keeping the Buffalo in its natural state. At times we literally jumped out of our chairs upon opening the morning papers. The *Harrison Daily Times* was especially prone to bring us out of our seats (May 17, 1969):

Amusement Park Complex Announced for Harrison Locale

Motion picture production along with a family fun park—a new and unique amusement complex—are scheduled for development soon in the Harrison area it was announced here over the weekend by Albert C. Gannaway of Little Rock, producer of motion picture, radio and television shows.

The amusement park development on more than 600 acres will be concentrated near the old Hurricane Cave [formerly owned by Jim Schermerhorn], 16 miles south of Harrison on U.S. 65. The cave has been renamed Music Mountain Caverns which will be the focal point for the amusement park activities.

The project is expected to take five years. Gannaway, a native of Arkansas and graduate of the University of Arkansas, has returned to the state after a number of years in motion picture, radio and television production.

"For so long we haven't been able to see the forests for the trees in our own state," he said. "Now we're going to develop the region to take advantages of the untold natural resources."

Gannaway here over the weekend said that Music Mountain would contain a family fun park which he referred to as "Ozarkland."

"Unlike Disneyland, Ozarkland will have a dual purpose," Gannaway explained. "Our Ozarkland will serve as a family fun park, and it will also serve as a location for filming movies and television shows."

The seriousness of the purpose was reflected Saturday with the arrival of Mr. and Mrs. David Dortort as Gannaway's guests. Dortort is NBC's executive producer and creator of the "Bonanza" and "The High Chaparral" programs. ...

"Ozarkland will actually be an outgrowth of Music Mountain. The Park will be built around five distinct themes: Westernland, Frontierland, Jungleland, International Land, and Movieland, all in the valley below," Gannaway explained. ...

The Movieland area will also contain a complete Sound Center, which will have all the facilities necessary for sound and video recording. Tourists will be able to tour this area and see recordings being made and TV shows being taped and filmed.

In conjunction with the Sound Center, there will be touring companies that will go out into the local communities to entertain, and to cultivate local talent for possible future use in Ozarkland's TV originations or recording sessions. ...

Other expected outgrowths of Music Mountain, Arkansas, include a Pilot's Paradise. This will be a housing development especially designed for the 50,000 pilots in this country that pilot their own aircraft for leisure and business. The project will create homesites where they can fly in and taxi right up to their own houses. Gannaway stated that these pilots would never have a need for a car, if they so choose, since they will be within walking distance of golf courses and fishing spots—all readily accessible by golf carts—when completed.

Another outgrowth expected is a housing development which would contain homes for motion picture personnel who would be in the area to make movies.

Also to be constructed are summer camps for boys and girls, a fishing center on the White River, a dude ranch, and a professional broadcasting school where students could learn all facets of the broadcasting industry to prepare for employment in that field. ...

He went on to explain that two waterfalls will flow over the side of Music Mountain near the entrance to the cave by rerouting of underground flow of water within the cavern.

He said visitors to the caverns would find streams and a waterfall within the cave with fish in the streams and ponds.

To be opened at the same time as the cave will be an authentically reconstructed ghost town near the entrance of a replica of an old zinc ore mine. Present plans call for a mining car ride through the vicinity, as well as burro rides and foot trails through the region. Also to be constructed is a museum building to house the many Indian artifacts that have been discovered in the cave and nearby regions. ...

The area known as Jungleland will be just that—complete with live, wild animals in jungle settings, Tahitian and Polynesian villages, and jungle boat rides on a simulated tropical river.

International Land will be a complete international shopping center. Streets from various countries will be contained in this area, and as with all five of the areas, it will serve the purpose of being a complete movie set for on-location shooting, as well as serving the tourist trade with fine shops and restaurants. ...

The Music Mountain Cavern was not immediately adjacent to the Buffalo River, but the ravine in which it was located opened on the river valley a short distance from the cavern. The acreage involved in Bonanzaland or Ozarkland would include a significant section along the river itself and was therefore subject to the 1967 disclaimer clause from the Department of the Interior. That may have served to put a damper on the visionary project, but, whatever the cause, no dirt was turned in its construction. It was obviously a top-heavy scheme dreamed up to out-Disney Disneyland—a pipe dream destined to dissipate after a few puffs, much to the relief of us frazzled conservationists.

Remains of the great sweet gum, beech, sycamore, and birch trees that bordered the Buffalo above Roark Bluff, destroyed by the management of Valley-Y Ranch, April 1965.

The Horse Ranch

On that first canoe ride with Justice Douglas down the Buffalo from Ponca to Erbie in 1962 we were privileged to pass under splendid arcades of huge beech and sweet gum trees that lined the banks from the low-water bridge on down to the eastern limits of the beech woods above Indian Creek. A few small fields back from the river had been cleared and farmed by the original settlers, but the shoreline was by then still intact. But on subsequent trips during the next few years we would observe evermore shocking destruction of that arcaded riverine passageway.

In the winter of 1963 driving down the road to Steel Creek we rounded a corner, and there sat the biggest bulldozer and earth mover we had ever seen. What they were doing there was attested by the scraped away valley floor and the jumbled remains of the woodland round about. Then in 1965 after passing Bee Bluff on the river we came upon a scene of sylvan slaughter. Great root wads of the old beech grove that once lined the left-hand bank lay scattered about, their butchered tops piled up in the background for burning later on.

The show barn and other buildings at Valley-Y Ranch. Note the exposed Buffalo River and canoe, *left of center*, at the present-day Steel Creek launching site, April 1970.

The "channelized" Buffalo River at Steel Creek, Valley-Y Ranch, April 1965. Ken Smith in foreground.

The big beech root wad left by Yarborough's dozer crew lies under Big Bluff as a monument to the owner, April 1971.

And one day in 1971 upon approaching towering Big Bluff we noted that it came into sight sooner than it should have. The reason—the old sweet gum and beech trees that formerly framed the scene were gone. There on the gravel bar at the foot of the great precipice lay a five-ton root ball of a big beech, gouged out of the riverbank, a fitting monument to the doer of these depredations. He hadn't liked the natural ground cover of the gravel bar where the O.W.W.C. had camped in '61 and we with Judge Douglas in '62. He therefore had it scarified away and planted with fescue, its borders decorated with piles of brush and dismembered tree tops pushed off into the river.

He wasn't satisfied with the natural arrangement of the river either. Therefore, on a float down that section in 1965 we were astounded to see a wall of gravel standing fifteen feet above the water, put there by man and his machinery and not by any violent superflood. That was just one manifestation of what was happening on five miles of the most scenic stretch of the upper Buffalo, then a part of Valley-Y Ranch. We were witness to what some called "progress." That section of the river was being *channelized*. It was a style of river management then in vogue, possibly inspired by some of our agricultural agencies. But from whomever the idea, the owner of the Valley-Y Ranch had adopted it with enthusiasm. He had ordered his big bulldozers into the bed of the river itself to ream it out like a drainage ditch to facilitate, so he (or they) thought, the passage of flood waters after heavy rains. When those flood waters did come, they picked up those artificial embankments and distributed them far and wide, leaving them wherever the tide allowed. Regrowth quickly took over the bare shore line, and today copses of common sycamore thirty feet tall and growing rapidly stand where the silvery barked beech once graced the stream. Even observant canoe paddlers would not today suspect that they were not passing by undisturbed riverbanks. Only a few of us who were there so briefly before the days of progress know the truth of its original beauty.

What was happening in that section of the valley below Ponca was eulogized by the official magazine of the Carroll Electric division of the Arkansas REA. In its July 1969 issue it was reported that:

> One of Arkansas' most distinctive ranches, it could well be in the prettiest location. Valley "Y" Ranch is in Carroll Electric's service area near Ponca, west of Jasper on highway 74, bordering the Buffalo River for miles—all 1,500 plus acres of it.
>
> The ranch is not a dude ranch. It's a working ranch, and both the ranch-hands and stock do their work better than average because the general manager is a pretty brunette with grey-blue eyes who has a way of getting things done. Degrees in sociology and philosophy, along with majors in art and music, somehow seem to give her the ability to convey the message to man and beast. ...

Her father had always talked of wanting a ranch—not only as a working ranch but a comfortable retirement home. When he flew over parts of Missouri and the Buffalo River country of Arkansas during World War II, he located the area that looked right.

A realtor in the Jasper area told him the section he was interested in was so remote it was practically inaccessible. But this did not stop Mr. Yarborough.

He purchased some small acreage below Big Bluff, fronting the Buffalo river and another area near the bridge at Ponca. These two segments were miles apart as the river winds through the mountains. He methodically set out to contact all the owners of this land, and after 18 years, the Valley "Y" Ranch is now intact. And with much hard work clearing and building a fine road off of highway 74, it is no longer remote.

Miss Yarborough has been managing the ranch for five years, and it is a smooth running operation. Her parents drive from Kansas City about every two weeks and stay awhile in the all-electric, natural stone ranch home—exquisite in its setting against a green hillside facing a meadow and the bluffs of Buffalo River.

One of the most attractive ranches in the State of Arkansas, it looks more like a movie setting than a working ranch. With such a background, what could be more fitting than raising show horses? The beautiful Arabians were chosen for the original stock.

Operated as a business and capable of taking care of itself, Miss Yarborough said, "It is doing just that."

After the recommendation of S.B. 855 for favorable action by the Senate Interior Committee, one would have thought that the operators of the Valley-Y Arabian Horse Ranch would have slackened their activity until after the hearing on Congressman Hammerschmidt's bill in the House. Instead they redoubled their effort, bulldozing off the mountainsides adjacent to the river, proceeding with the channelizing of the stream and initiating an ambitious building program. The stone building that came with the property was refurbished and large structures erected to accommodate those thoroughbred Arabian horses. The main show barn was almost as large as the one in the state fairgrounds in Little Rock.

What else the owner had in mind was no less expansive than the Bonanza Dreamland down river. Its scope would be laid out before the House Committee on Interior Affairs later on.

The owner of Valley-Y Ranch maintained a curious, self-defeating stance in opposition to the national river proposal throughout. His efforts to make the Valley-Y into a real "working ranch" became ever more labored and unprofitable. His financial base (an S&L-type venture in Kansas City, Kansas) experienced simultaneous difficulty—similar to what many S&Ls nationwide were to be subjected to twenty years later.

Fame and Fortune on the Line

We had known for some time that the Ozark Wildlife Club near Mount Judea was on the rocks. We had heard that it had been broken up and gone to the bottom but then came news that a salvor was on the scene to retrieve what could be had from the wreck. J. P. Odom started out as a successful small-town businessman in eastern Arkansas. He then founded an eminently profitable insurance company in Little Rock, becoming a millionaire as a result. Not satisfied with just that, he next appeared in Harrison where he became the financial sugar daddy of floundering Dogpatch. He soon had big things going there—an annual Sadie Hawkins Day, man and boy chase by the local women-folk, a new convention center, Swiss chalets for those who could afford them, and even a ski lift. He was overcome with the dream of an expanding theme park empire and was ready for a big addition. The Ozark Wildlife Club was his for the asking. He would raise that leaky vessel, and it would become (what with all those wild boars, wolves, and mountain goats ranging those fourteen thousand acres) the biggest theme park ever. Just as spectacular would be the people running it. They would have name recognition that would surely guarantee success.

Those of us Buffalo River battlers, then relaxed in smuggery after the S.B. 855 decision, were due for more than just a shock—the *Arkansas Democrat* shivered our timbers with a little item, June 16, 1969:

Sportsmen's Club Purchased by Odom

HARRISON—Jess P. Odom of Little Rock, chairman of the board of Dogpatch, U.S.A., today announced that he had purchased a 14,000-acre sportsman's club formerly known as Ozark Wildlife, Inc., between Harrison and Jasper.

The club will be named National Wildlife, Inc.

Former Gov. Orval E. Faubus has been named president of the enterprise. These duties will be in addition to his current responsibilities as president of Dogpatch. Other officers of National Wildlife will be Jack Beason, executive vice president; Dowell Naylor and Michael J. Odom, both vice presidents; J. Ted Blagg, secretary and treasurer, and Raymond Pritchett, comptroller. Ford Maggard has been named business manager.

Odom said National Wildlife will offer almost 14,000 acres of hunting, fishing, camping, horseback riding, trailer sites, archery and rifle ranges and swimming.

Currently in existence is a 32-unit motel with adjoining restaurant and swimming pool, a two-story bunkhouse, 1,100 homesites, daily locations for travel trailers and complete hookup facilities and two lakes stocked with fish. The area is also stocked with wild boar, deer, bobcats, wolves, wild goats and other animals. ...

About 2,100 members from the Ozark Wildlife Club have been carried over into the new club.

Faubus said he has invested in the club and disclosed that the club was the location of his recent honeymoon. Faubus said, "There is a need for such a facility that will provide activity for the business and professional world."

The newly named National Wildlife Club did not obtrude upon the boundary of the proposed Buffalo National River, but it was in close proximity and if fully developed would have had a definitely disturbing effect on the national park. There would have been noisy in and out traffic on newly built roads and on the airstrip. Clearing and construction on residential and commercial sites would ruin the naturalness of the landscape, and sewage from the project would contaminate Cave Creek and the nearby Buffalo. All of that had already happened at Dogpatch, and we were glad that there wouldn't be another such a short distance down river. But with the resurrection of the Wildlife Club with Orval Faubus at the helm, we were mortified, dismayed, and thunderstruck. Considering his popularity with the average man and his ability to accomplish big and unexpected things, there was no telling how far the "National Wildlife Club" might go.

But he had been through personal upheavals, a famous politician out of a job, a divorce and a new wife, and a big, expensive Fay Jones house in Huntsville. He had been offered a top position by Arkansas's most successful businessman, at that time, and like a big fish he took the lure. But he was not stupid. He soon detected that he was involved in a visionary venture that was bound to fail. He quietly exited the undertaking, leaving it to a strange and acrimonious course that would extend over the years to come.

At one point Dogpatch, by then having passed through the hands of receivers a time or two, found itself under the administration of a radio-television evangelist. Intending to transform it into some sort of religious Mecca, he renamed the place *Godpatch*. But it did not go over well, and mercifully it reverted back to old Al Capp's original label.

Hallucinations from the Aginners

At some point dreams may become hallucinations, and so it was with the BRC&RC-ALA-BRIA. As their opposition to the national river became ever more labored and frenetic, a curious and disturbing hodge-podge of unworkable ideas emerged from various points of the compass. This conglomeration they chose to call the Pastoral Plan for the Buffalo River.

Harry Pearson, still with us at the *Pine Bluff Commercial*, produced in the July 2, 1969, issue of that paper a good summary of that so-called plan:

"Pastoral River" Idea for Buffalo Gaining Support

As an alternative to the National Park Service's plan for saving the Buffalo River, the "pastoral river" idea, now being promoted by those who originally wanted to dam the river, has begun to gain support.

That support is coming from certain commercial interests, notably real estate dealers and resort owners in the Buffalo River area, who like the idea of keeping the Buffalo River under private ownership. ...

In informal interviews with residents of the Buffalo River area this week, it was learned that the pastoral river concept has, at least at first glance, provoked a favorable response among some businessmen who formerly leaned toward the national river concept.

The pastoral river plan, as presented by the American Landowner's Association, is not likely, however, to sway the conservation movement since it has at least three very controversial features:

1—The plan seems to call for damming the upper reaches of the river in order to make the river floatable year-round. The presumption behind such an idea is that water could be discharged from upstream reservoirs at a rate sufficient to make the river floatable all year. ...

2—Under the plan, the upper section of the river would be placed in a permanent preservation zone, administered by the National Forest Service. The lower sections of the river would be designed for mass recreation—and this does not rule out the possibility of a reservoir.

There would be no eminent domain exercised in either the resource-oriented zone (the upper section) or the user-oriented zone (the lower and middle sections). The user-oriented zone would, incidentally, be administered by a recreation cooperative which would determine the best uses for the river.

3—The conservationists have argued that the entire river is worthy of preservation in its natural state, and the National Park Service has agreed that the entire Buffalo is a national asset. Putting two-thirds of the river into a mass recreational and commercial development area (one administered by a co-op with the approval of the forest service) would destroy its wilderness value and, in the eyes of conservationists, its uniqueness.

In view of the unpredictability of the course of events in matters such as this, national park supporters were concerned as to the outcome. It may not have sounded unreasonable to the casual observer. In it there was something for everyone, or so it seemed, and we might be left looking like a bunch of mean and remorseless agents of federal tyranny.

But the Pastoral River Plan was of questionable parentage. No agency on a state or national level would own it, and there was no private benefactor to take it under wing. Having nowhere to go with this patchwork proposition, its begetters settled back into their well-established routine of negativism.

But in spite of a shaky takeoff, the Buffalo River Conservation and Recreation Council and its pastoral river cargo would manage to stay aloft until long after the establishment of the Buffalo National River. That caused much delay and confusion in the park service program, but we knew that there was dissention in their ranks in the fall of 1969, and we were most happy to welcome from them a major defector.

A Visit to Viorel, A Comedy of Errors

This recital of the trials and tribulations of two determined river savers is told in the words of Joe Marsh Clark. It records the lengths to which we were sometimes obliged to go in this unending contest and marks our final contact— or lack of it—with Harry Pearson:

George Viorel, Chairman of the Buffalo River Landowners Association, telephoned Dr. Neil Compton one day during the last part of December 1969 telling him that he was leaving the country for several months. He had sold his land bordering the river. He had become disenchanted with Thompson of The American Landowners Association and wanted some of us to come over to his home near Hasty and get his correspondence. He especially wanted to talk to Neil and to Harry Pearson. He wanted Harry to do an article on Thompson which he would sign. Apparently he thinks the people are being taken by Thompson.

Because of the bad weather, no attempt could be made to see Viorel until Friday, January 2. On that day Neil Compton met me at the Holiday Inn, Springdale, shortly after 1 P.M. We left for Hasty in his Scout, which had ignition trouble about 10 miles west of Harrison at the sports shop which is on the south side of the highway. No one was in the sport shop, so we checked at the trailer house nearby. No phone. We were told that there might be one at the next farm house nearby to the west.

We went back to the Scout, and a Mr. Vicks stopped with a pickup. Said he was going to Harrison and agreed to take us in. The area in front of the passenger seat was pretty well taken up with a case of beer and other packages. We crowded in extending our legs out over these packages. Mr. Vicks asked us if we'd "like a drink," but we said we were only interested in getting to Harrison and getting the car taken care of. We were afraid Vicks had been drinking, but apparently had not though I, sitting next to him, kept an eye on his driving. I offered to drive if he cared to do any drinking. He runs a country store near Harrison and said he had gotten fed up being snowed in and had gone to Green Forest to get some liquor so he could cheer things up. We found that Vicks originated near Witt Spring on upper Richland Creek and knew the folks we had been associating with in the area.

Mr. Vicks let us out at the Clark Implement Company, which is the International Harvester agency in Harrison. We began thinking how I was to get out to the

Intersection of 65 and 123 to meet Harry Pearson, who was supposed to be there at 4 P.M. Neil thought of Tommy House. Tommy graciously agreed to drive me to the intersection. I saw a house nearby so went to it and arranged to wait there and use the telephone. Harry was not there, but it was only a few minutes after four. I told Tom that he might just as well go back into town. As he left, I returned to the home of Mr. Foster Canady. As I got to the front door, a car drew up at the intersection and stopped a moment. It looked like a Mustang, and I thought that it must contain Harry. I waved, was not noticed, and the car went on toward Harrison.

I called in to the Clark Implement Co. and asked for Neil. They said he had driven to town trying out a new Scout but that the car was about ready to roll. I told them to tell Neil where he could find me if I was not at the intersection.

In a short time the car I thought was a Mustang bearing Harry came back, and I vigorously waved again from Canady's front porch. The car turned down 123 toward Hasty. I assumed that it carried Harry, would end up at George Viorel's, and that he'd be there when we arrived.

I called back to Clark Implement to make sure that Neil had been told that I would be waiting at the corner as it had originally been planned that I would go on to Viorel's with Harry. I was told that Neil had left in a borrowed Scout. I was afraid there might be a mix-up so I hung around the corner most of the time with it getting darker and colder. My down coat had been left in Neil's Scout because at that time it was not foreseen just how we would operate. It took Neil about thirty minutes to get there, and I was waiting when he showed up at 5:30.

I got aboard and we headed for Hasty, keeping our eyes open for a Mustang that might be heading west.

We finally started down a long hill, went through an iron gate and on down to George Viorel's. We were greeted by George and two big hounds. The house had a bed of straw on the porch surrounded by cardboard boxes.

We spent at least two hours hearing of George's disillusionment with Thompson and his Pastoral Plan. Neil and I broke away at about 9 P.M. Harry was not there when we arrived and never showed up, so we were concerned that he might have lost his way, got hung up in a ditch, and that he might be found the next morning in a block of ice. George agreed to call Neil between 10 A.M. and 12 noon the next day so he could give information on meeting Harry in the event that Harry had not driven up. It was agreed that if Harry could not come to Hasty the next day that he might meet Viorel at Conway as Viorel progressed out of the hills on the start of his hegira.

We got back to Harrison and at about 10 P.M. we called our wives and reported in. We then ate, got Neil's repaired Scout at Clarke's home and drove into Springdale where I had left my car. It was really frosted over, and I had to clean the windshield and the windows before I dared to get on 71. It was slicker than owl butter on the highway in front of the Holiday Inn, and I thought maybe I'd better work my way back and check in for the night. Good judgment did not prevail and the road cleared up and I got home at 1:30 A.M. Neil had promised to call me if he found out anything I should know as he was going to call Harry. He was really going to be concerned if he didn't get him. If he'd ever get him, it would be this time of morning. I couldn't wait so called Neil after reaching home. I called again and again, getting the busy signal, so I knew that Harry was safe even before Neil called me at 2 A.M. Neil said Harry was laid up after too much New Year's and couldn't make it. At least we did not have to worry about him being lost out in the wilderness.

In taking leave of George Viorel, he presented us with several large boxes of written material, correspondence between him and Thompson and reams of typewritten material from Lucile Hannon. Upon reviewing it later on, we found very little that was of value to us in the continuing contest for the river.

We were never to see George Viorel again, but we were pleased to have met him, a good example that in any argument one's adversaries may sometimes display insight. We had hoped that with his departure the BRC&RC would dry up and blow away, but there was more to come.

The Arkansas Farm Bureau Hits a Lick

Into Viorel's shoes stepped a former Texan, now a citizen of Dogpatch, G. L. Hutchingson, a respected citizen of that place, whom we would never meet. We had reason to fear that he would prove to be more effective than Tudor and Thompson, and in that we were not in error.

About a month after our visit with Viorel, Evangeline Archer's sister, Joy Markham, who owned property on the upper Buffalo, received a form letter from the BRC&RC. Its masthead stated: BUFFALO RIVER COUNTRY IS THREAT-ENED WITH A FEDERAL LAND GRAB.

It stated that wealthy, well-organized special interest groups were behind the national park proposal, that eminent domain would be applied to all in the area, that a bill to that effect had already passed the Senate and was pending in the House, that all landowners should support the pastoral river plan as a compromise, and that it would function through a Rural Recreation Cooperative (whatever that was), and that they needed membership, requesting five dollars annual dues.

One of the first moves by Hutchingson was to persuade the Arkansas Farm Bureau, an undeniably powerful political and financial entity (and of which he was an important member) to oppose the Arkansas Stream Preservation bill in the state legislature. As a result, that legislation which we sponsored was defeated, a sobering fact since the Farm Bureau at the same time stated its opposition to the Buffalo National River proposal then in Congress. That was probably as serious a setback as we had had since the first hearing in Marshall.

At the Ozark Society spring meeting at Wilhelmina State Park in April 1970 George Wells, a new reporter for the *Pine Bluff Commercial,* in an interview on the subject quoted commentary by me which came out in print in such a way as to cause a reaction in Dogpatch. We, of course, did not know exactly who was doing what in Dogpatch, Pruitt, and Jasper. The reporter merely recorded my concern, which was justified after that turn of events in Little Rock, and which may have been overly harsh in regard to officers of the BRC&RC.

Very shortly I received a hot potato in the mail, a letter from the irate wife of the chairman of the BRC&RC chastising me most severely for implying that the officials of the BRC&RC might be speculators, her husband most notably.

In addition came similar chastisement from Mrs. Don Gronwaldt, treasurer for the BRC&RC who had formerly been a member of the Ozark Society and a victim of the unpleasantries of the upper Buffalo on Memorial Day 1965.

Not being one to relish the wrath of womankind anytime, anywhere, about anything, it was my duty to rectify the gaffe if it was at all possible. My lengthy reply to Mrs. Gronwaldt took the form of a general policy statement of the position of the Ozark Society as it was at that stage of the confrontation and ended with the following observation:

May 19, 1970

… Had it not been for Ozark Society, the Buffalo River would today be dammed from Lone Rock, three miles above its mouth on White River, to the community of Pruitt. By now plans for the construction of the third dam at Pruitt would be well advanced, and you would be in the way of soon losing your property to the Corps.

As evidence of their determination to eventually dam all natural streams, I would like to quote the words of Major General Carrol H. Dunn, Deputy Chief of the Corps of Army Engineers, who spoke to an engineering society in Little Rock recently. In the Arkansas Gazette for Sunday May the 10th, 1970, General Dunn is reported to have said:

"Water Resource Development must not only be continued but materially accelerated. It will be necessary to at least double the present reservoir capacity."

There is but one agency with the experience and capability of maintaining the Buffalo River in anything like its natural state of loveliness and the know-how to direct its recreational potential. That agency is the National Park Service. Any other plan that we have yet heard of cannot possibly contain the ultimate aims of the Corps of Army Engineers and will prove ineffectual against the multitude of lesser threats now rising against this unique section of beautiful America.

Many of these destructive activities, less sudden in their effect than big dams, but ultimately just as ruinous are already in progress. Among these are: soil stripping for landfill resulting in heavy mud pollution, channelizing of the stream bed, denudation of adjacent steep slopes by the bulldozer and chain-saw, marginal mining operations, the broadcast of tree killing spray, organic pollution from human waste and chicken litter, and unwise and unsightly commercialization.

If we are successful in bringing about the establishment of the Buffalo National River in our time, our grandchildren will wonder why there was ever any opposition to it. If we fail, we will be remembered by the next ten generations for our greediness, lack of vision and ineptitude.

Yours very truly,
Neil Compton, M.D.
President, Ozark Society

Whether or not that epistle had any effect in Dogpatch I never knew, but at least I felt better having written it.

In the face of these burgeoning schemes from the private sector, the Ozark Society in 1969 was able to maintain a positive position in the controversy.

That summer Doug James arranged for the American Ornithological Union to come to Arkansas and float the Buffalo. They didn't see many birds at that season but enjoyed the trip on the river immensely (*Arkansas Audubon Society Newsletter,* December 8, 1969):

The AOU at the UOA

Never before had so many ornithologists come to Arkansas as during the first week of September when the AOU (American Ornithologists' Union) met at the UOA (University of Arkansas) in Fayetteville. Almost 350 scientists from 37 states including Alaska, from 3 provinces of Canada, and from such places as Panama and Ethiopia, attended the week-long annual meeting. The subsequent response in praise of the success of the meeting has been tremendous in scope, and rather overwhelming to me. Letters upon letters have arrived, many exclaiming it was the best meeting ever attended. Apparently everyone was enamored of the Ozarks, particularly the Buffalo River. Many people plan to return, some already have, and others are expected in the near future.

Among those present was Roger Tory Peterson, no less, with whom we Ozark Society guides had the pleasure of fraternizing. While it didn't have the impact of the W. O. Douglas trip, it nevertheless was a big boost to the save-the-Buffalo idea at that stage of the game.

Then came a letter from Ken Smith updating us on the status of his book:

... I am pleased to tell you that we are nearly sold out of the second printing of THE BUFFALO RIVER COUNTRY. We have sold around 7,000 copies of the first and second printings, and there remain, as of a few days ago, only about 350 hardcover books and 100 paperbacks. ...

Another happy note: Within the past month or so, we reached the break-even point, so that now we have covered all expenses—printing, postage, even the author's royalties—for the first and second printings of the book. Any revenue from the sale

Left to right: Dr. Fran James, H. Charles Johnston, Jr., and Roger Tory Peterson, August 1969, at Tony Bend.

of the remaining copies, for example, will remain in the Publishing Fund. This "surplus" will be plowed into printing the second edition of THE BUFFALO RIVER COUNTRY ... and sales of those books will bring the money back to finance new projects ...

That marked an end of all the hand wringing and uncertainty as to whether or not we should be involved in publishing ventures. The stage was set for the organization of the Ozark Society Foundation in the years to come.

A Ghost Town for Sale

Senate Bill 855 had been passed with such rapidity and ease that over-optimism permeated our ranks. Just about everyone knew that it was all over and in our favor.

But for those who paid attention, there were words of warning. As early as March 1969 the National Wildlife Federation learned that the Nixon administration had proposed a thirty-million-dollar cut in the Land and Water Conservation Fund, the major source of money for acquisition and development of new parks. That being the case, there was little point in authorizing new areas and adding to the backlog. That included the Buffalo, and those of us most concerned deduced that there was little likelihood of a companion bill passing the House in the current session of Congress, but we would agitate for introduction of the House bill anyhow.

At this stage of affairs the urgency for protective legislation on the Buffalo was not limited to that scenic upper section in Newton County. All of it figured in, but there was one site on the lower reach of the river that was of import, not so much because of its natural attributes but because of what had been done there by men looking for zinc ore. Rush in 1969 was a ghost town as interesting as any out west in Nevada or Arizona. There had been several surges of mining activity there, the first near the turn of the century when the ore was moved out to Batesville by flatboat or in wagons drawn by mule or ox teams. In one instance Rush was serviced by a small river steamer, the *Dauntless,* which unfortunately struck a log and sank in the big hole at the mouth of the Buffalo.

During the First World War the place achieved its greatest boom, reaching a fluctuating population of nearly three thousand. Enormous amounts of rock and ore were dug out of the hills along that narrow valley leaving caverns to be marveled at by later visitors.

There had been another flurry of mining during and after the Second World War, but by 1969 Rush was bereft of inhabitants except for Gus Setzer and Fred Dirst, an old miner who conducted tours into the mines for wandering visitors and who sometimes rented canoes to those needing them and charged a small fee for launching and taking out canoes at Rush.

Many of the old store buildings and houses along the tortuous road greeted the traveler with windows out and doors ajar—lonesome, forlorn, and forgotten. It was a place to inspire one's sentiment for those bygone days, a place to imagine the hustle and bustle, the rowdy crowds of miners, their women and children, and the preachers that came and tried to save their souls.

Those of us concerned for the fate of the river and its interesting nooks and crannies were not surprised by a headline in the *Arkansas Gazette,* June 3, 1970:

Old Ghost Mining Town for Sale;
Mines, Former Postoffice Included

The advertisement for the property reads: "This is an unusual piece of property with a colorful and historical background. An ideal location for recreational development. Some houses, old postoffice, store building, mine buildings and several old mines. ..."

Bob Baker, owner of the Development Company, said he thought the price was reasonable, if not too low. ...

The area for sale was formerly owned by Morning Star Mining Company of Pittsburgh, Pa., which found "Jumbo," the largest single piece of zinc carbonate found. It weighed 12,750 pounds. ...

No one has offered to buy the land. Baker said there had been some "promoters, bankers and lookers, but no doers."

That news report was not without its irony. We had known Bobby Baker since 1961 when he was a student at the U of A in Fayetteville. He was against the doings of the BRIA and helped keep us informed as to much of what they were up to. He was from Yellville, not far from Marshall, but knew all of the big dam pushers over there. We had lost track of Bobby, but in 1968 I ran into him by accident. He was out of school and had gone into the real-estate business in Yellville. Signs for his Buffalo Land and Development firm were conspicuous in the area.

That day he said to me: "Doc, we sure appreciate all that publicity that you people have generated for this part of the country. It sure has helped the real-estate business."

We already knew that we were doing just that, but there was nothing else to do except to continue the all-out effort to see the Buffalo River and Rush Creek in the jurisdiction of the National Park Service.

Rush Creek was not sold at that time, but later it did come into the hands of an industrialist who intended to make a big tourist development out of it. But he sold it to the National Park Service instead. A final disturbance came when vandals went in and removed most of the good boards in some of the old buildings. But some of it did remain. The park service has stabilized it and marked it off as a historic district in the national park.

Those of us who knew them can still go to that ghost town, there to muse upon the backbreaking toil, the occasional baptizings in the river, and the frolic of picnics and country dances that was the lot of such diggers as Fred Dirst and George Jones.

TWENTY

CONSERVATION—
A POLITICAL CHESS GAME

By 1969 those of us in the Ozark Society had learned something about the game that we were playing. It was environmental chess. If we were going to win, we would have to move our kings, bishops, and pawns with care and forethought. The board for this match of wits was the body politic upon which moved our elected elite. We had not at first realized how complicated would be the moves, with our favorite measure being kicked about in two houses in Washington and occasionally two more in Little Rock.

We understood from the start the importance of presidents, governors, and judges. If they were not in the right squares, we were subject to checkmate.

At that time the executive and legislative takeover by the various levels of the judiciary had barely begun. Had we been obliged to fight it out in court, we might have become bogged down in litigation and consumed with legal fees. Charley Johnston's suit against the Corps was mercifully brief but, fortunately, was not decisive. Wellborn Jack, himself a lawyer, in his court battle to halt construction on Gillham Dam on the Cossatot held up that project for two years, thanks to careful planning, sympathetic judges, and assistance from the Environmental Protection Agency.

As it was, we were courtesans to those decision makers, our elected office holders, to the various bureaucrats serving under them, and to that third estate, the media, in print and on the air. None of these could be expected to satisfy all of our conservationist coworkers all of the time. That became an increasing problem when Hammerschmidt's bill failed to move in the House. That impasse was best related by Harry Pearson in one of his final reports (*Pine Bluff Commercial*, October 30, 1969):

Buffalo Plan Has Rough Going in House

The prospects for creation of a Buffalo National River by the 91st Congress are not, at the moment, very bright.

Congressman Wayne N. Aspinall, Democrat of Colorado, chairman of the House Interior and Insular Affairs Committee, is presently bottling up, in his committee, legislation creating new national parks.

He will not, he says, hold hearings on the Buffalo National River proposal because of the lack of federal funds to create new parks. ...

Dr. Neil Compton of Bentonville, president of the Ozark Society, wrote Aspinall asking that a hearing be held.

Aspinall's reply said, in part:

"No action has been scheduled on this proposal to date and I regret to advise you that it is unlikely that action will be scheduled in the immediate future. In light of the present restrictions on funding and in light of the current backlog of existing parks and recreation projects, it would be inappropriate to hamper the current program with new authorizations which would cause an additional drain on the limited funds."

The anti-park forces are jubilant over Aspinall's letter. The chairman of the Buffalo River Conservation and Recreation Council, G. L. Huchingson of Pruitt, wrote to those campaigning against the national river:

"We are sure you are aware by now of the fact that S. 855 passed the Senate. This bill is now dormant and will not move unless the House abandons H.R. 10246 (the equivalent measure introduced by Congressman John Paul Hammerschmidt of Harrison), but this bill is not dead and will not become so until the end of 1970.

"Mr. Aspinall has announced that H.R. 10246 will not be called up for hearings this year, so there is only a very slim possibility of passage this year, however, we have good reason to believe that the House will push for passage by April of 1970."

If no action is taken by this session of Congress (which ends next year), then both bills will have to be reintroduced and those seeking to preserve the Buffalo River will have to go through the entire process again.

The Ozark Society has decided, for the moment, to do nothing. Compton, in a memo to society officials, wrote:

"From information we have it is believed that a letter writing campaign to Mr. Aspinall at this time might arouse his antagonism and it is requested that we hold this activity in abeyance for now."

On the other side of the fence, Huchingson advises his forces:

"These (delays) should be somewhat encouraging to you in that it gives us just a little more time, but let's not become complacent. Now is the time to really get busy.

"We encourage all of you to write to your Congressman, as well as Representative John Paul Hammerschmidt, Representative Wayne N. Aspinall ... Senator J. W. Fulbright and Senator John L. McClellan ...

"A personal letter in your own words can be a most effective aid to our project. We do ask that you stress three points: 1. opposition of the Park plan, 2. support of the pastoral river, and 3. request that your letter be made a part of the record of the House hearings when such hearings occur."

Hammerschmidt, asked for comment on Aspinall's decision, said he had no comment.

Then he added: "I am working on the matter and I still have hopes of getting the bill enacted in this session of Congress."

The situation described therein would hold true for nearly two more years. It was difficult for many to believe that the galling delay was due to a congressman from Colorado, Wayne Aspinall, who obviously could know little about Arkansas and its Buffalo River. But when it was finally over, his reasons as stated in this news release were valid. Taking into account the moves that had to be made in Congress, it is clear that we had little right to expect any more any sooner.

Hammerschmidt on the Griddle

Being confronted by this stalemate, there began to appear within our ranks the specter of political polarization. Most of our membership was of liberal or Democratic party leaning. Their target for the delay had to be someone closer to home—that Republican lumber dealer from Harrison, John Paul Hammerschmidt.

We began to hear such things as:

Wasn't his election a "fluke"?

How could a greenhorn freshman congressman know how to get a bill through the House?

Being in a minority party, how could he expect to enlist support from the Democrats?

Weren't the Republicans always against conservation measures?

He was a pawn of the Corps of Engineers and would double cross us when he got a chance!

Everybody knows that he was for finishing the dam on the Cossatot!

An increasing number of letters and phone calls began to plague Hammerschmidt and his staff, stating those charges and more, all originating within our ranks and not from the BRC&RC. In the end, Hammerschmidt would hold his pro-park position as surely as Jim Trimble had held out for the dams. But at this point in the game I was, for one, fearful that these attacks on Hammerschmidt's integrity and resolve might actually cause him to reverse his stand. I felt obliged to preach to our band of believers the doctrine of toleration, respect, and good will toward *all* politicians, Republican or Democrat, conservative or liberal. In the record of preservation and conservation of America's natural grandeur, neither party was preeminent. Both had played their part and would continue to do so. In an address to our membership assembled at our fall meeting in Hot Springs, October 15, 1969, I discussed this subject at length. In closing, a brief statement as to the influence of the Ozark Society on our country's lawmakers was made:

> In reflecting upon our present stature I recall a statement made by a member of one of our state agencies a year or two ago. He said: "The Ozark Society is a powerful organization."
>
> I would like to add the following observation: We are certainly not powerful in numbers, for our numbers are few and scattered. Neither is it our financing, for we operate on peanuts.
>
> The truth is, the Ozark Society has some powerful ideas. Ideas are the only real power that we shall ever have, and if we wish to be influential in politics or any other field of human endeavor, we must continue to generate and champion them.

Special Report on a Meeting with Congressman Wilbur Mills

In those years the most powerful man in Congress was Wilbur Mills of Arkansas's Second District, Chairman of the House Ways and Means Committee. At our fall meeting in Hot Springs, William J. Allen, field representative of the Wildlife Management Institute and a valuable technical advisor for our program, Dr. Joe Nix, and a few others discussed the possibility of arranging an interview with Congressman Mills. On November 29 that was done.

Attending this meeting in addition to myself were Clayton Little, Joe Nix, Harold Alexander, and Bill Allen. We found that arrangements had been made by the simple expedient of Dr. Nix having written a letter to Congressman Mills asking for an audience.

At 9:30 A.M. Mr. Mills received us in his office in the basement of the post office building in Searcy. He appeared to be interested in the subject and informed us that he had, at the request of Congressman John Paul Hammerschmidt, already spoken to Congressman Wayne Aspinall. He informed us that Congressman Aspinall was not positively antagonistic toward this legislation and that he was, in fact, concerned about conservation measures such as this. He said that there was a large backlog of such legislation in Congress, and this might delay a hearing, but that he believed that such a hearing might be held early in 1970. He informed us that he was on good terms with Congressman Aspinall and that he believed that the hearing would be arranged as soon as the proper legislative machinery could be set in motion.

Some discussion was held as to whether or not Senate Bill 855 should be considered or whether Congressman Hammerschmidt's House Bill should be given priority. Congressman Mills felt that the latter should be the one.

Mr. Mills was quite attentive and gave due consideration to what each of us had to say. He displayed considerable knowledge of problems concerning artificial reservoirs, aware of the fact that sport fishing had declined in many of them and that in some cases the hatch of young trout had been disappointing due to the presence of manganese in the water. This was a fact that Dr. Joe Nix had discovered in his research on water chemistry, and it was gratifying to find that Congressman Mills was already aware of Dr. Nix's work.

The desirability of the Buffalo National River project was discussed, and it appeared that Congressman Mills was enthusiastic concerning this project. His knowledge of the subject, however, was not as thorough as it might have been, and we came prepared with a number of items for his review. Included in these was a copy of Ken Smith's book, *The Buffalo River Country,* and also a copy of the Stream Preservation Report prepared by Harold Alexander. In addition, I had brought along thirty glossy 8-by-10 prints showing the natural beauty of the Buffalo River area and also showing some of the abuses that were being perpetuated along it, such as the highway department activities and the bulldozing and logging operations in some of the side canyons.

The subject of opposition to the Buffalo National River was also considered, and it was pointed out that most of the recent opposition was coming from out of state operators such as Mr. Charles Thompson of Harpers Ferry, West Virginia. The matter of real-estate values also was discussed, as was the concern over the development of commercial enterprises along the stream.

In addition to discussing the Buffalo River, it was also brought out, mainly by Harold Alexander, that other streams in the area were being considered for wild river status to take care of the overflow traffic that would ensue on the Buffalo in case it was made a national river. The ones that were especially mentioned in this case were the Mulberry and the Kings River.

Also the matter of the Blanchard Spring Cavern was discussed, and it was noted that Mr. Mills had a very large black-and-white print of this cavern in the anteroom of his office. He described to us some of the details that had taken place in efforts to obtain money for the opening of Blanchard Spring Cavern.

The next subject discussed was that of the Soil and Water Conservation Fund. This fund was set up by Congress to furnish monies with which to purchase new national park lands and public recreational area lands. Mr. Mills mentioned the fact that this fund had been cut already under the Johnson administration and that further cuts had been made in it by the Nixon administration. He reminded us again that the Nixon administration was not for any new starts in this field. This again emphasized the fact that the Buffalo River legislation would have to wait until other such similar legislation cleared Congress. He thought, however, that if such legislation did pass that it eventually would obtain its share of whatever was left of the Soil and Water Conservation Fund.

At the end of the discussion, it was agreed by all that a well-balanced recreational complex in northern Arkansas would include such features as the Blanchard Spring Cavern, the Buffalo River, and various reservoirs that have already been constructed. The discussion session ended on a most friendly and hopeful note, and we took leave of Mr. Mills very much encouraged in the prospects for the eventual enactment of legislation for the protection of the Buffalo River under the National Park Service plan.

Bushwhackers on Roark Bluff

A few months after our visit with Congressman Mills I was pleased to receive from Joe Nix a request from the congressman for more photographs showing environmental damage along the Buffalo. We had word that Senator John McClellan also wanted some of the same. That called for immediate action on my part to obtain some good shots of the ongoing damage in addition to those already on hand, which involved for the most part heavy-handed highway construction on Arkansas #21 and 43 in Boxley Valley.

The most serious environmental injury was being committed on the Valley-Y Ranch, and there I repaired in company with Larry Burns one day in May 1970. We did not regard it wise to barge in at headquarters at Steel Creek with cameras, lenses, and tripods to shoot up the place in full view of the manager, carpenters, and dozer drivers. However, some of the steep slopes recently bulldozed and now eroding were visible from Highway 74 east of Ponca. We took advantage of that and of the skinned-off hilltop at the main entrance of Valley-Y. The rest would be more ticklish. We wanted pictures of the big show barn and other structures now abuilding at Steel Creek. There would be a vantage point from the mountainside above Roark Bluff to the north. To get there we drove up Highway 43 to the first bench where an abandoned wagon road led off to the east around the steep mountainside. There we left our car to go skulking through the woods like a couple of revenuers searching for a moonshine still. But we couldn't see much. Spring was well along, and the trees were in full leaf, hiding the ranch complex down below. There was nothing else for me to do but pick out a likely tree and climb up to where one could see all of that activity down in the valley. While not as shocking as they would have been from closer up, the pictures were acceptable and were sent off to Washington with some satisfaction. Whether or not we were on Yarborough's property up there I do not know, and whether or not it was the thing for an honest conservationist to do, others will have to decide. It is told here to illustrate the lengths to which those who play this game must sometimes go.

We then journeyed over the end of Gaither Mountain and down to Erbie where, on the north side of the river below the old church, we stopped to photograph as dismal a thing as could be wreaked by the hand of man. There a tract of nearly mature timber had been sprayed the year before with 2,4,5-T. Many of those tall hardwoods stood half dead with clumps of leaves hanging irregularly, with others stark and dead all the way down. From the understory shrubbery sprouts had come up this season, creating an impenetrable barrier of briers, brush, weeds, and vines. The pictures of all that were good ammunition to have in the battle for the Buffalo. Shortly, we were to learn that the man responsible for that spray job was the nephew of our own Dick Murray. He was an absentee owner, a major in the army, living in Hawaii.

On the way home that day we saw the means by which this environmental atrocity had been accomplished. Sitting in the yard of a farmhouse over near Rule in the Osage valley was a small helicopter. It was the property of a returnee from Vietnam who had learned to fly a helicopter in the air force and was now in the business in this Ozark back country selling his ability as a timber killer to misguided ranchers and farmers for miles around.

Such doubts as one might have had concerning the reception that those hard-to-come-by photos received in Washington was dispelled by a letter from Congressman Mills:

June 3, 1970

Dear Dr. Compton:

Thank you very much for your letters of May 26 and 29 forwarding to us the pictures along the Buffalo River. They are just what I suggested to Dr. Joe Nix I would like to have and I am most appreciative of your making them available.

For your confidential information at this point, the Chairman of the Subcommittee of the Committee on Interior and Insular Affairs advised me this week that he thought before long they could make what he referred to as a "field survey" of this project that may or may not involve hearings but, at least, it is a step in the direction of bringing about the enactment of the legislation by the House of Representatives.

We are placing the material you gave us in the hands of the Chairman of the Subcommittee, the Honorable Roy Taylor, of North Carolina.

With kindest regards and best wishes, I am

Sincerely yours,
Wilbur D. Mills

His disposition of them and the impending visit to the Buffalo by Congressman Roy Taylor was as exciting as any news that we had had in the prolonged contest. Surely now we were in its final stages.

More Yet for John Paul

But many of our members continued to radiate gloom and doom. Hammerschmidt had to be the fly in the ointment. That belief was voiced by Joe Nix in a letter to me concerning possible principal speakers at our fall meeting in Fayetteville that year:

> I have been thinking about the problem for the annual meeting of the Ozark Society. … How about asking one of our congressional delegation? This might have two purposes, an open discussion on the Buffalo River Bill, and serve as some political pressure for the next try at the bill. …
>
> I am talking as if the Buffalo River Bill had already met its defeat, and I believe that, for all practical purposes, it has. I know how you feel about Hammerschmidt, but I think that you are wrong about him. I think he has sold us down the drain, and Wilbur Mills didn't help the situation either. Bill Allen pushed the Governor to send a telegram to Aspinall and Hammerschmidt a couple of weeks ago, and Harold did draft such a telegram. When it was submitted to Faulkner in the Governor's office, I understand from the grapevine, he (Faulkner) called Hammerschmidt and asked him if he wanted the telegram to be sent. He did not and it was not. We have again been put off, I think there is little doubt. …

That attitude permeated our ranks despite disclaimers by Congressman Hammerschmidt; witness a letter from him to Joe and Maxine Clark, both of whom were convinced of his deviousness:

> I understand your concern regarding the Buffalo River Bill. May I assure you, I am doing everything possible to bring about favorable action on the bill.
>
> There is absolutely no doubt of the support for the preservation of the Buffalo River, nor is there any doubt in my mind of the correctness of the cause.
>
> Please be assured you have my support, and by all means your gains will not be lost. For instance, the Sleeping Bear Dunes National Lakeshore Bill just now having hearings passed the Senate in 1963. The past action and support given to the river cause by the Ozark Society will ultimately accrue to final proper action by the Congress.

At our last general meeting Evangeline Archer had bowed out as secretary of the Ozark Society. She had been replaced by Lois Imhoff, wife of John Imhoff, professor of engineering at the University of Arkansas. The Imhoffs were long-time friends and coworkers and, like so many others, regarded Congressman Hammerschmidt with distrust.

My own discomfiture was stated in a letter to Lois, August 27, 1970:

> … As for Mr. Hammerschmidt's position we have no positive knowledge as to why he has taken the course that he has chosen to take. This also includes Congressman

Wilbur Mills. Both of them appeared to be very much for the Buffalo National River a few months ago. We have no real reason to believe that they are not for it now, but I am certainly disappointed that the legislation is not going through this session of congress. I can't help but believe that it will be harder to accomplish the next time. However, they do understand the workings of that body better than we do and perhaps are doing what is the best after all.

A Black Hat for NC

After the fall meeting of the Ozark Society in Fayetteville, our membership, dispirited because of the failure of Hammerschmidt's bill, gave way to grumbling and tail twisting, if not mutiny. As a result, I along with John Paul became an object of suspicion. It was best said by Allan Gilbert, Jr., Senator Fulbright's nephew. He was a reporter for the *Northwest Arkansas Times* in Fayetteville, a paper then owned by Senator Fulbright. While my personal relationship with Allan Gilbert was always cordial and friendly, he and many others sensed that my preference was the "other" party. It had been ever since FDR ran for his third term, but I never talked of it in their presence. My persistent defense of Hammerschmidt may have been the giveaway. Regardless of political attitude, I better than anyone knew that, by 1970, had it not been for John Paul, there would have been a 240-foot-high dam on the Buffalo at Gilbert.

Allan Gilbert had written to Fulbright's principal aide, Lee Williams, for an updating on the House bill and received an interesting reply:

> Since returning and since receiving your letter, I have been doing a little gumshoeing on the Buffalo River Bill. To refresh your recollection, prior to the Congressional recess, the Bill was scheduled to be taken up along with other measures, but it was not included in the measures that were considered, and a staff man tells me confidentially that it was withdrawn from consideration at that time because there was not unanimity among the Arkansas Congressional (House) delegation. My supposition is that JPH did not want the Committee to act on it prior to the election because it would have given him another issue to deal with which he wanted to avoid.
>
> … I have discussed it with the Senator, and when we can pin down Chairman Aspinall the Senator is going to find time to walk over to the House side and visit with him about the Bill in the hope that he can persuade him to take action on it during the bobtailed session. There is always the possibility that JPH will now make his own attempt to get it through, and I have always felt that Wilbur Mills can be very influential if JPH asks him to intercede to get the Committee to act. We will do whatever we can and let you know if there are any significant developments.
>
> I do not wish to be in a position of charting your strategy from that end, but it could be that a lil ole column in the paper commenting on the present status of the Bill (*Northwest Arkansas Times*, of course) might stir up some JPH constituents to urge him to go ahead, but that is, of course, a matter for you to decide.

In regard to '72 and that seat, I think it is going to be ripe for the right man, and we should compare notes on that one day before too many heads crop up.

The last paragraph of that message reveals the inter-party intrigue that exists on Capitol Hill. They were looking for a viable opponent to oppose Hammerschmidt in the next election.

Allan Gilbert's reply contained more of the same:

As you instructed, Purvis called me regarding Buffalo River bill status with Aspinall's committee in House. His report was about what we'd heard from another member of committee (John Saylor). I then discussed the situation with Ozark Society folks, who asked me to find out what they could do to light a fire under John Paul Hammerslip. I asked the Senator and he suggested an end run, wherein the Ozark Society people would appeal to Alexander, Pryor and Mills, making it obvious that they were dissatisfied with the help they were getting from John Paul. The idea, the Senator said, would be to embarrass J.P. into doing something on his own hook for his own district.

I reported this to the Ozark Society, but Dr. Compton, who is now the bottleneck, insisted on talking the strategy over with Hammerschmidt, who assured the good Doctor that he would personally take the matter up with the committee as soon as Congress reconvenes. He also told the good Doctor that he would do what he could to get the state delegation in line behind the bill, but cautioned him that one serious obstacle to approval in the House was the lack of endorsement of the plan by the Park Service.

The Ozark Society has now endeavored to contact the Park Service to find out where they stand on the bill, so I presume John Paul has passed the buck off to that agency for the time being. I'm writing to fill you in from here, and Purvis, and to ask if you can ascertain if the Park Service opinion is crucial, vital, significant or impera-tive for House action. I hadn't peered very deeply into the Buffalo bill before, but now that I have it appears clear that Dr. Compton believes whatever JPH tells him, and is willing to settle for that, and JPH is determined to stall as long as he can in the belief that nobody will get too upset if nothing but a lot of study and discussion takes place.

I have offered myself up for full-time help for any reasonable candidate for Third District in 1972 against JPH. There's a question, of course, whether my services (like Trumbo's) are asset or liability.

However far off Allan Gilbert may have been in his "bottleneck" statement, he was for certain in error in saying that Hammerschmidt was passing out information that the National Park Service did not endorse the Buffalo River plan. That plan was their baby, and they never wavered in strong support of it.

In offering himself up as a full-time warrior to defeat Hammerschmidt in 1972, Allan Gilbert was doomed to disappointment. JPH had just demolished Donald Poe in the 1970 election and would continue in his office without serious challenge until the present day. He was never a prevaricator. That he was not a hard-core conservationist we all admit, but he was not a scheming, conniving politician or developer either. He was astute enough to recognize the Buffalo National River as a good thing for the state and nation and abided by that throughout.

Joe and Maxine Clark were then convinced of Hammerschmidt's duplicity, and in order to verify that Joe wrote to Wilbur Mills for evidence. Mills' reply, December 7, 1970, stated in a few words the reason for the delay, bringing us back again to the basic obstacle—the chairman of the House Committee on Interior and Insular Affairs:

Dear Mr. Clark:

Your Congressman, the Honorable John Paul Hammerschmidt, showed me the Autumn 1970 issue of your Bulletin and I thought perhaps it would be wise for us to get the record straight so far as Mr. Hammerschmidt's efforts to pass H.R. 10246 are concerned.

On several occasions he and I both have talked to the Chairman of the Committee on Interior and Insular Affairs, the Honorable Wayne Aspinall of Colorado, about this legislation. We urged him to conduct a hearing and allow the bill to be favorably reported to the House of Representatives, where I am certain we will have no trouble in passing it.

Mr. Aspinall said he had a great number of questions about the project that would have to be answered and that the schedule of the Committee was such, as well as his own schedule of leaving the Congress to campaign before Congress actually recessed for the election, that it was not possible to report the bill at that time.

Since we have reconvened following the election, Mr. Aspinall has been absent and, apparently, his Committee is not reporting additional legislation during his absence.

We are still hopeful that he will return and that we can pass the bill before the end of this session, but if not you may be assured we will all be pressing hard for its enactment in the next Congress.

With kindest regards and best wishes, I am

> Sincerely yours,
> Wilbur D. Mills

The contents of this letter became known and was commented upon in the editorial page of the *Northwest Arkansas Times,* December 21, 1970:

Flotsam and Jetsam

The Buffalo River bill is dead for this session of Congress, according to Rep. Wilbur Mills of Kensett, congressman from the state's Second District.

... why is Congressman Mills, of the Second District, explaining the demise of a matter that pertains to the Third District? The Buffalo River is in Republican Rep. John Paul Hammerschmidt's Third District, and Hammerschmidt is author and sponsor of the House version. It is unusual for someone besides the bill's sponsor to explain its disposition. Or, if Mills is speaking for the Arkansas delegation, why aren't Rep. Bill Alexander of the First District and Rep. David Pryor (who has evidenced more concern for the bill even than Mills) in on the denouement? ...

The Buffalo River bill didn't have the fullest support of all its congressmen a couple of months ago. Which raises the question of what its support will be next year, when President Nixon may very well have decided that economies, in Interior affairs, are even more a matter for careful review than during his first two years in office.

We can only conclude that Congressman Mills' letter was written with an eye toward comforting the disappointments of those who have worked so long and hard on the Buffalo's preservation, including many in his own district. Or, he might have written it as a favor for his colleague from the Third District. That, in itself, would be a curious thing, too, though, don't you think?

The *Northwest Arkansas Times,* like many other papers, was baffled by the message from Wilbur Mills. "The lack of unanimity in the Arkansas delegation" posed a quandary. They would have liked for the laggard to have been Hammerschmidt—or was he really Alexander? or Pryor? As far as we ever knew, there was none. Those congressmen didn't declare themselves since the bill was not coming out, and there was no need at that time to do so.

Some Last Moves—
The Beginning of the End

Nineteen seventy-one, the last full year of the contest, was ushered in by a welcome exchange between two of Arkansas's influential people in national government. On January 4 Senator Fulbright dispatched a note to Senator McClellan announcing his reintroduction of the Buffalo National River bill under the bill number S-7 and requesting McClellan's cosponsorship. In a brief reply McClellan stated that he would be glad to act as cosponsor.

From those two heavyweights, we could not have asked for more.

Meanwhile, those of us here at the grassroots level were busy planning our next move as best we could, at board meetings, by frequent telephone calls, and through the mail. Ken Smith was kept advised by letter:

... I did have a long talk with John Paul Hammerschmidt a few days ago on the telephone. He was emphatic in stating that nothing should be done right now. It seems to me that he is a little touchy on the subject for some reason and for what we do not know. Joe Nix who was at our meeting yesterday thought that perhaps it was due the activity that has been taking place on the Cossatot River. We do know that some people down that way did write him some rather nasty letters, and it could be that since the project was already under construction he was put out by such goings on. Other members of the Society have written him some rather bad letters about the Buffalo River also …

I have been trying to get everyone to soft pedal such opinions. There is no point in calling him a liar and other worse terms since I am sure that he is sincerely interested in seeing the bill pass eventually. …

I have had to absorb an awful lot of blame for nothing being done since I was try-ing to keep people from badgering either Hammerschmidt or Mills. I certainly do hope that our hot-headed and emotional members will not indulge in any more unwise remarks because if we are going to get this thing passed we are going to have to be on the very best terms with Hammerschmidt and Mills. … We feel, however, that Mills and Hammerschmidt are pretty close and that Mills has a personal interest in him for some reason even though he is a Republican. …

Governor Bumpers has agreed to speak to the Ozark Society on the occasion of its spring meeting 1971 at Petit Jean on March 27th. I certainly hope that he keeps his promise. We wouldn't expect much out of him, but to have him be there with even a short message would be a tremendous help to us. Of course, we are going to get the Buffalo River Bill passed someday, but we must have the help of people like him and the honest efforts of people like Mills and Hammerschmidt before anything can ever be done. …

As of right now we do not have any large, well-organized opposition to the Buffalo National River project. We still have the Buffalo River Conservation and Recreation Council formerly headed by G. L. Huchingson. Also we know that there is well-financed opposition from Mr. Yarborough, the developer of the Valley-Y Ranch. Recently he had a movie made, and it was shown on national television about what he was doing to the Buffalo River, which is absolutely ruinous but which was played up to be a conservation project.

The importance of keeping in touch with Ken Smith cannot be overstated. Although stationed at Grand Coulee in Washington State early in the year, he would be transferred to Washington, D.C., in July 1971, where he would play an important, though unofficial, role in our last and best efforts to gain national recognition for the Buffalo River.

At that meeting on March 27 there occurred an event which significantly rein-forced our advantage in this drawn-out chess game.

After that meeting, with the tide running so obviously in our favor, harsh words for Hammerschmidt and his apologists were heard less and less.

The New Governor Takes a Stand

It had long been our policy to contact candidates for pivotal public office so that in the event of their election they would at least know something of the national park proposal for the Buffalo and perhaps even endorse it. In doing so, we were to meet Dale Bumpers, who would be a prime example of the importance of persuasion.

During the campaign for governor in the summer of 1970, most of us sensed that Bumpers was the front runner, a man whom we should not neglect.

Joe Nix, who was good at making arrangements, was able to schedule a meeting with Dale Bumpers at the Holiday Inn in Fayetteville that September. There Joe and I spent an hour describing the beauty of the Buffalo River and its urgent need for protection.

Mr. Bumpers, who listened with patience and good attention, had frequent important questions and pertinent comment concerning the proposal about which he was heretofore unfamiliar. At the end of the interview, he followed us to the door and said: "I want you fellows to know that I lean your way."

That was reason for us to feel good about our effort and about his subsequent election in November. Then we were genuinely elated when he accepted our invitation to appear at the Ozark Society meeting on March 27, 1971, to say something about his position on the Buffalo. At 1:30 P.M. Governor Bumpers and several of his staff walked into the dining hall in Mather Lodge where two hundred or more of our partisans were gathered. As we all held our breath, he walked up to the podium, held up his hand, and said: "I want you people to know that I am for the Buffalo National River!" Never before or since has such a cheer gone up from the heights of Petit Jean as was heard that day.

Dale Bumpers was to continue as the recipient of our applause in the years to come, especially in his role as senior senator from Arkansas. In that office he has repeatedly championed funding for the Buffalo National River when it was urgently needed and not forthcoming otherwise. He has been the man whom the park service turned to above our other representatives. Then in 1984 he, in company with Ed Bethune (Arkansas's only bona fide conservationist congressman), shepherded the Wilderness Bill through the legislative labyrinth in D.C. That added 117,000 acres in the Ozark and Ouachita national forests to full protection by the federal government from any form of human exploitation.

Joe Nix Goes to Washington

The next important move was the rerun of the new Senate bill, designated S-7, which was identical to S.B. 855. It was presented on January 25, 1971, by Senator Fulbright (for himself and Senator McClellan). We were advised by good authority that its passage was assured. The hearing for S-7 was set for April 22. From reliable sources we were advised that we would not need to send as many delegates to this hearing as previously, but a few reliable spokesmen would be in order. That being the case, Dr. Joe Nix volunteered to represent the Ozark Society along with Dick Broach from the Arkansas Game and Fish Commission. Events transpiring were described by Dr. Nix in a special report written at the time:

… On Wednesday evening, I met Dick Broach at the Little Rock airport, and we flew to Washington on the same plane. … [W]e also met John Fleming at the airport, and he accompanied us to Washington. …

On Thursday morning we went to the Wilderness Society where Ken Smith had made arrangements for one of their personnel to spend the day with us. Mr. Arthur Wright helped us a great deal finding the places that we needed. He also had some very good suggestions and comments as how to get the Buffalo River legislation through. …

Our next stop was in Senator McClellan's office. His legislative assistant, Mr. Emon Mahoney, met with us and informed us that both Senators would be giving statements. He also said that one of them would be introducing Mr. Broach and me. The Senator joined us, and we walked to the hearing room with him. He made the statement to me, "We have got to get the Buffalo River saved before I get out of Congress, and we will unless I get out before I expect."

For those of you who attended the first Senate hearing, this one was in the same place. Senator McClellan gave the first testimony followed by Senator Fulbright. Both were very strong statements which I am sure you have read some about in the newspapers. I will mention one thing regarding Senator McClellan's statement. Repeatedly he said, "We cannot save all of the rivers, but we must the Buffalo." Keep this in mind with specific regard to the Mulberry.

At the close of his speech, Senator Fulbright introduced Mr. Broach and myself. Senator Bible asked that he be allowed to hear the testimony of the Park Service before ours. Senator Bible and Senator Anderson (N.M.) were the only members of the committee present. Senator McClellan left immediately after his testimony, but Senator Fulbright stayed through the testimony of the Park Service.

The testimony from the Park Service was given by Mr. Hartzog with Bernie Campbell by his side. The general scope of the plan was reviewed, and up-to-date cost estimates were given. We were all taken a little when they reported that the cost of the project had jumped from around 9 million to 12 million. Somewhere around 1.5 million of the increase can be accounted for by the new Uniform Relocation Act of 1971. This act provides for relocation cost of persons displaced by the project. We should remember this when we talk to persons about the proposed park. …

Just as he did at the first hearing, Senator Bible brought up the question of the transfer of the state park lands to the national park system. Mr. Hartzog answered that he had no commitment from the State of Arkansas regarding this matter. At this point I realized that Mr. Surles was not present. It was rather embarrassing that there was not someone representing the Governor at the meeting. …

Since my statement was rather long and since the whole hearing seemed to have the air of a formality, I briefly summarized it, reading only selected portions. The entire statement will be included in the record.

Tom Kimball followed me, then Art Wright from the Wilderness Society. This ended the hearing.

Tom Kimball spoke with Mr. Hartzog before he left. Tom told me that Mr. Hartzog said (confidentially) that Mr. Mills is the one who is the key figure in the House bill and that we must get to work on him. (Remember this.)

… I asked Mr. Pryor about the Buffalo, specifically, is Mr. Mills going to help us on it this year. …

Mr. Hammerschmidt was working on the floor of the House so we met him in a small room just off of the House Chamber. We talked with him for about 30 or 40 minutes. First of all, we know that he is listening to Waymon Villines and the group who is asking that some concessions be made on land ownership policy. …

I asked him about Mr. Mills and how he stood on the bill, and he did not think there was any trouble here. He told me that the bill would, in all probability, be introduced within the next week or two. The setting of the hearing date would be determined after introduction, but apparently he has some assurances that he can obtain an early hearing. …

As expected, S-7 passed the Senate without dissent, and we eagerly awaited something as good from the House. On May 16, 1971, the local news media announced the submission of House Resolution 8382 by Congressman John Paul Hammerschmidt to establish the Buffalo National River in Arkansas. It differed from the Senate bill in some respects, being more liberal to tenants within the proposed park and providing greater tax relief to counties involved. The date for the hearing was not then set, but this development was a signal for us to make plans for the final move.

The BRC&RC Makes a Move

One of the factors favoring the designation of the national park along the Buffalo was the virtual absence of permanent human habitation along it. The population in those rough and by now impoverished counties of Newton, Searcy, and Marion had peaked in 1900, at which time a sustained exodus set in. It continued on, accelerating after the end of the Second World War. In the decade from 1950 to 1960, Arkansas lost 250,000 people and one congressman as a result. Then there remained in Boxley Valley 60 or 70 residents of all ages and no more than that in and around Woolum, the only two places with enough good bottomland to sustain more than a few farms. As for towns along the course of the Big Buffalo, there were none. Jasper, with 450 people, was on the Little Buffalo five miles away. Ponca, something less than a village, with 40 or 50 people, and Gilbert with about the same were closer, but both had been excluded from the proposed park boundary.

This left Pruitt where Highway 7 crosses the river with about twenty individuals living immediately adjacent to the stream. It was as far from being an urban community as a place could be, although there might have been forty or more residents a mile or more back from the river. But there at the bridge, residing on a lot leased from Pearl Holland, was Lucile Hannon, the would-be nemesis of the national park. There also was Marvin Sherman, Hilary Jones, and a few more in search of a means to blockade the legislation now in Congress.

Early in the year they came up with a plan which would be an aggravation to all and an embarrassment to some. For our newly sworn-in governor they had a proposition: they requested that Governor Bumpers ask Congressman Hammerschmidt and Senator Fulbright to exclude the town of Pruitt from the proposed Buffalo National River.

In a short epistle to Senator Fulbright, Governor Bumpers provided us with a good example of the need for us all to keep an eye on our servants in public office, May 21, 1971:

> Enclosed is a copy of a letter I received from the Mayor and Recorder of Pruitt, Arkansas, which is self-explanatory. It seems their request is entirely appropriate.

A perfect politician would surely be one who could accommodate all of the people all of the time; but there is no such in a democracy where ideologues of every sort abound, all armed with the power of the vote. Governor Bumpers, still not as seasoned as some in politics, aspired to that perfection and took the "City of Pruitt" bait without looking for the hooks in it or asking for consultation from any other source.

Senator Fulbright, with his long years of experience in public controversy, first requested advice and an opinion from that agency soon to be entrusted with the management of the Buffalo River. From the Department of the Interior, on July 22, 1971, he received this statement:

> The city of Pruitt was incorporated in 1969 to include an area of 1,933.56 acres, of which 1,571.24 acres lie within the boundaries proposed for the Buffalo National River. The population center of this incorporation, according to our survey, contains 14 permanently occupied structures, plus a few seasonal or temporary occupancies, mostly within a 1-mile radius of the river crossing. The remainder of the incorporated area is agricultural or forested lands with scattered farms, permanent and seasonal homes.
>
> The city of Pruitt, if excluded, would subtract more than 1,500 acres from the proposed national river and would completely bisect it. Also, it could create an intrusive and discordant development which would not be in keeping with the purpose of the

legislation. Moreover, the legal act of incorporation has not, in itself, altered the character of the properties, which, we believe, should be protected within the national river boundaries.

That report from the Department of the Interior should have been the end of the Pruitt charade, but its begetters kept it going for a couple more years.

Ken Smith, upon learning of this, dug up some interesting facts about the newborn town:

1. The population of the entire 1,950 acres within the city limits was 70.
2. Those developments near the river are especially unattractive and shoddy (which those of us who had been there already knew).
3. The originator of the scheme was Marvin Sherman who was elected (?) mayor, started a subdivision, and as mayor approved of his own subdivision.
4. Sherman presented but did not sign the original petition for incorporation.

After the hearing on H.R. 8382, Hammerschmidt's new bill for the national river, on October 28 and 29, Governor Bumpers received counseling from authorities in Washington and on the local level admonishing against his support of the Pruitt plan to bisect the national river. It was necessary for him to crawfish out of that situation, which he did in a letter to Lucile Hannon:

> Thank you for sending me a copy of your letter to Mr. Roy A. Taylor, Chairman of the Sub-Committee on National Parks and Recreation. I have read with interest your reasons for opposing a national river on the Big Buffalo. We are deeply sympathetic regarding your views and have given careful thought and study to this controversial problem. We have concluded that we must preserve our most precious areas in the state and do everything possible to protect our natural resources. We did ask Congressman Hammerschmidt to take the town of Pruitt out of the bill.

That was unacceptable in the BRC&RC camp and left us with a bad taste as well, although relieved that Bumpers again leaned our way. The effort to pass the buck to Hammerschmidt accomplished nothing. Miss Hannon expressed her feelings concerning the preservation of our "most precious areas" in a reply to the governor the next day in which she expressed her regret at having voted for him for governor. She is reported to have said, "We've got our fighting pants on … we're not going to sit down and let them take the river."

By the end of the year the hassle over Pruitt had about run its course. Senator Fulbright had, no doubt, had his fill of it. In a letter December 7, 1971, to Hannon his frayed patience as a public servant is visible:

> I would like to acknowledge the receipt of your several mimeographed letters concerning the proposed establishment of the Buffalo National River.
> In checking my files, I determine that my first letter from you was dated May 12, 1971, well after the hearing on this legislation by the Senate Interior Committee, April

22. In that letter, on behalf of the City of Pruitt, you thanked me for sending you copies of the bill and requested further information about specific aspects of the bill.

I wrote to you on May 17 and at the same time requested the assistance of the Department of the Interior in obtaining the information. As you will recall, the Senate bill on the Buffalo passed unanimously on May 21.

On May 27, I wrote you again, enclosing an interim reply from the Director of the National Park Service. On June 30 I wrote you once more, enclosing certain materials you had requested and advising you as to how to locate other information.

Shortly after the passage of the bill in the Senate, I received a letter from you and Mayor Marvin Sherman of Pruitt, dated May 19, in which you said you did not desire to add to the controversy, but requested the exclusion of all incorporated towns and cities from the boundaries of the National River. Soon thereafter I received a letter from Governor Dale Bumpers enclosing a copy of a similar letter you had written to him.

On June 7, after contacting the National Park Service about your request, I wrote to Mayor Sherman, acknowledging your letter. On June 28, I again wrote Mayor Sherman, sending along some of the information you requested.

Finally, on July 26, I sent you a copy of a letter to Mayor Sherman along with a lengthy letter and further information about your suggested exclusion of Pruitt from the boundaries of the Buffalo National River. For your information, I am again mailing you a copy of this letter, since it dealt in some detail with this question.

I hope this will help set the record straight. I regret that we are not in agreement about the proposed Buffalo National River, but I have tried to be attentive to your requests for information and prompt in answering your letters.

I might also mention that in your mimeographed letter dated November 27, you include a copy of a letter from Representative Sterlin Hurley in which he indicates that he received no reply from me to his letter of April 12, 1971. I have checked my files and find that I did respond to Mr. Hurley at some length in a letter of April 29, and am sending him a copy of that letter.

This is a matter which has been under consideration for a number of years. At the Senate Interior Committee hearing on April 22, no opposition to the bill was expressed. I remain convinced that the establishment of the Buffalo National River will ultimately be to the advantage and benefit of all concerned.

The above letter resulted in a lengthy tirade too labored to present here in full but which ended in a clear-cut threat to the senator's political well-being.

The Horse Ranch Has a Visitor

By 1971 the Valley-Y Ranch was well along on the owner's plans for an extensive commercial and agricultural development on his holdings between Ponca and Kyle's Boys' Home.

P. W. Yarborough had witnessed the developing controversy with understandable concern but had refrained from becoming much of a participant until late in the game. Perhaps he believed that the national park proposal would never succeed.

Just before the hearing on Hammerschmidt's bill, the owner of Valley-Y was to receive a surprise visit from Senator Fulbright. We were no less astonished upon learning of it but were certainly not as taken aback as he, since he was absent during the senator's unscheduled visit.

Had there been any doubt as to the rightfulness of his pending legislation in the mind of Senator Fulbright, it would have been dispelled by the sight of the building complex then imposed upon the breathtaking scenery under Roark and Harvest bluffs at the mouth of Steel Creek.

SOME FINAL MOVES—PRO AND CON

By January 1971 we had arrived at the last full year of the contest. We enjoyed almost every advantage on the board, and our opponents were obliged to resort to whatever harassment that they could dream up. That would be varied, curious, and downright aggravating.

The Firing of Delos Dodd

The first episode had had its beginning the year before. It involved the Buffalo River State Park and its parent, the State Parks, Recreation and Travel Commission, of which Edward Tudor was then a member. He had enough leverage with other commission members to set up a couple of attempted blockades before being outvoted by newcomers on the commission.

That former BRIA boys were still capable of unpleasantries was recorded by the *Arkansas Gazette*, January 10, 1970:

> **Buffalo Park Chief Asked to Step Down**
>
> Delos Dodd, superintendent of Buffalo River State Park, has received a request from the state Parks, Recreation and Travel Commission that he resign, but he said Friday he did not plan to. ...
>
> "The man should never have been fired without a chance to defend himself," said Orville Richolson of Newport, a member of the Commission. "It's a disgrace." He said it was humiliating to Dodd, who had given 14 years to the Buffalo River park. ...
>
> Ray G. Cooper, director of the Commission, notified Dodd in a letter dated

Wednesday that the Commission had requested his resignation. It said Edward Tudor of Marshall, a member of the Commission, had directed him to write the letter.

Those of us long-time friends of Delos Dodd were angered by this low blow to a good park manager. We were able to generate significant support for him, the result of which was reported by the *Arkansas Gazette,* February 21, 1970:

Parks Panel Rescinds Resignation Request, Orders Spending Probe

The State Parks, Recreation and Travel Commission rescinded Friday its action of last month in requesting the resignation of Delos Dodd, superintendent of the Buffalo River State Park, but it ordered a complete investigation of the Park's operations after one Commission member said Dodd had paid state money to persons working in his son's restaurant.

At that time Tudor made charges that Dodd was a party to other vague infractions, none of which he was ever found guilty of. At the same time, the Parks, Recreation and Travel Commission itself was found by the accounting firm of Ernst and Ernst of Little Rock to be guilty of poor administration of its funds. The commission was presented with a report from the accounting firm to set up a completely new accounting system, which it voted to do without reservation.

Delos Dodd was permitted to retire on his own terms a year or two later.

Following that spat over the management of the Buffalo River State Park, Ed Tudor attempted to use his influence with the new governor, Dale Bumpers, to gain the chairmanship of the Publicity, Parks, and Travel Commission but without success.

In spite of setbacks, Commissioner Tudor still harbored a compulsion to stop or disrupt the national park plan, if he could, by denying the donation of the state park lands to the national park system. How Commissioner Tudor fared was reported by a number of newspapers, the *Harrison Daily Times* being an example (August 30, 1971):

Parks Commission Gives Buffalo River State Park to Federal Government

FORREST CITY—The State Parks and Tourism Commission voted Friday to give the Buffalo River State Park to the federal government if Congress designates the Buffalo as a national river.

This action, ending years of indecision on the matter by the Commission, eliminates one barrier to passage of the Buffalo National River Bill later this year. The bill has passed the Senate and will have a hearing this fall before the Park and Recreation Subcommittee of the House Insular Affairs Committee.

Also part of the package will be the Lost Valley State Park, a 280-acre undeveloped park in Newton County that was acquired by the Department in 1966.

Earlier this summer, the state Game and Fish Commission voted to give land that it owns along the Buffalo to the Park Service if the National River bill becomes law.

The vote to offer the parks was 7 to 0. Edward Tudor of Marshall, who has persistently opposed the national river concept, argued against it but abstained on the vote, along with three other commissioners.

May Not Be Able to Give Park Away

LITTLE ROCK (AP)—Atty. Gen. Ray H. Thornton Jr. said in an opinion today that there is no specific statutory authority for state park lands to be given away or donated to the National Park Service.

The opinion went to William E. Henderson, director of the state Department of Parks and Tourism, who had asked whether any such authority existed.

Thornton said, however, that state-owned riparian lands could be conveyed to the National Park Service if the federal government agreed to maintain the project. ...

This action removed, for all practical purposes, any remaining influence that the Tudors had on the course of events. The *Harrison Daily Times'* notice of the attorney general's ambiguous statement reflects their pro-dam, anti-park stand in the matter. Mr. Thornton, ambitious in politics, was only trying to say something soothing for both sides of the argument.

The month before, the Arkansas Game and Fish Commission had agreed to donate two thousand acres of their sixteen-thousand-acre Buffalo River Wildlife Management Area below Hasty to the National Park Service if they would allow hunting in the area involved. Hunting privileges in the area would not be denied by the park service and no obstacles to the Buffalo National River now remained on the state level.

Bell Foley and the Saline Dams

The Corps of Engineers, being an agency of the government, is obliged to abide by federal law. With firm legal status of the Buffalo National River now at hand, the Corps recognized that fact, and from the highest recess of the pentagon there issued as curious a statement as we had ever expected to see (*Arkansas Gazette*, June 13, 1971):

Corps Endorses Plan for Buffalo River

The Army Engineers, which for years proposed building a dam on the Buffalo River, now are on record in Congress as endorsing the plan to make the 132-mile stream a national river to preserve its wild and natural character.

"The Corps of Engineers supports the wild river formally and officially," Col. William C. Burns, Little Rock District Army Engineer, said last week. …

Colonel Burns said that by supporting the national river bill the Engineers were carrying out the recommendations of a comprehensive plan for the development of the water resources of the White River Basin prepared in 1967 and 1968 by representatives of 14 federal agencies and the states of Arkansas and Missouri.

The plan, among other things, called for preserving the Buffalo, Eleven Point and Current rivers as national free-flowing streams. It also recommended the construction of several dams on other rivers in the Basin. …

Please re-read the last sentence.

If they were to be denied the Buffalo, the Corps of Engineers' obsession to dam something was not lessened thereby. There had to be, if not a river, then a rivulet still flowing somewhere, a threat to mankind in flood and a boon for motor boaters if dammed. Tucked away in Arkansas's eastern Ozarks they found what they were looking for, the Strawberry River. It originated in Izard County near Union, flowed across Sharp County near Poughkeepsie, entered Lawrence County near Jessup, and thence into the Black River and finally the White. It was a small, clear-water, gravel-bottom stream flowing through low hills—as unknown a stream as any could be in the Ozarks. Most of us had never heard of it until its fate was revealed by the *Arkansas Gazette*, April 25, 1971:

George Goff, 77, to Lose His Land to Bell Foley Dam

POUGHKEEPSIE—Sharp County residents have spent more than 30 years waiting for a dam that promises to change their lives. …

The currently proposed dam site is northeast of Poughkeepsie and would cover 18,000 acres. The storage was increased in order to maximize potential of the site, which has not been approved.

The Bell Foley project is one of the first such projects in Arkansas and possibly in the nation that has been designed since Public Law 8972 was passed allowing the Army Engineers to design recreation facilities in a project with the stipulation that a portion of the cost be shared by a local government agency. …

The tentative estimated total cost of the modified project is $25.4 million and it is estimated that the local share of the costs allocated to recreation and fish and wildlife enhancement will be approximately $3.15 million. The benefit-to-cost ratio is estimated at 3.7 to 1.

On the Bell Foley project the Corps lacked an important element of support. There was no big dam-pushing newspaper editor to sell the plan on the local level. Then there was widespread uncertainty among politicians great and small as to whether or not dam building was still the answer to tenure in office. The surprising results of the Buffalo River war had done much to generate that

(Courtesy of George Fisher)

uncertainty. Bell Foley would be evaded by governors and would lack the support of congressmen and senators and after a few years would simply fade away. The Ozark Society and other conservation organizations were prepared to intervene against the Bell Foley dam had it become necessary. There were too many drowned valleys in the Ozarks already.

But the engineers had yet another card up their sleeve. There was an undammed river, the Saline, that arose in the eastern Ouachitas and flowed out onto the coastal plain below Benton and Little Rock. From there its course was almost directly south past Rison and Warren to enter the Ouachita a short distance from the Louisiana line, a rather lengthy waterway. It was not as scenic or swift as our upland streams but neither was it as muddy as those bayous in the nearby delta. Regardless of that, it did have recreational and aesthetic value, good swimming, fishing, and boating, with an interesting border of hardwood and cypress. Near Benton there was a possible dam site that the Corps had not overlooked. The time had arrived for them to make a move for the Benton dam. Some of the businessmen of the city of Benton were enlisted in its support, and a number of local politicians and legislators followed suit. Their main goal was to obtain the endorsement of the Benton dam by Dale Bumpers, then the governor.

His response to the Benton dam promoters was, at first, favorable but became vague after being subjected to opposition by the members of the conservation coalition.

As it turned out, Dale Bumpers' state of uncertainty didn't matter in the end.

It developed that much of the area to be inundated by the Benton dam was more or less built up with human habitation. It was a sort of suburb of Benton. In the face of a real forced exodus of many well-established residents, the Benton dam plan has simmered on for years, refusing to die in spite of that.

Parting Shots and a Few Last Words

The Cossatot affair ran consecutively with the final stages of the Buffalo fight. Most of us sensed that we were not going to be able to stop the partially finished Gillham dam. But having set out to do just that, we were morally obligated to carry on. Construction would be held up for two years on that utterly needless project by Wellborn Jack's court battle. Reaction against that effort to deprive them of their dam had aroused the ire of many local citizens. They would show us how to settle the argument.

Shocking news of how that was done came to us from a newspaper from far outside the theater of action (*Houston Chronicle,* May 4, 1971):

On Canoe Trip
Beaumont Scouts Shot At—Leader

DE QUEEN, Ark. (AP)—Fred Madenwald, an adviser for Explorer Scout Post 12 at Beaumont, Texas, said that about 11 shots were taken at his group of Explorers as they floated down the Cossatot River near here.

Madenwald said Monday that no one was injured and that the shots Sunday went across the bows of the five canoes carrying his 10-member scout group. Madenwald said one shot splashed water into a canoe but that most of the shots hit about five feet from each canoe.

The shooting apparently is an outgrowth of a feud between landowners along the scenic Cossatot and conservationists. Landowners are upset about the effort by conservationists to keep the Cossatot a free-flowing stream through the Ouachita Mountains in Southwest Arkansas. The conservationists have filed suit seeking to keep Army Engineers from building a dam on the river.

Federal Judge G. Thomas Eisele has halted construction on the dam until Army Engineers file a complete "environmental impact" statement on the project. Notice of appeal of Eisele's decision has been filed by both sides.

The float Sunday was called the second annual Cossatot Pilgrimage and attracted more than 225 persons. ...

Madenwald said he did not see the shooting himself but said two of his Explorers told him they saw a man behind some brush signal another man on a cliff about 100 feet above the river to start shooting. Madenwald said there appeared to be five or six men involved altogether but that only the man on the cliff did any shooting.

No arrest has been made in connection with the shooting.

According to Mr. Madenwald, who has spent a great deal of time taking the

Explorers on trips which have included Canada (twice), Mexico, Florida, and other places, the above article is quite accurate. However, the last sentence contains a great deal more meaning than it appears to.

When this incident was reported to the sheriff, he accused the scouts of causing trouble. He was very uncooperative, did not want to take down any information, stated that he could not receive the complaint, stated that his deputy was too busy to look into it but might later on. One can only wonder what his actions would have been had the scouts been firing rifles from the canoes at citizens on the banks. One could only wonder also what sort of lesson a group of impressionable boys learned about respect for law enforcement officers.

From that episode we were offered a view of the evil side of human nature and a reminder of the precarious position that we conservationists sometimes occupy in attempting to stay the destruction of a part of the world by men determined to do it.

That June there had occurred an event about which we local partisans knew nothing but which we were overjoyed to learn of through the *Yellville Mountain Echo*, June 17, 1971:

Group Floats Buffalo River Sunday

A group of national and state officials floated the Buffalo River Sunday from Maumee to the state park, apparently wanting to see for themselves the river over which there has been so much controversy in congress in recent years.

The group included Rep. Bill Alexander, D-Ark.; Rep. Roy Taylor, D-N.C.; and Rep. Harold T. Johnson, D-Calif., members of the Parks and Recreation Subcommittee of the Interior and Insular Affairs Committee; George Hartzog, Jr., director, National Park Service; William E. Henderson, director, Arkansas Department of Parks and Tourism; and Bernie Campbell, director, Hot Springs National Park.

There was a total of 20 persons in the group. Seven boats were used to float them.

The group landed at the Mountain Home airport Saturday and toured Blanchard Spring Caverns that afternoon. They were guests at a catfish fry at the Gaston resort near Bull Shoals dam that night.

Congressman Roy Taylor, chairman of the House Subcommittee on Parks and Recreation, had intentionally requested no publicity or assistance from us on this trip. It was felt that the presence of instigators of the national park plan would be improper. Disgruntled big dam backers were not wanted either, which was fair enough. In spite of the fact that newspaper reporters were not wanted, the *Harrison Daily Times* learned of the trip somehow and sent out a man to interview the inspection group. Other local papers picked up the story and passed it on, a morale-boosting revelation for our side, to say the least.

TWENTY-ONE

THE JUBILEE BUS

Through the summer and early fall of 1971 we had waited expectantly the announcement of the crucial hearing on H.R. 8382. Finally, the *Pine Bluff Commercial,* September 23, 1971, was able to provide us with the welcome news:

House Hearings Set on Buffalo Bill

Congressman John Paul Hammerschmidt of Harrison announced yesterday in Washington that House hearings had been set for October 28 and 29 on his bill to make the Buffalo River in North Arkansas a national river. ...

Hammerschmidt's bill, which he called "the latest best estimate" by the National Parks Service, would appropriate $16.1 million for acquisition and more than $12.1 million for development. He said he thought that perhaps even these higher figures were "conservative" estimates.

October 28 was scarcely a month away, and although we had anticipated the event since January and had put forth more than one directive to our membership, we were going to be cramped for time. The mechanics of the thing was going to require greater effort than anything we had been through up until now.

The Ozark Society Goes to Washington

The first action was to issue an appeal to all members to write to the chairman of the subcommittee offering personal endorsement of H.R. 8382. Immediately thereafter all conservation agencies and organizations concerned were contacted with the same request.

The next consideration was the matter of sending delegates to the hearing. We knew that it would not be wise to swamp the committee with a huge number of people, but, on the other hand, we did not want to appear with a scanty crowd. The Arkansas Game and Fish Commission had offered the use of its airplane which was to carry delegates from the commission and also from the

451

Publicity, Parks, and Travel Commission and the Wildlife Federation. It could carry only five individuals besides the pilot. It goes without saying that our contingency fund, which had been set aside especially for this occasion, did not contain enough to fly more than a few more people to Washington, D.C. To obtain any considerable number of representatives through the use of private automobiles also was obviously not a practical method to use. We did not want to leave matters to pure chance, trusting that whoever could go would go and perhaps that we might come up with a respectable number.

In considering the matter, I recollected that the Upper Eleven Point Association had on more than one occasion sent delegations to Washington on a chartered bus. At first this did not seem like a good idea, but while in Little Rock in early October contact was made with Everett Bowman, who had on several occasions in the past arranged for the chartered bus to conduct the Ozark Society bus tours in the spring and fall. Everett was on good terms with the Continental Trailways officials in Little Rock, and very shortly he had received information from them that they would furnish a forty-passenger bus from Little Rock to Washington, D.C., and return for the surprisingly low figure of $1,200. This service would also include moving the delegation about in the city of Washington while we were there. This made it appear to be an even more attractive idea. Also while in Little Rock I received a long distance call from Ken Smith who was again, by an act of fate, on an assignment in Washington. Ken had been considering how best to accommodate us while in Washington, D.C., and had discovered that the management of the dormitory complex where he stayed, the McLean Gardens, would accommodate us for $5.25 per person per night if we had more than twenty in our party. This was also a surprisingly reasonable rate and made it possible for the Ozark Society to guarantee lodging for delegates taking the time and trouble to make the trip. Without Ken's help, what turned out to be a very efficient operation could have been a frustrating adventure in the big city.

Some of our members had suggested that the bus should start from Fayetteville, picking up delegates as it went along across the state of Arkansas. This seemed an excellent idea, and Everett Bowman was able to find that Continental Trailways would initiate the trip at Fayetteville for an extra three hundred dollars or a total of fifteen hundred dollars for a round trip from Fayetteville to Washington, D.C., and back with a total driving time one way of about twenty-five hours. This, plus the dormitory fee for thirty or thirty-five people for two nights, was still within the limits of our emergency fund, and a definite commitment was made to charter the bus. Harold Cox, our bus driver, who was aware of the rate, remarked that we had obtained a bus for an exceptionally low amount. It had been figured on the basis of sixty cents per mile

The Jubilee bus loading for Washington, D.C., October 26, 1971. *Left to right:* Evangeline Archer, the Herb Fowlers, Ilene Gipson, Jo Wilson, Niki Kimball, Maxine Clark, Lois Imhoff, Nancy Jack, Joe Clark, Neil Compton, Joe Carver, Tommy Jenkins, and Willie Priff.

with charges usually running around seventy-five cents per mile. In discussing this matter with Everett Bowman, a conclusion was made that perhaps the Continental Trailways Company, having given us service before for our sightseeing trips and knowing of the Ozark Society and its activities, was anxious to be of some assistance since it considered our cause a commendable one.

Early in the afternoon of October 27, we rounded Front Royal, Virginia, and soon thereafter arrived at McLean Gardens without error.

It had been decided that we would have a briefing session in the main lobby of Gadsden Hall. We were joined during the evening by others of our delegation who had flown in, and among these were Bob Apple of the Arkansas Wildlife Federation and Dr. Joe Nix from Arkadelphia and Nancy Jack and Dwayne Kelly from Kansas City. Before the meeting was much more than underway, Ken Smith arrived on the scene with a copy of H.R. 8382. He suggested that it would be a good idea to go over the bill line by line and to explain all points not understood thoroughly by all concerned. Before this was done, however, assignments were made from those present for visits to all of the offices of all the members of the Committee on Interior and Insular Affairs. We knew that the hearing would most likely adjourn at about noon on the first day, and these visits were to be made during the afternoon on Thursday the twenty-eighth. We realized that conditions were in our favor, and it was possible that the bill might well have proceeded without such maneuvering; but since we were there, bent upon securing its passage, it would have been improper not to have made an effort to obtain personal interviews with all the congressmen concerned.

Ken Smith then read each line of the bill, and discussion was held concerning each provision of it. This was time consuming, but when the session was terminated, there was no doubt in anyone's mind but that we were in much better order to make an appearance than we had been before.

On the morning of Thursday, October 28, we were all up and aboard the bus by 8:30 and were off across the city, bound for the Longworth Building. Before long, the hearing room on the third floor was full and overflowing with persons interested in the legislation at hand, most of whom were on our side, but we at once discerned Mr. P. W. Yarborough and his lawyer sitting on the front row in the delegate section. Later Hilary Jones and three or four others arrived from the Jasper-Pruitt area, but as far as we could tell, these were the only ones opposing us at this meeting. At the beginning of the hearing as many as twelve of the congressmen on the subcommittee were in attendance and which we understood was an unusually large number to be present, a fact which we accepted as being favorable for our side. The hearing was called to order by Congressman Wayne Aspinall, something else which was, according to our information, unusual since he did not always make his appearance on these occasions. Mr. Aspinall was the chairman of the Full Committee on Interior and Insular Affairs, with Congressman Roy Taylor being the chairman of the Subcommittee on Parks and Recreation. Mr. Taylor was to take over later when Mr. Aspinall departed.

Mr. Aspinall in his preliminary remarks stated that he was aware that there had been a great deal of interest in Arkansas in favor of the proposed Buffalo National River, implying that he had been receiving much mail and other inquiries expressing concern for the passage of this legislation. He then asked if there was anyone present from Arkansas who was of that opinion and, if so, would they please hold up their hands. A whole forest of hands went up, and Mr. Aspinall laughed and said, "That's enough."

There could have been no better omen under which to start the hearing than this simple display of support. It was almost as if the act had been pre-ordained. Those two words from Chairman Aspinall were ample justification for all the trouble and expense to which we had gone to organize our delegation and to charter the bus in order to be present on this day.

The first witness to speak before the committee was Congressman John Paul Hammerschmidt, author of H.R. 8382. He gave a clear, well-balanced, and reasoned summary of his purpose in presenting the measure. There should surely have been no doubt as to his convictions regarding it after his statement.

He apologized for the absence of Congressman Wilbur Mills, who had already gone on record in favor of the bill.

Then Congressman Bill Alexander, who had introduced H.R. 9119, an identical bill to H.R. 8382, delivered a very emphatic and forceful endorsement of the proposal. What he had to say was as well conceived as anything that any of us who had for so long fought for the preservation of the Buffalo River could have said.

We had noted with interest at the beginning of the hearing that John Fleming of the *Arkansas Gazette* in company with H. Charles Johnston, Jr., occupied prominent seats in the press section. After bidding us good-bye at the bus station in Little Rock, they had flown to Washington to participate in the hearing. During the recess that occurred later that day, an opportunity presented to discuss their activities and John offered information that they had spent most of the evening before with Congressman Alexander and that they had, in fact, written his speech for him.

Following Mr. Alexander's statement, Congressman David Pryor delivered another very effective speech in favor of the proposal. He said, "There are very few rivers in the United States in a free-flowing state having contiguous land area not yet developed and that have water that is relatively pollution free. The Buffalo River is one of them."

The next speaker in behalf of the measure was Nathaniel P. Reed, assistant secretary for Fish, Wildlife, and Parks of the Department of the Interior. Mr. Reed explained some of the mechanics that would ensue providing the bill was passed, and from his testimony it was regarded that the Department of the Interior was on record in favor of a quick beginning on land acquisition, which he estimated could be done in five years.

Following Reed's statement, William E. Henderson, director of Parks and Tourism of the Arkansas Parks Publicity and Travel Commission, testified. Shortly before this hearing was set, Jimmy Driftwood, the chairman of the State Parks, Recreation and Travel Commission, had obtained from that body approval of the proposal to transfer the Buffalo River State Park and the Lost Valley State Park to the National Park Service in the event that the bill passed. Jimmy Driftwood also appeared at this time, as did Harold Alexander of the Department of Planning of the State of Arkansas and Dick Broach of the Arkansas Game and Fish Commission.

These gentlemen all represented the state of Arkansas, with Mr. Henderson speaking for Governor Bumpers. It was obvious that Arkansas was represented unanimously in favor of H.R. 8382.

Next on the agenda was Mr. P. W. Yarborough, the owner and operator of the Valley-Y Ranch. Mr. Yarborough did not speak but was represented by his attorney, Thomas E. Allen, of Kansas City, Missouri.

George Smith, writing in the *Baxter Bulletin,* and which reporter was in position to hear comment not audible to the rest of us, offered significant comment by Mr. Yarborough. At the beginning of the hearing in his preliminary remarks, Chairman Roy Taylor had said, "I visited the Buffalo last summer, and I can understand why it should be preserved. It is truly a beautiful river." Mr. Yarborough, upon hearing this, threw up his hands and said, "Well, the hearing is over."

Mr. Allen was permitted to speak at length and proceeded to give an elaborate account of the activities carried out at the Valley-Y Ranch. He described the policies which he or Mr. Yarborough considered to be good conservation practices. It appears that these policies had been set by the Soil Conservation Service, however, and that they did in fact involve channelizing the stream, a fact that we are all familiar with already.

Then a most surprising plan was revealed in which an entire new program for Mr. Yarborough's holding was related. It was declared that he had determined the feasibility for a ten-million-dollar development that would include a shopping center, 25 vacation and retirement homes, 250 condominium units, a golf course, a landing strip, many new roads, and other developments. Everyone in the room was shocked by this revelation except Mr. Yarborough and his lawyer. Following the meeting, one congressman was overheard to say to a colleague, "Why in the world did he mention THAT? That is what we are trying to avoid."

At first, many of us assumed that his intent was to establish a claim for inflated value of his property and to have it on record before the subcommittee. In later consideration of the situation, however, it appears that this was not the most likely motive and that Mr. Yarborough really did believe that the revelation of such plans to the subcommittee would have a beneficial effect for his contention that H.R. 8382 should not be passed.

Whatever his motives, the release of these plans for the total commercialization of the best part of the Buffalo River constituted the real clincher in the argument for its acquisition by the National Park Service in order to remove forever the likelihood of such unnecessary schemes for its transformation into something which it should never be.

It was my lot to follow Mr. Yarborough, and the initial part of my statement was similar to that which had been delivered before the Senate subcommittee two years before. But, in addition, I included commentary concerning the role of physicians in conservation as voiced by the American Medical Association in reference to the need of an unpolluted and unspoiled environment for people's good health.

In the years during which I have been identified as a conservationist involved in an effort to salvage some of the remaining natural beauty of my part of the country, my position has been questioned by a few as to why a physician should be involved in such matters. Some believe that I should be concerned with human physiology, pathology and the treatment of various ailments and not with the problems of the world in general in which I live. First, I would like to say that I am not alone. Many doctors of medicine have been involved in this very subject and will continue to be as long as our profession exists. I believe that all of us engaged in the treatment of human illness are aware of the fact that body, mind and emotions are all intertwined in such a way that they cannot be separated. Almost all human afflictions are definitely affected by tension, anxiety and frustration, which factors become increasingly a part of our daily lives as technology and artificial existence become compounded. Thus, more and more we yearn to seek out the undisturbed, remote and beautiful places away from the crowds of people, the unending miles of pavement and heavy traffic and the hurly-burly of urban living which has practically engulfed us all. Along the Buffalo River in its canyons and on its gravel bars, we may still seek spiritual, emotional and physical reconstitution, and here those of us who know it seek to turn from the stress of our daily lives. The need for such places for all of us has been most eloquently stated by the doctors of America, speaking as a group, and I would like to call your attention to a statement appearing in the October 8th issue of *Life* Magazine in the form of an advertisement by the American Medical Association.
In it they say:
We are doctors of medicine. Naturally, we care about the health of your body.
But we care, too, about the health of the body of land you live on.
America's body is not well. Its condition is critical. Suddenly the words "ecology" and "pollution" are in the air. People seem worried. But we hope their worry isn't just another passing fad.
A sick environment can make people sick. It can undo everything a doctor works for.
In fact, disease induced by the environment now costs us $38 billion a year. That bill will go up if we don't nurse our environment back to health.
How to do the job? We at the American Medical Association worry about it. But we want everyone to worry. Because everyone is going to have to help get the job done.
There's no use our trying to keep people well in a dead land.

Mr. Taylor indicated that enough had been said. Immediately thereafter, he called upon succeeding speakers to give a very brief statement for or against, and this was complied with by those who were to follow. Most of them took thirty seconds and certainly not longer than a minute. That included Sturgis Miller, Dr. Joe Nix, Henry Shugart, Mrs. Jimmy Brown, Tom Foti, Tommy Jenkins, Niki Kimball, and finally, Herbert K. Fowler, who did not have time to begin his statement at noon when Congressman Taylor called an adjournment until the following morning.

The next day several members of the opposition had an opportunity to speak. These consisted of individuals from Pruitt and Jasper and who were in no way allied with Mr. Yarborough. The first of these was Mr. Hilary Jones, who repeated much of what he had had to say at the Senate hearing two years before. He managed to inject a new theme, however, and that was the national park proposal would cause a great influx of criminals into Newton and Searcy counties with the establishment of hippie communities and other dreadful things. Mr. Taylor stated in reply to him that if he thought this would be the case that he, Mr. Taylor, would resign from Congress. He stated that a study of the situation revealed that the increase in crime in national park areas had not increased any more than it had in the country in general over the last several years. As an example, he stated that there had been only fourteen arrests in Hot Springs National Park, Arkansas, in 1970.

Following this, Mr. Oxford Hamilton, the county clerk of Newton County, Arkansas, made a statement. He said that Newton County would not be able to cope with the situation because they would not be able to hire enough deputy sheriffs or other peace officers to take care of the crime wave that would ensue. He also brought up the same subject of revenue loss to the county. Upon stating this, Mr. Taylor asked him how many miles of the Buffalo River were located in Newton County. Mr. Hamilton did not know. Mr. Taylor then asked him how many acres of Newton County would be involved in the park project. Mr. Hamilton did not know this. He then asked him how many dollars would be lost in Newton County, and he knew not the answer to that question either.

After this, a Mr. James E. Carter, a recently arrived landowner from Cozahome, made a statement in which he described his intention to build a house on the banks of the Buffalo. The point of his argument remained unclear to us and to the members of the committee. Following this, Mr. Wendell McCutcheon, who was the principal of schools at Mt. Judea, Arkansas, made a statement not unlike that of his friends. He dwelt upon the loss of revenue from taxes and stated that Jasper and Mt. Judea would become ghost towns if this subject were enacted.

Other speakers appearing in the affirmative on the second morning were Bob Apple, executive director of the Arkansas Wildlife Federation; Nancy Jack, speaking from the standpoint of a landowner; Col. Jack Diggs, representing the

Joe and Maxine Clark and Neil Compton before boarding the bus for home after the hearing.

National Sierra Club; Jim Gaston, representing the Arkansas Travel Council and the Ozark Playground Association; Herb Fowler from the Department of Architecture, University of Arkansas; Bob Ferris of the Tulsa Canoe and Camping Club; Tommy Jenkins of the University of Arkansas Chapter of the Ozark Society; Tip Davidson of Shreveport, Louisiana; W. L. Pope of Kansas City; Virginia Ferguson of Conway, another landowner; H. Charles Johnston, Jr., of Little Rock, Arkansas, speaking for the "Gallinule" Society; and Joe and Maxine Clark, speaking for the Ozark Society. Jo Wilson and Leonard Heman wound up the appearances before the committee. All of this was accomplished by 11:30 A.M., and we immediately proceeded to our rendezvous with the bus, which was awaiting with the engines running on the street below, ready for the trip home.

We carried with us a sense of deep satisfaction and a feeling of accomplishment for what had happened in the halls of Congress. There was no doubt that after ten long years of ups and downs and disappointments that the hopes for seeing a national park along the Buffalo River were now a very positive possibility.

Harold Alexander running to catch the waiting "Jubilee Bus" for the trip home after the hearing in Washington, D.C., on H.R. 8382.

Later, in reporting the trip home, Oxford Hamilton, our opponent from Jasper, lamented in the *Harrison Daily Times* that they had no "Jubilee Bus" to ride home on. This indeed it was, and Mr. Hamilton should be thanked for his graphic description of this marvelous machine that carried us along through Tennessee and across its Buffalo River to Memphis and to Little Rock.

The rest of the return trip up I-40 and Highway 71 was quite uneventful with many of us resting and napping along the way. Arrival time at Fayetteville was 5:00 P.M. with the sky overcast and a light rain falling. The great expedition to Washington, D.C., to save the Buffalo River from any and all forms of exploitation was over. It seemed almost impossible that we could have arrived at this favorable point in the long contest to preserve some of our Ozark scenery for posterity. This event should stand as an example to all of us from this time onward, demonstrating the fact that our present system of government does permit the expression of the desires of its citizens and the accomplishment of their will if they so apply themselves to the achievement of their convictions. Henry Shugart of the Audubon Society summarized it as we rode along between

Memphis and Little Rock. He said, "This goes to show that we can do anything that we want to if we will work at it."

While the hearing before the Subcommittee on Parks and Recreation was not the last act in the establishment of the Buffalo National River, it was, for all practical purposes, the clincher. The next steps would be more or less automatic, barring a veto by President Nixon. That he was not about to do to important legislation sponsored by the first Republican congressman from Arkansas since Reconstruction.

This hearing, therefore, was the most decisive event in the prolonged controversy and is, of all the action that took place, the most important historically. A list of those who made the trip to Washington is, for that reason, in order:

DELEGATES ON CHARTERED BUS

Harold Alexander, Conway, Arkansas Department of Planning
Everett Bowman, Little Rock, President Pulaski Chapter, Ozark Society
Mrs. Milton Brown, El Dorado, Arkansas Audubon Society
Joe Carver, Fayetteville, University of Arkansas student
Joe M. Clark, Fayetteville, Editor, Ozark Society Bulletin
Mrs. Joe M. Clark, Fayetteville, Assistant Editor, Ozark Society Bulletin
Neil Compton, Bentonville, President, Ozark Society
Marvin Demuth, West Memphis, Bluff City Canoe Club
Mrs. Marvin Demuth, West Memphis
Mrs. Hubert Ferguson, Conway, Landowner, Buffalo River
John Ferguson, Conway, Senior, Arkansas State Teachers College
Robert A. Ferris, Tulsa, Oklahoma, Tulsa Canoe and Camping Club
Tom Foti, Pine Bluff, Jefferson Audubon Society
Herbert K. Fowler, Fayetteville, Architecture Department, University of Arkansas
Leonard Heman, Kansas City, Missouri, Landowner, Buffalo River
Nancy Jack, Kansas City, Kansas, Ozark Wilderness Waterways Club
Dr. Frances James, Fayetteville, Arkansas Academy of Sciences
Tommy Jenkins, Fayetteville, President, University of Arkansas Chapter, Ozark Society
Mrs. Don Kimball, Fayetteville, Science Teacher, Elkins
Elston Leonard, Little Rock, Soil Conservation Service
Dr. Jewel Moore, Conway, Arkansas Division, American Association of University Women
Henry Shugart, El Dorado, South Arkansas Audubon Society
Mrs. Henry Shugart, El Dorado
David Strickland, Muskogee, Oklahoma, Oklahoma Scenic Rivers Association
J. William Wiggins, Little Rock, Professor of Chemistry, University of Arkansas at Little Rock

Mrs. Jo Wilson, Fayetteville, Educator speaking as a mother
Mrs. Marcia Wood, Fayetteville, League of Women Voters
Mrs. Ilene Gipson, Fayetteville, Arkansas Environmental Research
Tip Davidson, Shreveport, Louisiana, Girl Scouts, Shreveport (return trip only)

DELEGATES TRAVELING VIA AIR OR AUTO

Robert E. Apple, Dardanelle, Executive Director, Arkansas Wildlife Federation
Dick Broach, Little Rock, Arkansas Game and Fish Commission
Mrs. Bryant Davidson, Shreveport, Louisiana, Shreveport Chapter of Ozark Society, Sierra Club, Louisiana Environmental Society
Col. Jack E. Diggs, Fayetteville, Sierra Club for the State of Arkansas
Jimmy Driftwood, Timbo, Chairman, Arkansas Parks, Recreation and Travel Commission
Jim Gaston, Lakeview, Arkansas Travel Council
William E. Henderson, Little Rock, Arkansas Department of Parks and Tourism
Mrs. Wellborn Jack, Shreveport, Louisiana, Southwest Chapter, Sierra Club
Dwayne Kelly, Kansas City, Missouri, Ozark Wilderness Waterways Club
Sturgis Miller, Pine Bluff, Arkansas Legislature
Dr. Joe F. Nix, Arkadelphia, Arkansas Stream Preservation Committee
W. L. Pope, Kansas City, Missouri
George Smith, Mountain Home, Chamber of Commerce, Mountain Home
Buddy Surles, Little Rock, Director, State Parks, Arkansas Parks, Recreation and Travel Commission

Acknowledgements Where Due

Immediately upon termination of proceedings in the Longworth Building, a memorandum was sent to the chairman of the Full Committee on Interior and Insular Affairs:

TO THE HONORABLE WAYNE ASPINALL

It is reassuring to note that Congress is concerned and deeply interested in the rapidly increasing problems created by the incredible technological progress of our country during the last generation. That these problems must be dealt with by innovative legislation is clear to us all. We are anxious that such legislation be just and fair to all concerned. After participating in the hearing for H.R. 8382 and H.R. 9119 before the Subcommittee for Parks and Recreation, we are reassured that we have in our country the best possible means to accomplish such laws for the betterment of its citizens.

Neil Compton, M.D.
President, Ozark Society

As soon as possible after that similar acknowledgments were sent under my signature to those most deserving recognition for what had been accomplished. To John Paul Hammerschmidt, who walked the politico-legislative tightrope without wavering:

November 5, 1971

Dear John Paul:

I would like to take this means to express to you the sincere appreciation of all members of our organization, and especially of those of us who were present at the recent hearing on H.R. 8382, for your tactful and efficient handling of this piece of legislation. Your well prepared and reasonable statement to the Subcommittee on Parks and Recreation was but one of several demonstrations of ability made by you during the sometimes tedious course that the proposal for the Buffalo National River has had to take.

It was obvious during this hearing that one of your most effective decisions was the one to have some of the members of the Subcommittee to actually see the Buffalo River and to have floated it last summer.

Then there was the patience which you had to display during the sometimes trying period last fall when some of our overzealous members were demanding a premature showdown on this bill. ...

I trust that you will have the honor of bringing this piece of legislation to reality. It will be a precedent setting accomplishment which should significantly increase your stature as a public servant and would, I hope, help to assure your tenure in this office, or even a more responsible one, for as long as you should desire to serve our country.

On December 6, 1971, an AP news release revealed the favorable decision of the House Subcommittee on Parks and Recreation. The decision of the Full Committee on Interior and Insular Affairs was delayed until after the Christmas holidays. That decision was made known by the media throughout the state. The first paragraph of a long report by the *Pine Bluff Commercial*, February 8, 1972, told us what we wanted to know:

Buffalo Park Bill Passed by House

WASHINGTON—The Buffalo National River bill was approved by the House of Representatives yesterday on a unanimous voice vote, according to the office of Congressman John Paul Hammerschmidt of Arkansas.

A spokesman for Hammerschmidt said the congressman was delighted at the passage and hoped it would be approved quickly by the Senate and signed in the near future by President Nixon.

Passage of the bill by the House virtually assures the creation of the park this year, since the Senate has already passed a similar bill and the President has included $4.6 million for the park in his budget. ...

The Conference Committee to iron out any differences between the Senate and House bills acted favorably without delay. The finished legislation package then went to the Oval Office. President Nixon's signature brought to an end the long, laborious, and acrimonious struggle to place a federal guarantee upon the integrity of the Buffalo River in Arkansas.

Some interesting sidelights of that ceremony were described in the *Arkansas Gazette*, March 12, 1972:

Buffalo River Centennial Item for NPS

The Buffalo National River became a reality on March 1. A little noted angle of the initial successful hurdle was the fact that President Nixon signed the bill just exactly one hundred years from the day that President U. S. Grant signed the bill that created Yellowstone National Park. ...

The Buffalo National River will be the fifth Park Service area in Arkansas. Hot Springs National Park was designated as a federal reservation as early as 1832, but it didn't become a National Park until March 4, 1921. It was the nation's seventeenth Park. The other three National Park Service areas in the state are the Pea Ridge National Military Park in Northwest Arkansas, the Fort Smith National Historic Site in Western Arkansas and the Arkansas Post National Memorial in Eastern Arkansas.

Formal recognition of every one of those several thousand people who contributed to this victory on the Buffalo was beyond the capabilities of our limited staff. But to those most obvious champions, these words of thanks:

March 2, 1972

Ed Freeman III
Pine Bluff Commercial

Dear Ed:

I wish to offer sincere appreciation to the *Pine Bluff Commercial* for the outstanding support that it has afforded the conservation cause in Arkansas since the appearance of the Pearson series on the Buffalo River controversy. From the very beginning, those of us involved realized that we could not long survive without publicity. We did not at first have any outstanding personalities or prime politicians on our side. We were in the possession of nothing but a good idea. Without some method of presenting it to the public, no one would have ever heard of it. It is true that important early events were recorded by some of the newspapers in the state, but these usually came in the form of routine news releases and were much the same wherever printed. We knew that what we needed was special reporting about the people and the events that were taking place. The press in general approved of our program with a number of excellent editorials appearing from time to time. But first hand, in-depth and sustained reporting on this developing controversy was lacking until the *Commercial* appeared on the scene. I have read your paper carefully since that time and have come to the conclusion that the *Commercial* is one of our few live newspapers. Over the years if we wanted to know the latest about the Arkansas legislature, Orval Faubus or the Cossatot River, it was to be found in your pages. We have noted that many, many times the other papers in the state, even up here in Northwest Arkansas, turn to the *Commercial* for information concerning current happenings, especially in reference to conservation. You are to be commended for good journalism in this day of canned news and canned opinions coming over the television, the radio and on the wires of the press.

You are also to be commended for the fact that the *Pine Bluff Commercial* presents both sides of the subject. All of us are politically opinionated in some way or another, and it is discouraging to note the intentional censorship that some papers exercise in reporting current events. But the *Commercial* is read with respect by those who oppose its position as well as those who support it because it is an honest forum.

As for the success of the Buffalo National River, you are as entitled to credit for its accomplishment as any one individual or any organization. Without the diligence of George Wells, Tom Parsons, Kathy Gosnell and Harry Pearson and others on your staff, we would still be floundering. I can assure you that all of us in the Ozark Society will never forget the timely reporting that you provided when the going was rough.

We know that you will keep up the good work.

To the senator who made it all possible:

March 16, 1972

Dear Senator Fulbright:

The Ozark Society and its many thousands of friends extend to you our sincere appreciation to the senator who made it all possible for your efforts on behalf of the newly created Buffalo National River. To you goes the distinction of being our first public official to recognize its value and to use the influence of your office to its ultimate establishment. Had it not been for the grant for the National Park Service to conduct its original survey of the Buffalo River in 1961, there would have been no alternative to the proposal to build the high dams which were scheduled for it.

You will always be remembered for the part that you have played over these several years in achieving what we believe is one of the finest pieces of federal legislation ever to be accorded the State of Arkansas. We deeply appreciate your support for both bills for this legislation as they appeared in the Senate and want to reassure you that your good judgment in so doing has given a tremendous boost to the much needed improvement of the conservation practices here in Arkansas.

The next fall we were to have a final contact with Senator Fulbright. He had accepted an invitation to speak at the Ozark Society general meeting in Little Rock, the only occasion at which we would be so honored.

That night in the Jeff Banks building on the University of Arkansas Medical School campus he said to us: "Of all the legislation that I have authored during my time in the Senate, I take more pride in the bill for the Buffalo National River than all the rest"—a statement that I shall never forget.

And to a dam builder who changed his mind:

March 17, 1972

Dear Senator McClellan:

Those of us who have been so intimately concerned with the program for the establishment of the Buffalo National River wish to commend you for your very valuable participation in the enactment of this legislation. We sincerely appreciate your endorsement of this measure upon its appearance in the Senate. We believe that your recognition of the importance of this pivotal conservation measure will enhance your image as an understanding and progressive member of our delegation in the National Congress.

And to Hammerschmidt once more:

March 16, 1972

Dear John Paul:

The announcement of President Nixon's signing of the bill to create the Buffalo National River was especially gratifying to the several thousand of us in this part of the country who had been actively interested in this legislation and who have worked in various ways for its passage. It was just as satisfying as well to many thousands of other citizens of Arkansas, Oklahoma, Louisiana, Missouri and Tennessee who have sympathetically followed this proposal since your introduction of it made it a definite likelihood. As the years pass, literally millions of people will have an opportunity to see and enjoy the beauty of an outstanding Ozark stream as it was created by nature.

This has been made possible by your wisdom in making the decision to support this measure and your untiring efforts to secure its passage. I hope that all will recognize the fact that you were the key man in this campaign. Only those who were on the scene and engaged in the controversy prior to your election can really know how true this is.

Now we must all work for the orderly and proper development of this exceptional recreational development.

To Orval E. Faubus we extended the honor of being the principal speaker at the spring meeting of the Ozark Society in Arkadelphia, March 25, 1972. On that occasion, he provided us with further insight regarding a fact that we already knew—why Orval Faubus was the right man in the right office at the right time (1965) (*Pine Bluff Commercial*, March 26, 1972):

Faubus Discusses '65 Decision on Buffalo Dam Block

ARKADELPHIA—Former Governor Orville E. Faubus of Huntsville said yesterday that he had decided to block a dam on the Buffalo "two years or more" before he made a formal statement.

Faubus told the spring meeting of the Ozark Society during a meeting at Ouachita

Baptist University here how he came to make his decision to block the dam and preserve the Buffalo as a free-flowing stream. ...

Faubus said that a combination of things had developed at that time that made his decision meaningful. One of these was the increasing interest by conservationists and other people in preserving free-flowing streams. ...

Faubus said he waited two years or more to make his formal statement "and kept my own council" because, if he had not, the Engineers might have waited until he left office and the next governor might have approved the project. ...

Faubus said that he had made his decision basically on "those intangible things that touched the hearts and minds of people," rather than for political reasons. ...

Another political factor, Faubus said, was that some of his biggest critics were for preserving the river. "I caught unshirted hell from the Pine Bluff Commercial and the Arkansas Gazette for months that ran into years," Faubus said, but they agreed with him on the issue of the Buffalo. ...

"The Buffalo could not be preserved by leaving it alone," Faubus said. "Those who thought it could were indulging in a mirage." Faubus said that "fortunately" the Engineers had given lower priority to dams on the Buffalo than to other such structures in the White River basin. "There was time for numbers of people to become aware of the beauties of the river," he said. ...

He said that he met at length in his office with representatives of the Little Rock district of the Engineers and once traveled to Washington to discuss the dam with the chief of the Engineers.

"If I had approved it, it would have been half finished before I left office," Faubus said. He said that he then supported the National River project because "the river could never have remained the same" without some form of protection.

The public must maintain a vigilant attitude toward the river, he said, to insure that its beauty is not destroyed.

A few days before that meeting in Arkadelphia, newspapers throughout the state carried a front-page account of an event that seemed to be more than coincidence. If the bell had tolled for Lone Rock and Gilbert, it now sounded for the prime supporter of those great dams as well (*Marshall Mountain Wave*, March 16, 1972):

James Trimble Dies at Age 78

James W. Trimble, aged 78, of Berryville, who, as the state's Third District congressman, brought dams and reservoirs and rural electrification and millions in federal money into Northwest Arkansas, died Friday at a hospital in Eureka Springs.

He was regarded as the state's most liberal congressman.

As a member of the House Rules Committee, he wielded power that helped him get many projects for Arkansas.

To become a member of this Committee, Mr. Trimble gave up his seat on the House Public Works Committee, where he championed Arkansas waterway projects.

Judge Trimble's stand for more dams helped bring on his unexpected defeat in 1966 by John Paul Hammerschmidt, the conservative Republican lumberman from Harrison.

Mr. Trimble believed that to complete his dream of water control on the White River of North Arkansas, dams were needed on the Buffalo River.

But opposition from conservationists and sportsmen who wanted to keep the Buffalo free flowing led to concerted action to beat him. He got 72,635 votes to Hammerschmidt's 80,495.

He was not bitter. He said of his defeat: "It's part of the game." And with the graciousness that characterized him in Washington, he offered Hammerschmidt his help. ...

After his defeat, he and Mrs. Trimble returned to their home at Berryville. His death came less than two weeks after President Nixon signed the bill to make the Buffalo River a National River, free of dams.

From that obituary we derived no satisfaction. Although he had been an opponent, he was not an enemy. We, too, honored him for being what he was, a gentleman to the end.

EPILOGUE

After ten years of wrenching controversy the Ozark Society in 1972 stood victorious, its goal achieved: the Buffalo River was now a significant part of the nation's park system. But what now could the United States Park Service, Congress, the federal courts, and the citizens of Arkansas and the nation do with this "narrow rope of land"—this "gander-necked park"? Its development and management will, like any other federal project, be subject to the whims of Congress and ultimately to the will of the voting public. For that reason it must be constantly monitored by level-headed conservationists and defended from exploitation when need be.

Such action has already taken place. The Pindall landfill project of the 1980s was defeated by a coalition of established conservation groups, along with a spontaneous uprising against it by local citizens. Other such confrontations are bound to come about. We and those who come after us must be on guard.

The Buffalo National River is not like Congressman Trimble's house with an "unfinished roof." The narrow, lengthy configuration is but a foundation upon which we hope future generations will build. There was so much of immediate importance that has been left out, now crying out for protection; the "Penitentiary" next to Lost Valley, the Beech Creek gorge above Boxley, the Boat Mountains, Cecils Cove, and much more. These scenic marvels, if not someday added directly to the Buffalo National River, could achieve protection from important state and private agencies, now well established.

A protective attitude by human inhabitants for the entire watershed of the Buffalo River will be mandatory if it is to survive as a beautiful clearwater stream of national significance. That will mean restrictions on industry and certain types of agriculture in the area. Such modalities we must learn to accept and live with if there be places on this earth where our descendants can know and understand the wonders of creation.

Don Winfrey, geologist and cartographer.

David Strickland, member of the National Park Advisory Committee, at Pecos Ruins, New Mexico.

Jim Schermerhorn, speleologist.

L. B. Cook, resort owner at Theodosia on Bull Shoals Lake.

Dr. Douglas James, ornithologist at the University of Arkansas.

INDEX

Markham, Joy, 419
Marshall, 79, 86, 91, 120, 127, 139, 142, 150, 155,
    159, 161, 164, 168, 180, 181, 184, 185, 200, 202,
    205, 207, 214, 218, 232, 234, 301, 304, 308, 322
Marshall Mountain Wave, 64, 68, 80, 81, 91,
    121, 122, 127, 131, 132, 137, 139, 143, 144, 146,
    149, 153, 163, 167, 184, 234, 252, 272, 276, 279,
    299, 301, 303, 304, 312, 316, 328, 362, 363, 467
Marvin, Horace, 93
Mather Lodge, 28, 167, 372
Mather, Stephen, 28, 29
Maumee, 161, 188, 191, 329, 330, 331
Maumee Mines, 160, 182
Maynard, Charles, 155, 156, 183, 306, 326
McBroom, James T., 181
McClellan, John L., 9, 15, 130, 184, 189, 199,
    203, 291, 306, 378, 392, 393, 394, 430, 436,
    438, 466
McClellan Kerr Seaway, 306
McCutcheon, Jack, 117, 288, 345, 347, 405, 406
McCutcheon, Wendell, 458
McKinley, R. L., 309
McMath, Sid, 241
McPherson, Harry, 58, 345
McPherson, Virginia, 123
McRaven, Charles (Charley), 166, 167, 184,
    199, 200, 212, 236
Mena, 32
Meeman, E. J., 120
Memphis Commercial Appeal, 120
Memphis Press Scimitar, 120
Meramac River, 12
Micropterous Dolomieui Society (Smallmouth
    Bass Society), 160, 161
Millard, Tom, 45
Mill Creek, 285, 287
Miller, Sturgis, 458, 462
Mills, Joe, 207, 209, 210
Mills, Wilbur, 228, 306, 428, 429, 430, 431, 436
Missouri River, 6, 77
Mississippi River, 17
Mizell, Leonard, 330
Monte Ne, 18
Moore Creek, 351, 353, 354, 355, 377
Moore Creek Bluff, 42. *Also* Sikes Bluff, 351.
Moore, Dwight, 82

Morgan, B. B., 156
Morgan, Tillman, 154
Mosley, Dick, 97
Mossville, 84, 373, 375
Mount Gaylor, 34
Mount Judea, 34, 346, 414
Mount Magazine, 41
Mount Nebo State Park, 28
Mount Sherman, 83
Mountain Home, 12, 64, 93, 128, 276, 334
Mt. Hersey, 205, 207, 208, 211
Muir, John, 26, 159
Mulberry, 77
Mulberry River, 34, 35, 96, 387, 388, 429
Murray, Dick, 43, 183, 202, 326, 330, 333, 334,
    355, 364, 377, *385*, 386, 387, 431
Mussel Shoals, 4

Narrows, the, 208
National Association of Campers and Hikers,
    279, 280, 340
National Audubon Society, 121, 401
National Council of the Arts, 272–73
National Grange, 200
National Scenic Rivers bill, 119
National Science Foundation, 347
National Wildlife Club, 415
National Wildlife Federation, 26, 30, 152, 323,
    326, 380, 400, 422, 452
Natural Heritage Commission, 260
Nature Conservancy, 73, 86, 340
Naughton, Mike, 119
New York Times, 184
Newark (New Jersey) *Star-Ledger*, 272
Newman, John, 299, 367
Newport, 23, 273
Nix, Joe, 250, 388, *389*, 400, 428, 429, 438, 439,
    453, 458, 462
Nixon, Richard, 461, 464
Norfork, 79
Norfork Dam, 14, 22, 23
Norfork (North Fork) River, 9, 11, 12, 13, 14, 15,
    16, 135
Norman, Dean, 94
North Little Rock Times, 313
North Pole Knob, 287

NEIL ERNEST COMPTON was born August 1, 1912, at Falling Springs Flats in northwest Arkansas. He graduated from Bentonville (Arkansas) High School in 1931 and from the University of Arkansas in 1935 with degrees in zoology and geology. He received his doctor of medicine degree from the University of Arkansas School of Medicine in 1939 and worked for the Arkansas Board of Health in Bradley and Washington counties until joining the U.S. Navy in 1942. He served with the Naval Medical Corps in the South Pacific and continued in the Naval Reserve until his retirement in 1972 with the rank of Captain.

Compton was married to Laurene Putman and they had three children: Ellen, Edra, and Bill.

Compton was actively engaged in conservation work in Arkansas, particularly involving the Ozarks region. He was the first president of the Ozark Society to Save the Buffalo River (later the Ozark Society) and served in that office until 1974. He was active in securing the Arkansas Wilderness Act of 1983 and continued to champion preservation efforts involving stream protection.

Compton was the author of three books of text and photographs about the Ozarks: *The High Ozarks: A Vision of Eden* (1982); *The Battle for the Buffalo River: A Twentieth-Century Conservation Crisis in the Ozarks* (1992); and *The Buffalo River in Black and White* (1997). *The Battle for the Buffalo River* was nominated for the National Book Award and, in 1993, received the Arkansas Library Association's Arkansiana Award.

For his conservation work Compton received numerous awards, including the American Motors Conservation Award (1964), the Distinguished Arkansas Conservationist Award (1966), the designation of Honorary National Park Ranger (1987), the first annual congressional Teddy Roosevelt Conservation Award (1990), the National Wildlife Federation Achievement Award (1992), and the Arkansas

Outdoor Sportsman's Hall of Fame (1992). Compton died in Bentonville, A̶
on February 10, 1999, at age eighty-six, having hiked in the Arkansas Ozarks a ̶
before he died.

The Ozark Society Papers and the Neil Compton Papers are housed in the Special Collections Department of the University of Arkansas Libraries, Fayetteville, Arkansas.

KENNETH L. SMITH has been involved in conserving Arkansas's Lost Valley near Boxley and the Buffalo River since 1956. He holds an engineering degree from the University of Arkansas and a master of science degree in Natural Resources Information from the University of Michigan. He has worked across the country for the National Park Service, particularly with the Buffalo National River in Arkansas, where he is a designer and construction supervisor of the Buffalo River Trail along the entire length of the river. He is currently involved with the Buffalo River Land Trust.

Smith is the author of four books: *The Buffalo River Country in the Ozarks of Arkansas* (1967), *The Illinois River Country* (1977), *Sawmill: The Story of Cutting the Last Great Virgin Forest East of the Rockies* (1986), and *Buffalo River Handbook* (2004). He has also written text and designed layouts for two Buffalo National River maps for Trails Illustrated of the National Geographic Society.